DISCRETE MATHEMATICS

DISCRETE MATHEMATICS

By

Sudhir Dawra

ANMOL PUBLICATIONS PVT. LTD.

NEW DELHI - 110 002 (INDIA)

ANMOL PUBLICATIONS PVT. LTD.
4374/4B, Ansari Road, Daryaganj
New Delhi - 110 002
Ph.: 23261597, 23278000
Visit us at: www.anmolpublications.com

Discrete Mathematics

First Published, 2003

PRINTED IN INDIA

Published by J.L. Kumar for Anmol Publications Pvt. Ltd., New Delhi - 110 002 and Printed at Mehra Offset Press, Delhi.

Contents

Preface

Computers are extensively used in business and industry to assist managers for information processing and decision making. With the help of Discrete Mathematics Problems can be automatically solved. The objective of this book is to train the students of BCA, MCA, B.Tech., M.Tech. and O, A Level courses to accomplish the objectives of mathematics and problems can be decomposed.

This book has been written for the beginner. Prior some knowledge of mathematics is good enough to go through this book.

The most important stage in the process of learning is the Synthesis of the knowledge. The Discrete Mathematics with the help of Computers can be effectively used for assisting a programmer only if the knowledge of Mathematics can be synthesized with the skill in Program writing.

Different Techniques, methods and Exercises have been included in this book to make it more comprehensive.

In addition to the coverage of fundamentals of discrete Mathematics, solved examples and theory con various topics have been carefully selected and applied in Red life situations.

The manuscript have been reviewed by a numbe of eminent editors. I am highly thankful for Mr. Joshi for his valuable suggestions for this book. I shall be failing in my duty if I don't acknowledge, the contributions and blessings of Mr. J.L. Kumar and Mr. J.K. Dawra. Any suggestions for the improvement of the book are welcome and added in the subsequent revised Edition.

—Sudhir Dawra

1

Sets and Relations

Whenever a word of term becomes too used, it finds its usages in many areas. As a result of this, it becomes very difficult to given an universally accepted meaning to the word. Therefore, the word gets its contextual meaning. For example, network, communication, input etc. *Set* is one of the word among them. We have studied how to write a program in a programming language. We have used variables. Before using a variable, we have declared that variable of a type.

Once we have declared a variable of a type, say, integer (int) the variable can be assigned only an integer value. What I am trying to explain to you is that we can always make collection of certain type of objects and give a name to that collection to refer it thereafter by that name.

1.1 Sets and its Representation

Set

"A set is a well-defined collection of any type of objects, things or numbers without repetition".

A set is known by its elements. A set is well-defined if it can be *determined* that a given object is a member of the set.

Examples of sets:

$D = \{0, 1, 2, 3, 4, 5, 6, 7, 8, 9\}$

This set is collection of all digits. In this example all the elements are placed inside '{'&'}'. The collection {0, 1, 2......, 9} is represented here as *'D'*. This is a way to represent a set. The same set can be written as:

$D = \{x \mid x \text{ is a digit}\}$

Here, we have not listed all the objects, but general name of the objects.

Yet another way to represent a set is to use a diagram. e.g.

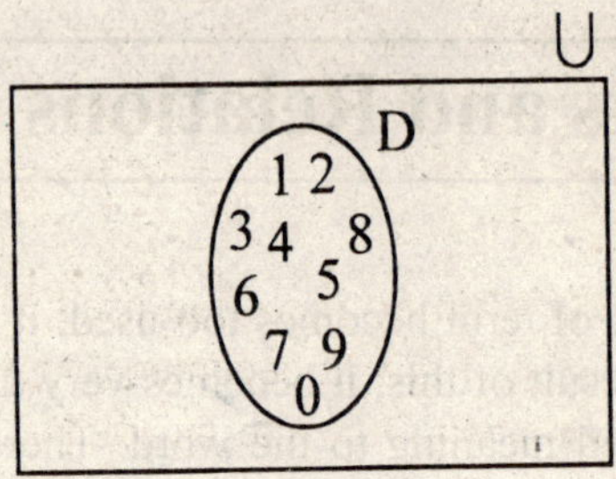

Fig. 1.1.

Membership of a Set

Here, 0, 1, ...9 are called members or elements of the set D. If an object 'x' is a member of a set 'D', then we write

$x \in D$

and it is read as 'x belongs to D', i.e. $\in$ stand for 'belongs to'. If an object 'x' is not a member of a set 'D', then we write

$x \not\subset D$

and is read as 'x does not belong to D'. The symbol '$\in$' is also called 'membership operator'.

Some Standard Examples of Set

1. Set of Integers

   $Z = \{-\infty, ...-3, -2, -1, 0, 1, 2, 3......\infty\}$

2. Set of Positive Integers

   $Z_+ = \{1, 2, 3, 4, ...\infty\}$

3. Set of Negative Integers

   $Z = \{-\infty, ...-4, -3, -1\}$

4. Set of Natural Numbers

   $N = \{0, 1, 2, 3, ...\infty\}$

5. Set of Rational Numbers

$$Q = \{x \mid x = p/q \;\; p \in Z \; \& \; q \;\; \in Z+\}$$

6. Set of Real Numbers

$R = \{x \mid x$ is a distance of point on a line from origin$\}$

7. Set of Alphabets

$$\Sigma = \{a, b, c, \ldots z\}$$

Set Representation

A set is constituted by its members. Members, also called elements, of a set is denoted, in general, by lowercase alphabets like *a*, *b*, ...*z*. The set as a whole is denoted in general, by uppercase alphabets like *A*, *B*, *C*, ...*Z*. A set is represented in following three ways

- Enumerated way/tabulation method
- Symbolic way/phrase structure method/set former method
- Diagrammatic way

Enumerated way: In this method of representation, all members of a set are listed explicitly e.g.

$A = \{1, 2, 3, 4\}$; set of positive integer ≤ 4

$B = \{a, e, i, o, u\}$; set of vowels of English alphabets of lowercase

$C = \{0, 1\}$; set of bits

This method of representation is also called **tabulation method** of set representation.

Symbolic way: While representing a set in Symbolic way, elements are described by giving a statement of properties of the elements. e.g.

$Z = \{x \mid$ is an integer$\}$; this is set of integers

$P = \{x \mid$ is a prime number$\}$; this is a set of prime numbers

$C = \{x \mid x$ is serial no of PC's manufactured by HCL$\}$; this is a set of serial numbers of all personal computers manufactured by Hindustan Computer Limited.

$A = \{x \mid x^2 \leq 16\}$; is another example of a set represented in symbolic way. List the elements of *A*. The symbolic representation of set is also called **'set former method'** or **'descriptive phrase method'**.

Diagrammatic way: This method uses two geometrical entities circle and point to represent a set. Circle is used to represent a set and points for elements of that set. e.g.

$A = \{1, 2, 3\}$

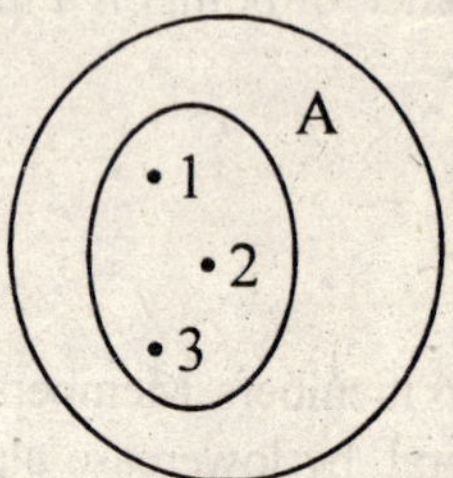

Fig. 1.2.

there will be as many points as many elements in a set. More about this we will study in next section of this chapter.

1.1.1 Types of Set

Nature of elements determines nature of sets, e.g. set of students, set of courses etc. The number of elements in a set determines what type of set we are dealing with.

Finite Set

A set is finite, If it has a finite number of elements. (i.e. elements of such a set can be counted by a finite number). The number of elements in a finite set A is called *cardinality* of A and is denoted by $n(A)$ or $|A|$

Example 1: Let $F = \{a, b, x, z, 0, 18, p\}$. Then $n(F) = 7$, or $|n| = 7$

Example 2: Let $S = \{x \mid x$ is a floating point operation executed in sec$\}$

Example 3: Let $C = \{x \mid x$ is a name of TV channels broadcast over a cable$\}$

Infinite Set

If a set has an infinite number of elements, it is an infinite set. The elements of such a set cannot be counted by a finite number.

Example 1: $R = \{ x \mid x$ is a natural number$\}$

Example 2: $\{x \mid x$ is a grain of sand on the sea shore$\}$

Point Set

A set of points along a line called *Point* set.

____.____.____.____.____.____.____.____.____

-3 -2 -1 0 1 2 3 4

A Point set is an infinite set. If it is considered in computer science in graphics display context, it is a finite set.

Example 1: $P = \{x \mid x$ is a pixels aling a line on VDU$\}$ (finite)

Example 2: $I = \{x \mid x$ is a point on a any geometric line$\}$ (infinite)

Null or Empty or Void Set

A set containing no elements. A null set is denoted by (phi) ϕ

Example 1: $A = \{x \mid x$ is a married bachelor$\}$

Example 2: $B = \{x \mid x$ is a register capable of storing infinite number$\}$

Example 3: $C = \{x \mid x$ is real and $x^2 = -1\}$

Singleton or Unit Set

A Set containing only one elements

Example 1: $A = \{a\}$

Example 2: $B = \{x \mid x$ is a integer which is neither negative nor positive$\}$ $\{o\}$

Universal Set

The Universal set is the totality of elements under consideration. It is the set of all possible objects relevant to a particular application. It is denoted by Ω

Example 1: $\Omega = \{$head, tail$\}$

$\Omega = \{0, 1\}$

$\Omega = \{x \mid x$ is an ASCII code$\}$

$\Omega = \{x \mid x$ is roll no. of student of MCS$\}$

Disjoint Set

Two sets are said to be disjoint if they have no element in common.

Example 1: $A = \{x \mid x$ is an odd numbers$\}$
$B = \{x \mid x$ is an even number$\}$

In this example, A and B have no element in common, therefore these are disjoint.

Example 2: $M = \{x \mid x$ is a pointing device of a computer system$\}$
$C = \{x \mid x$ is a keyboard$\}$

Here M & C are two sets having no elements in common.

Overlapping Set

Two sets are said to be over-lapped if they have some element in common.

Example 1: $A = \{x \mid x$ is a name of student pursuing DM course$\}$
$B = \{y \mid y$ is a name of student pursuing UNIX course$\}$

It is possible that a few students may be pursuing both the courses in the same semester.

Example 2: $A = \{x \mid x$ is a character represented by 7 bits$\}$
$B = \{y \mid y$ is a character represented by 8 bits$\}$

Here, A & B are overlapping sets, because characters represented by 7 bits are also represented by 8 bits.

Subset and Superset

A set A is said to be subset of another set B, if every member of A is a member of B. It is written as $A \subseteq B$.

Example 1: Let $A = \{a, b, c\}$, and $B = \{a, b\}$. Since all elements of B are also elements of A, we say that $B \subseteq A$.

Example: $x = \{1, 2, 3\}$ $y = \{1, 2, 3\}$

Example 2: Let Z_+ be the set of positive integer and Z be the set of all integer, then $Z_+ \subseteq Z$.

Whenever $A \subseteq B$, we say that A is contained in B or B contains A. Similarly, if $A \subseteq B$ i.e. if A is subset of B, is called **Superset** of A, and is written as $B \supseteq A$

In the above examples find which set is superset of other one.

Proper Subset

A set X is said to be proper subset of another set Y if

- all elements of X are in Y, and
- there is at least one element in Y such that, that element is not in X

We represent proper subset as $\subset$. Thus X is a proper subset of Y is represented as $X \subset Y$.

Similarly, Y is said to contain X properly and is represented as $Y \supset X$.

Example 1: Let $A = \{a, b, c\}$ and $B = \{a, b\}$. In this example all elements of set B are also elements of set A, and $c \in A$ is not an element of B, therefore $B \subset A$.

Note: 1. Every set A is a subset of itself.
2. A null set is a subset of every set.

Equal or Identical Set

If every elements of one set is also an element of another set then the two sets are said to be identical (equal). Equal sets have exactly the same elements (may be possibly in different order).

If $A \subseteq B$ & $B \subseteq A$ then $A = B$.

Example 1: Let $A = \{x \mid x$ is letter of word 'equal'$\}$, and $B = \{e, q, u, a, l\}$; since $A \subseteq B$ and $B \subseteq A$ {please. verify) we can write $A = B$

Example 2: Let $A = \{x \mid x$ is a digit$\}$ and $B = \{x \mid x$ is a non-negative integer and $x < 10\}$. Verify that both the sets are equal.

Example 3: Let $A = \{a, b, 2, 3\}$ and $B = \{2, 3, a, b\}$. Here also, both these sets are equal.

Equivalent Set

Two sets A and B are said to be equivalent if they have same cardinal number, i.e. number of elements in both the sets are equal. Nature of elements are not important.

Example 1: Let $A = \{1, 2, 3\}$ and $B = \{x, y, z\}$, then both A and B are equivalent sets but not identical sets.

Example 2: Let $A = \{x \mid x$ is an even integer$\}$ and $B = \{y \mid y$ is an odd integer$\}$, then both A and B are equivalent sets.

Note: Prove that, if any two sets A and B are identical then they are equivalent but reverse is not always true.

Power Set

Till now we have studied only set of elements. **Sometimes** a set also becomes a member of a set. If all the elements of a set are sets, then we call the set as **'set of sets'.** If some elements of a set are sets and some are basic elements then we call that set as **'mixed set'.**

Example 1: Let $A = \{\{1\}, \{1, 2\}, \{3\}\}$. Since all the elements of A are sets, we call A as set of sets.

Example 2: Let $A = \{1, 2, \{3\}, \{2\}, \{1\}\}$. Since some of the elements of A are basic elements like 1, 2, and some are sets like $\{3\}$, $\{2\}$, $\{1\}$. A is a mixed set.

Now, consider a set $A = \{1, 2, 3\}$. Let us construct all distinct subsets of A. Since ϕ is a subset of every set, $\phi \subseteq A$. Also A is subset of itself. So $A \subseteq A$. Other distinct subsets are $\{1\}$, $\{2\}$, $\{3\}$, $\{1, 2\}$, $\{1, 3\}$, $\{2, 3\}$. The set of these subsets is

$$\{\phi, \{1\}, \{2\}, \{3\}, \{1, 2\}, \{1, 3\}, \{2, 3\}, \{1, 2, 3\}\}.$$

This set is a set of all possible subsets of the set A. We call the set as **Power set** of A and is represented as either $P(A)$ or 2^A. Thus **Power set** of a given set can be defined as a collection of all possible distinct subsets of the given set.

$$\therefore P(A) = \{\phi, \{1\}, \{2\}, \{3\}, \{1, 2\}, \{1, 3\}, \{2, 3\}, \{1, 2, 3\}\}.$$

Example 1: Let $A = \{a, b\}$ then $P(A) = \{ \phi, \{a\}, \{b\}, \{c\} \}$

Example 2: Let $B = \{x\}$ then $P(B) = \{ \phi, \{x\} \}$.

From above examples we are able to observe that if $|A| = n$ then $|P(A)| = 2^n$, i.e. if number of elements in A is 1, then its power set will have $2^1 = 2$ elements. Similarly, if number of elements is A in 3, then its power set will have $2^3 = 8$ elements.

Now let us co-relate this power set concept with bit combinations, which we have studied in computer fundamental courses. If we fix the position of elements in a given set and represents its various subset by a bit-string indicating the absence of an element by '0' and presence by 1, we can have a bit representation of every possible subsets of the given set. The length of each such bit string will be equal to the number of elements in the given set. To illustrate it, let us consider the following example.

Example 1: Let $A = \{1, 2, 3\}$. Its all possible subsets, its bits representation and remarks are given below:

Sr. No	*Subsets*	*Bit's representation*	*Remark*
1.	ϕ	000	None of the 1, 2, 3 is present
2.	{1}	100	Only 1 is present
3.	{2}	010	Only 2 is present
4.	{3}	001	Only 3 is present
5.	{1, 2}	110	3 is absent
6.	{2, 3}	011	1 is absent
7.	{1, 3}	101	2 is absent
8.	{1, 2, 3}	111	All are present

We know that if we have 3 bits, we can have 2^3 combinations. Similarly if a set has 3 elements it can have $2^3 = 8$ possible subsets. In general if we have a set of n elements its power set will have 2^n elements.

Multi-set

We have studied that a set is a collection of distinct objects. In real life, we may encounter many situations when collection are not distinct. For example, collection of names of students in a class. Possibly, we may have two or more students having same name.

Now, if we make a set of names of all students in a class, we may not have all the elements distinct. Such a set in which an element may appear more than once, we call it a **multi-set.**

Example: Let $A = \{a, a, a, b, b, d\}$. Since 'a' appears three times in A, 'b' appears two times, we call A as multi-set.

Number of times an element appears in a multi-set is called **multiplicity** of that element in multi-set. In the above example, multiplicity of a is 3, that of b is 2 and multiplicity of d is 1.

Note: A set is a special case of multi-set in which the multiplicity of every element is one.

The cardinality of a multi-set is the number of distinct element in the multi-set.

1.1.2 Operations on Sets

Binary set operators: Union (∪), Intersection (∩), Difference (–), Symmetric difference (⊕), Cartesian Product (X) and unary set operator: Complement (~) are used to know either combined, common or relative properties of set.

Union A ∪ B:

The union of two sets A and B is the set of all elements belonging to set A or to set B or to both A and B i.e $A \cup B = \{x \mid x \in A$ or $x \in B$ or $x \in$ both A and $B\}$

Example 1: Let $A = \{x \mid x$ is rational number$\}$, and $B = \{x \mid x$ is irrational number$\}$ then

$$A \cup B = R = \{x \mid x \text{ is real}\}.$$

Example 2: Let $A = \{x \mid x$ is instruction to be executed when if condition is true$\}$, and

$B = \{x \mid x$ is instruction to be executed when if condition is false$\}$, then

$$A \cup B = \{x \mid x \text{ is instruction in if-else body}\}$$

Example 3: Let $A = \{1, 2, 3\}$ and $B = \{2, 3, 4\}$ then $A \cup B = \{1, 2, 3, 4\}$.

Intersection A ∩ B:

The Intersection of two sets A and B is the set of all elements belonging to both A and B i.e. $A \cap B = \{x \mid x \in A \text{ and } x \in B\}$

Example 1: Let $A = \{1, 2, 3\}$ and $B = \{2, 3, 4\}$ then $A \cap B = \{2, 3\}$

Example 2: Let $A = \{x \mid x \text{ is a rational number}\}$, and
$B = \{x \mid x \text{ is an irrational number}\}$ then

$A \cap B = \phi$, since a number is either rational or irrational, it can never be both and hence there is no common element in A and B.

Set Difference A–B

The difference of two sets A and B is the set of all elements belonging to A but not to B i.e.

$$A - B = \{x \mid x \in A, x \notin B\}, \text{ and}$$
$$B - A = \{x \mid x \in B, x \notin A\}$$

Example: $A = \{a, b, x, y\}$, $B = \{c, d, x, y\}$, then $A - B = \{a, b\}$, $B - A = \{c, d\}$.

Symmetric Difference

The symmetric difference of two sets A and B is defined as the set of elements that belong to either A or to B but not to both A and B i.e.

$$A \oplus B = \{x \mid (x \in A \text{ and } x \notin B) \text{ or } (x \in B \text{ and } x \notin A)\}$$
$$= (A - B) \cup (B - A) \text{ or } A \cup B - A \cap B$$

Example 1: Let $A = \{a, b, c, d\}$ and $B = \{a, c, e, f, g\}$ then
$A \oplus B = \{b, d, e, f, g\}$

Complement of Set

Let A be a set and Ω be its universal set, then complement of A denoted by A' or A^c, is defined as the collection of all elements of Ω, which are not in A. i.e.

$$A^c = \{x \mid x \notin A\}$$

Example 1: Let $\Omega = \{x \mid x \text{ is an integer}\}$ and
$A = \{x \mid x \text{ is an even number}\}$ then
$A^c = \{x \mid x \text{ is an odd integer}\}$

Example 2: Let $A = \{1, 2, 3\}$ and $\Omega = \{x \mid x \text{ is a digit}\}$ then

$$A^c = \{0, 4, 5, 6, 7, 8, 9\}$$

Example 3: Let $A = \{0\}$ and $\Omega = \{0, 1\}$, then $A^c = \{1\}$. Similarly if $B = \{1\}$ then $B^c = \{0\}$. This property of complement conveys that if we have only two possibilities OFF(0) and ON(1) then complement of 0 is 1 and that of 1 is 0.

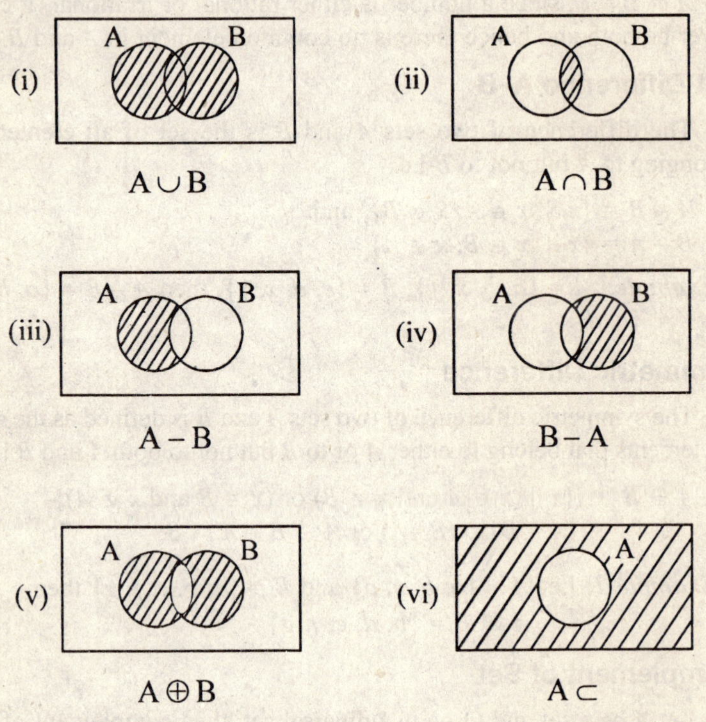

Fig. 1.1.

Cartesian Product

Let $A = \{a, b\}$ and $B = \{x, y, z\}$. Now let us construct pairs in such a way that first element of the pair comes from A and second element comes from B then, we get a set of pairs as shown below:

$\{(a, x), (a, y), (a, z), (b, x), (b, y), (b, z)\}$.

In this collection of pairs (u, v). $u \in A$ and $v \in B$. This collection of pair is product set of sets A and B. We also call it as Cartesian Product of sets A and B. It is denoted as $A \times B$, i.e.

$$A \times B = \{(a, x), (a, y), (a, z), (b, x), (b, y), (b, z)\}.$$

We can define cartesian product of any two non-empty sets A and B as collection of all ordered pair (x, y) such that $x \in A$ and $y \in B$ i.e.

$$A \times B = \{(x, y) \mid x \in A \text{ and } y \in b\}$$

Similarly, $B \times A = \{(x, y) \mid x \in B \text{ and } y \in A\}$

If A and B are defined as above, then

$$B \times A = \{(x, a), (x, b), (y, a), (y, b), (z, a), (z, b)\}.$$

Note:

- A pair (a_1, b_1) and (a_2, b_2) are said to be equal if and only if $a_1 = a_2$ and $b_1 = b_2$. Thus (x, a) need not be equal to (a, x) since x may not be equal to a. For example, in 2-D plane, points having coordinate (1, 3) and (3, 1) are different.
- In general $A \times B \neq B \times A$
- Cartesian Product is the basis of **Relation** on a set. Relation is a very important topic in Discrete Mathematical Structure. Therefore, students are suggested to understand the concept of Cartesian Product, thoroughly.

Laws of Operation on sets:

A. *Idempotent Laws or Laws of Inclusion or* Tautology

1. $A \cup A = A$
2. $A \cap A = A$

B. Commutative laws

3. $A \cup B = B \cup A$
4. $A \cap B = B \cap A$

C. Associative laws

5. $(A \cup B) \cup C = A \cup (B \cup C)$
6. $(A \cap B) \cap C = A \cap (B \cap C)$

D. Distributive Laws

7. $A \cup (B \cap C) = (A \cup B) \cap (A \cup C)$
8. $A \cap (B \cup C) = (A \cap B) \cup (A \cap C)$

E. Identity Laws

9. $A \cup \phi = A$ 10. $A \cap \phi = \phi$

11. $A \cup \Omega = \Omega$ 12. $A \cap \Omega = A$

F. Complement Laws

13. $A \cup A' = \Omega$ 14. $A \cap A' = \phi$

15. $(A') = A$ 16. $\Omega' = \phi$ 17. $\phi' = \Omega$

G. De Morgan's Laws

18. $(A \cup B)' = A' \cap B'$ 19. $(A \cap B) = A' \cup B'$

Readers are advised to prove the above mentioned laws. For illustration, proof of 7 and 18 are given below.

Proof of Law 7:

To prove the law 7, we have to proof the following two results:

$A \cup (B \cap C) \subseteq (A \cup B) \cap (A \cup C)$ ——————— (1)

and $(A \cup B) \cap (A \cup C) \subseteq A \cup (B \cap C)$ ——————— (2)

First we shall prove 1. Let x be any element of $A \cup (B \cap C)$, then

$x \in A \cup (B \cap C) \Rightarrow x \in A$ or $x \in (B \cap C)$

$\Rightarrow x \in A$ or ($x \in B$ and $x \in C$)

$\Rightarrow$($x \in A$ or $x \in B$) and ($x \in A$ or $x \in C$)

$\Rightarrow x \in (A \cup B)$ and $x \in (A \cup C)$

$\Rightarrow x \in (A \cup B) \cap (A \cup C)$

$\therefore A \cup (B \cap C) \subseteq (A \cup B) \cap (A \cup C)$

Similarly, we can prove that

$(A \cup B) \cap (A \cup C) \subseteq A \cup (B \cap C)$

Therefore, combining 1 and 2 we have

$A \cup (B \cap C) = (A \cup B) \cap (A \cup C)$

Proof of Law 18:

Let x by any element of $(A \cup B)^c$ [or $(A \cup B)'$], then

$x \in (A \cup B)' \Rightarrow x \notin A \cup B$

$\Rightarrow x \notin A$ and $x \notin B$

$$\Rightarrow x \in A' \text{ and } x \in B'$$
$$\Rightarrow x \in A' \cap B'$$

$\therefore (A \cup B)' \subseteq A' \cap B'$ ———————————— (1)

Similarly $A' \cap B' \subseteq (A \cup B)'$ ———————————— (2)

Therefore combining 1 and 2 we have $\mathbf{(A \cup B) = A' \cap B'}$.

1.1.3 Principles of Inclusion and Exclusion

Let us begin with an example of finding all integers between 1 and 100 that are divisible by 3 or 7. Let A and B be two sets defined as:

$A = \{x \mid x \text{ is divisible by 3 and } \leq x \leq 100\}$

and $B = \{x \mid x \text{ is divisible by 7 and } \leq x \leq 100\}$

Now, our problem is to find the number of elements in $A \cup B$ i.e. count of those numbers which are divisible either by 3 or by 7. Further, there are certain elements (numbers between 1 and 100) that are divisible by both 3 and 7, i.e. $A \cap B$ is not a null set. Hence, we can write

$$|A \cup B| = |A| + |B| - |A \cap B|$$

Since, while counting the numbers, the common elements of A and B have been counted twice, so once it has to be excluded. Therefore we are subtracting the count of numbers in $A \cap B$.

In the above equation, $|A|$, $|B|$ & $|A \cap B|$ can be computed as below:

$$|A| = 33 \quad \left[n = \frac{l-a}{d} + 1 = \frac{99-3}{3} + 1 = 33\right]$$

$$|B| = 14 \quad \left[n = \frac{l-a}{d} + 1 = \frac{98-7}{7} + 1 = 14\right]$$

$$|A \cap B| = 4 \quad \left[n = \frac{l-a}{d} + 1 = \frac{84-21}{21} + 1 = 4\right]$$

$\therefore |A \cup B| = 33 + 14 - 4 = 43$ Ans.

Example: At a university 60% of the teacher play tennis, 50% of them play bridge, 70% jog, 20% play tennis and bridge, 30% play tennis and jog and 40% play bridge and jog. What is the percentage of teachers who jog and play bridge and tennis.

Solution: Let T, B and J be the set of teachers who play tennis, who play bridge and those who jog respectively. Therefore, we have

$|T| = 60: |B| = 50: |J| = 70$ [Assuming there are 100 teachers in University]

$|T \cap B| = 20; |T \cap J| = 30; |B \cap J| = 40$

$|T \cup B \cup J| = 100$

$|T \cap B \cap J| = ?$

Now we can write,

$$|T \cup B \cup J| = |T||B| + |J| - |T \cap B| - |T \cap J| - |B \cap J| + |T \cap B \cap J|$$

or, $100 = 60 + 50 + 70 - 20 - 30 - 40 + |T \cap B \cap J|$

or $|T \cap B \cap J| = 100 - 90 = 10$

Thus 10% of the teachers jog and, play bridge and tennis.

Now we are in a position to know formal statement related to principle of Inclusion and Exclusion, which is based on the Cardinality of sets.

1. Let A and B are two sets then

   $|A \cup B| = |A| + |B| - |A \cap B|$

   If A and B are disjoint then $|A \cup B| = 0$

   $\therefore |A \cup B| = |A| + |B|$

2. Let A, B and C are three sets then

   $|A \cup B \cup C| = |A| + |B| + |C| - |A \cap B| - |A \cap C| - |B \cap C| + |A \cap B \cap C|$

   If all these three sets are mutually disjoint then

   $|A \cup B \cup C| = |A| + |B| + |C|$

3. In general, Let $A_1, A_2, \ldots. A_n$ be n sets then

$$|\bigcup_{i=1}^{n} A_i| = |A_1| + |A_2| + \ldots + |A_n| - |A_1 \cap A_2| - |A_1 \cap A_3| -$$

$$\ldots. |A_{n-1} \cap A_n| + |A_1 \cap A_2 \cap A_3| + |A_1 \cap A_2 \cap A_3|$$

$$+ \ldots\ldots + |A_{n-2} \cap A_{n-1} \cap A_n| + \ldots\ldots + (-1)^{n-1} |\bigcap_{i=1}^{n} A_i|$$

If all A_i's are mutually disjoint then

$$|\bigcup_{i=1}^{n} A_i| = |A_1| + |A_2| + \ldots + |A_n|$$

$$= \sum_{i=1}^{n} |A_i|$$

4. Let A and B be two sets then

$$|A \cap B| \le \min\{|A|, |B|\}$$

$$|A \oplus B| = |A| + |B| - 2|A \cap B|$$

$$|A - B| \ge |A| - |B|$$

$$|A \times B| = |A| \triangle |B|$$

Exercise 1.1

1. Determine whether each of the following statements is true or false. Briefly explain your answer.

(a) $\phi \in \phi$ (b) $\phi \in \phi$

(c) $\phi \in \{\phi\}$ (d) $\{\phi\} \subseteq \phi$

(e) $\{\phi\} \in \phi$ (f) $\{\phi\} \subseteq \{\phi\}$

(g) $\{\phi\} \in \{\phi\}$ (h) $\{a, b\} \subseteq \{a, b, c, \{a, b, c\}\}$

(i) $\{a, b\} \in \{a, b, c, \{a, b, c\}\}$ (j) $\{a, b\} \subseteq \{a, b, \{\{a, b\}\}\}$

(k) $\{a, b\} \in \{a, b, \{\{a, b\}\}\}$ (l) $\{a, \phi\} \subseteq \{a, \{a, \phi\}\}$

(m) $\{a, \phi\} \in \{a, \{a, \phi\}\}$

2. Determine the following sets:

(a) $\phi \cup \{\phi\} = \{\phi\}$ (b) $\phi \cap \{\phi\} = \phi$

(c) $\{\phi\} \cup \{a, \phi, \{\phi\}\} = \{a, \phi, \{\phi\}\}$ (d) $\{\phi\} \cap \{a, \phi, \{\phi\}\} = \{\phi\}$

(e) $\phi \oplus \{a, \phi, \{\phi\}\} = \{a, \phi, \{\phi\}\}$ (f) $\{\phi\} \oplus \{a, \phi, \{\phi\}\} = \{a, \phi\}$

3. (a) Let A and B be sets such that $(A \cup B) \subseteq B$ and $B \subseteq A$. Draw the corresponding Venn diagram.

(b) Let A, B, and C be sets such that $A \subseteq B$, $A \subseteq C$, $(B \cap C) \subseteq A$, and $A \subseteq (B \cap C)$. Draw the corresponding Venn diagram.

(c) Let A, B, and C be sets such that $(A \cap B \cap C) = \phi$, $(A \cap B) \neq \phi$, $(A \cap C) \neq \phi$, and $(B \cap C) \neq \phi$. Draw the corresponding Venn diagram.

4. Give an example of sets A, B, and C such that $A \in B$, $B \in C$, and $A \notin C$.

5. Determine whether each of the following statements is true for arbitrary sets A, B, C. Justify your answer.

(a) If $A \in B$ and $B \subseteq C$, then $A \in C$ (b) If $A \in B$ and $B \subseteq C$, then $A \subseteq C$.

(c) If $A \subseteq B$ and $B \in C$, then $A \in C$. (d) If $A \subseteq B$ and $B \in C$, then $A \subseteq C$.

6. Let A, B, C be subsets of U. Given that

$$A \cap B = A \cap C$$

$$\overline{A} \cap B = \overline{A} \cap C$$

Is it necessary that $B = C$? Justify your answer.

7. Given that

$$(A \cap C) \subseteq (B \cap C)$$

$$(A \cap \overline{C}) \subseteq (B \cap \overline{C})$$

show that $A \subseteq B$.

8. What can you say about the sets P and Q if
   (a) $P \cap A = P$? (b) $P \cup Q = P$?
   (c) $P \oplus Q = P$? (d) $P \cap Q = P \cup Q$?
9. (a) Let $A \subseteq B$ and $C \subseteq D$. Is it always the case that $(A \cup C) \subseteq (B \cup D)$? Is it always the case that $(A \cap C) \subseteq (B \cap D)$?
10. (a) Given that $A \cup B = A \cup C$, is it necessary that $B = C$?
    (b) Given that $A \cap B = A \cap C$, is it necessary that $B = C$?
    (c) Given that $A \oplus B = A \oplus C$, is it necessary that $B = C$?
    Justify your answers.
11. For $A = \{a, b, \{a, c\}, \phi\}$, determine the following sets:
    (a) $A - \{a\}$ (b) $A - \phi$
    (c) $A - \{\phi\}$ (d) $A - \{a, b\}$
    (e) $A - \{a, c\}$ (f) $A - \{\{a, b\}\}$
    (g) $A - \{\{a, c\}\}$ (h) $\{a\} - A$
    (i) $\phi - A$ (j) $\{\phi\} - A$
    (k) $\{a, c\} - A$ (l) $\{\{a, c\}\} - A$
    (m) $\{a\} - \{A\}$
12. Let A, B, C, be arbitrary sets.
    (a) Show that $(A - B) - C = A - (B \cup C)$
    (b) Show that $(A - B) - C = (A - C) - B$
    (c) Show that $(A - B) - C = (A - C) - (B - C)$
13. Let A, B, C be sets. Under what condition is each of the following statements true?
    (a) $(A - B) \cup (A - C) = A$ (b) $(A - B) \cup (A - C) = \phi$
    (c) $(A - B) \cap (A - C) = \phi$ (d) $(A - B) \oplus (A - C) = \phi$
14. Let A, B, C be two sets.
    (a) Given that $A - B = B$, what can be said about A and B?
    (b) Given that $A - B = B - A$, what can be said about A and B?

15. Let A denote the set of all automobiles that are manufactured domestically. Let B denote the set of all imported automobiles. Let C denote the set of all automobiles manufactured before 1977. Let D denote the set of all automobiles with a current market value of less than \$2000. Let E denote the set of all automobiles owned by student at the university. Express the following statements in set-theoretic (symbolic) notations:

    (a) The automobiles owned by students at the university are either domestically manufactured or imported.
    (b) All domestic automobiles manufactured before 1977 have a market value of less than \$2000.
    (c) All imported automobiles manufactured after 1977 have a market value of more than \$2000.

16. Let A denote the set of all freshmen, B denote the set of all sophomores, C denote the set of all mathematics majors, D denote the set of all computer science majors, E denote set of all students in the course Elements of Discrete Mathematics, F denote the set of all students who went to a rock concert on Monday night, G denote the set of all students who stayed up late Monday night. Express the following statements in set-theoretic notation:

    (a) All sophomores in computer science are in the Elements of Discrete Mathematics.
    (b) No student in the course Elements of Discrete Mathematics went to the rock concert Monday night. (The obvious reason is the long problem sets in the course Elements of Discrete Mathematics).
    (c) The rock concert was only for freshman and sophomores.
    (d) All sophomores who are neither mathematics nor computer science majors went to the rock concert.

17. Determine the power sets of the following sets.

    (a) $\{a\}$ (b) $\{\{a\}\}$ (c) $\{, \phi \ \{\phi\}\}$

18. Let $A = \{\phi, b\}$. Construct the following sets:

    (a) $A - \phi$ (b) $\{\phi\} - A$

    (c) $A \cup \, !P(A)$ (d) $A \cap P(A)$

19. Let $A = \{\phi\}$. Let $B = P(P(A))$

    (a) Is $\phi \in B$? $\phi \subseteq B$? (b) Is $\{\phi\} \in B$? $\{\in\} \subseteq B$?

    (c) Is $\{\{\subseteq\}\} \in B$? $\{\{\in\}\} \subseteq B$

20. Let $A = \{\phi, \{\phi\}\}$. Determine whether each of the following statements is true or false.

    (a) $\phi \in P(A)$ (b) $\phi \subseteq P(A)$

    (c) $\{\phi\} \subseteq P(A)$ (d) $\{\phi\} \subseteq A$

    (e) $\{\phi\} \in P(A)$ (f) $\{\phi\} \in A$

    (g) $\{\{\phi\}\} \subseteq P(A)$ (h) $\{\{\phi\}\} \subseteq A$

    (i) $\{\{\phi\}\} \in P(A)$ (j) $\{\{\phi\}\} \in A$

21. Let $A = \{a, \{a\}\}$. Determine whether each of the following statements is true or false.

    (a) $\phi \in P(A)$ (b) $\phi \subseteq P(A)$

    (c) $\{a\} \in P(A)$ (d) $\{a\} \subseteq P(A)$

    (e) $\{\{a\}\} \in P(A)$ (f) $\{\{a\}\} \subseteq P(A)$

    (g) $\{a, \{a\}\} \in P(A)$ (h) $\{a, \{a\}\} \subseteq P(A)$

    (i) $\{\{\{a\}\}\} \in P(A)$ (j) $\{\{\{a\}\}\} \subseteq P(A)$

22. Determine whether each of the following statements is true or false. Briefly explain your answer.

    (a) $A \cup P(A) = P(A)$ (b) $A \cap P(A) = A$

    (c) $\{A\} \cup P(A) = P(A)$ (d) $\{A\} \cap P(A) = A$

    (e) $A - P(A) = A$ (f) $P(A) - \{A\} = P(A)$

23. Determine A if the Ω is (a) positive integer, (b) the integer, (c) the real number, (d) the complex number

    (a) $A = \{x \mid x \text{ is of the form } 3y + 1 \text{ for } y \in u\}$.

    (b) $A = \{x \mid x \text{ is of the form } y - 4 \text{ for } y \in u\}$.

    (c) $A = \{x \mid x^2 + x - 2 = 0\}$.

    (d) $A = \{x \mid x^2 - x - 1 \leq 0\}$.

    (e) $A = \{x \mid x^3 - 2x^2 - 2x - 3 = 0\}$.

24. Write a program fragment (any language) to test whether x is an element of A, a set with n elements.
25. Assume that $U = \{a, b, ..., k\}$ and $A \subseteq U$. Write a program fragment to computer $\overline{A}$.
26. Assume that $U = \{1, 2, ..., 100\}$, and A and B are subsets of U. Write a program fragment to test whether $A \subseteq B$.
27. Assume that $U = \{1, 2, ..., 100\}$, and $A \subseteq U$. Write a program fragment to find all elements x such that $\{x, x + 1\} \subseteq A$.
28. A survey was conducted among 1000 people. Of these 595 are Democrats, 595 wear glasses, and 550 like ice cream; 395 of them are Democrats who wear glasses, 350 of them are Democrats who like ice cream, and 400 of them wear glasses and like ice cream; 250 of them are Democrats who wear glasses and like ice cream. How many of them who are not Democrats, do not wear glasses, and do not like ice cream? How many of them are Democrats who do not wear glasses, and do not like ice cream?
29. The 60, 000 fans who attended the homecoming football game bought up all the paraphernalia for their cars. Altogether, 20, 000 bumper stickers, 36, 000 window decals, and 12, 000 key rings were sold. We know that 52, 000 fans bought at least one item and no one bought more than one of a given item. Also, 6000 fans bought both decals and key rings, 9000 bought both decals and bumper stickers, and 5000 bought both key rings and bumper stickers.

    (a) How many fans bought all three items?

    (b) How many fans bought exactly one item?

    (c) Someone questioned the accuracy of the total number of purchases; 52, 000 (given that all the other numbers have been confirmed to be correct). This person claimed the total number of purchases to be either 60, 000 or 44, 000. How do you dispel the claim?

30. Out of a total of 130 students, 60 are wearing hats to class, 51 are wearing scarves, and 30 are wearing both hats and scarves. Of the 54 students who are wearing sweaters, 26 are wearing

hats, 21 are wearing scarves, and 12 are wearing both hats and scarves. Everyone wearing neither a hat nor a scarf is wearing gloves.

(a) How many students are wearing gloves?

(b) How many students not wearing a sweater are wearing hats but not scarves?

(c) How many students not wearing a sweater are wearing neither a hat nor a scarf?

31. Among 100 students, 32 study mathematics, 20 study physics, 45 study biology, 15 study mathematics and biology, 7 study mathematics and physics, 10 study physics and biology, and 30 do not study any of the three subjects.

    (a) Find the number of students studying all three subjects.

    (b) Find the number of students studying exactly one of the three subjects.

32. Seventy-five children went to an amusement park where they can ride on the merry-go-round, roller coaster, and Ferris wheel. It is known that 20 of them haves taken all three rides, and 55 of them have taken at least two of the three rides. Each ride costs $0.50, and the total receipt of the amusement park was $70. Determine the number of children who did not try any of the rides.

1.2 Relation

A relation is an *association* between two or more things. When a relation suggest a correspondence or an association between the elements of two sets it is a *binary* or *dyadic* relation. For three elements it is *ternary* or *triadic* and so on.

Most of the time, in this course, we shall be dealing with binary relation only. We have studied Cartesian product of the two sets. We also know what is an ordered pair. Now let us get down to the job of defining a relation.

Let A and B be any two non-empty sets. Then $A \times B$ is defined as:

$$A \times B = \{ (x, y) \mid x \in A \text{ and } y \in B \}$$

Let R be any subset of $A \times B$. Then R may contain zero or more (at the most all) ordered pairs of $A \times B$. This set R is called a relation from set A to set B. In general, we define a relation R from a set A to another set B as any subset of $A \times B$. If R is a subset of $A \times A$ i.e. R is a relation from set A to itself, then R is said to be a relation on A.

Example 1: Let $A = \{1, 2, 3\}$ and $B = \{x, y, z\}$ then

$A \times B = \{(1, x), (1, y), (1, z), (2, x), (2, y), (2, z), (3, x), (3, y), (3, z)\}$

We have defined a relation as any possible subset of $A \times B$. Thus $\{(1, x)\}$, $\{(1, x)\ (2, x)\}$ etc. are relations from A to B.

In fact, we have $2^9 = 512$ possible relations from set A to set B in this example. Verify this fact. In general, if $|A| = n$, $|B| = m$, then $|A \times B| = mn$, and so number of possible subsets of $A \times B = 2^{mn}$. This large number of possible relations from a set A to another set B forces us to define different types of relations according to the nature of ordered pairs it contains from $A \times B$.

1.2.1 Types of Relation

Let R be a relation from set A to set B. The product set $A \times B$ is the exhaustive list of possible ordered pairs. This set $A \times B$ will be used to define different relations R from A to B. Whenever we refer a relation R from set A to set B. A and B are non-empty sets.

A relation R from a set A to another set B is said to be *Universal Relation* if

$$R = A \times B$$

Example 1: Let $A = \{1, 2\}$, $B = \{a, b\}$ then $R = \{(1, a), (1, b), (2, a), (2, b)\}$ is a universal relation from set A to set B. Hence $R = A \times B$

A relation R is a universal relation on A of $R \Rightarrow A \times A$

Example 2: Let $A = \{x, y\}$ then $A \times A = \{(x, x), (x, y), (y, x)\ (y, y)\}$ is a universal relation on A.

Complement of a Relation

A relation R from a set A to another set B is called complement of another relation S from A to B if

$R = A \times B - S$ or, $S = A \times B - R$

So R and S are complement of each other. We can easily observe that if R and S are two relations from set A to set B, and if they are complements of each other. Then,

$R \cup S = A \times B$ and $R \cap S = \phi$

Example: Let $A = \{1, 2\}$ and $B = \{3, 4\}$ then $A \times B = \{(1, 3), (1, 4), (2, 3), (2, 4)\}$

Let $R = \{(1, 3), (2, 4)\}$, then its complement is $A \times B - R = \{(1, 4), (2, 3)\}$.

The complement of a Relation R is denoted as $\overline{R}$ or R^c or R'. Similarly, if R is a relation on a set A, then its complement R^c is given by $A \times A - R$.

Null, Void or Empty Relation

We know that any subset of $A \times B$ is a relation from set A to set B. We also know that ϕ is a subsets of every set. A relation R is called Null (or void or empty) relation from set A to set B if $R = f$.

Example: Let $A = \{x \mid x$ is name of males of a town$\}$
$B = \{y \mid y$ is name of females of that town$\}$

Let R be relation defined as "is sister of" from A to B.i.e. xRy (read as x is related to y) of $(x, y) \in R$ when $x \in A$ and $y \in B$. Since no male can be sister of any other person, $R = \phi$

If R is a relation on a set A, and $R = \phi$ then, R is a null relation on A.

Inverse Relation

Let R be a relation from set $A = \{0, 1\}$ to $B = \{F, T\}$ and is given as $R = \{(0, T), (1, F)\}$. then the inverse of R is given as $R^{-1} = \{(T, 0), (F, 1)\}$. To obtain R^{-1} from R we have just inverted the position of elements of ordered pairs of R i.e. $(0, T)$ to $(T, 0)$ and $(1, F)$ to $(F, 1)$. Further, we observe that $R^{-1} \subseteq B \times A$ i.e. R^{-1} is a relation from B to A whenever R is a relation from A to B. Thus if $R = \{ (x, y) \mid x \in A$ and $y \in B\}$ is a relation from A to B then inverse of R denoted as R^{-1} is a relation from B to A and is defined as $R^{-1} = \{(y, x) \mid y \in B$ and $x \in A\}$

In the same fashion, if R is a relation on a set A and $R = \{(a, b) \mid a, b \in A\}$ then its inverse $R^{-1} = \{(b, a) \mid a, b \in A\}$. R^{-1} is still a relation on A.

Now on-wards, we will be discussing relation defined on a set. In some books the following types of relations are placed under properties of relations, but in this book we shall continue to treat them as a type of relation.

We shall illustrate the following type of relation with set $A = \{1, 2, 3\}$ i.e. $A \times A = \{(1, 1), (1, 2), (1, 3), (2, 1), (2, 2), (2, 3), (3, 1), (3, 2), (3, 3)\}$.

Reflexive Relation

A relation R is said to be a *reflexive relation* on A if R contains atleast (x, x) for all $x \in A$ i.e $\forall\ x \in A,\ x\ R\ x$ (read as "for all x belonging to A x is related to x)

On the above set A, following relations are reflexive

$R_1 = \{(1, 1)\ (2, 2), (3, 3)\}$
$R_2 = \{(1, 1)\ (1, 2)\ (2, 2),$
$(3, 1), (3, 3)\}$ [Since $\forall\ x \in A, (x, x) \in R$]

the following relations are not reflexive.

$R_3 = \{(1, 1)\ (2, 3), (3, 3)\}$ [since for $b \in A$, $(2, 2) \notin R$]
$R_4 = \{(1, 1)\ (1, 2), (1, 3)\}$ [since $(2, 2), (3, 3) \notin R$]
$R_5 - \{(1, 2)\ (1, 3), (2, 3)\}$ [since b $(1, 1), (2, 2), (3, 3) \notin R$]

Irreflexive Relation

A relation R is said to be *irreflexive relation* on A if R does not contain any of the pair (x, x), i.e $\forall\ x \in A, (x, x) \notin A$ or $\forall\ x \in A$, $x \sim R\, x$ (read as "for all x belonging to A, x is not related to x). In the above example only R_5 is an irreflexive relation, because none of $(1, 1)$, $(2, 2)$ and $(3, 3)$ is an R_5.

Non-reflexive Relation

A relation R is called *non-reflexive* on A if R is neither reflexive nor irreflexive i.e. for some $x \in A$, $x\ R\ x$ and for some $x \in A$, $x \sim R\, x$. In the above examples R_3 and R_4 are non-reflexive.

Symmetric Relation

A relation R is called *symmetric relation* on A. If for $x, y \in A$ and $x \neq y$, if $(x, y) \in R$ then (y, x) also $\in R$. i.e

$(x, y) \in R \Rightarrow (y, x) \in R$

In the previous examples, R_1 is symmetric relation by default, since there is no pair (x, y) in R such that $x \neq y$, and hence there is no need to look for (y, x) in R.

R_2 is not symmetric because $(1, 2) \in R_2$ but $(2, 1)$ is not in R_2. Similarly R_3, R_4 and R_5 are not symmetric, verify it.

Another examples of a symmetric relation on A can be given as:

$R_6 = \{(1, 2), (2, 1), (3, 1), (1, 3)\}$
$R_7 = \{(1, 2), (2, 1), (1, 1)\}$

Asymmetric Relation

A relation R on a set A is called *asymmetric* if presence of a pair (x, y) in R, for $x, y \in A$ and $x \neq y$, excludes the presence of (y, x) in R. i.e. for $x, y \in A$ and $x \neq y$, if xRy then $y \sim R\, x$

Amongst R_1 to R_7 given as above, R_2, R_3, R_4, R_5 are examples of asymmetric relation.

Anti-Symmetric Relation

A relation R on a set A is said to be *anti-symmetric* if for $x, y \in A$ and $x \neq y$,

$(x, y) \in R$ and $(y, x) \in R \Leftrightarrow x = y$
i.e. if $x \neq y \Rightarrow$ either $x \sim R\, y$ or $y \sim R\, x$ or both

In the previous examples R_1, R_2, R_3, R_4, R_5 are all anti-symmetric relation. Observe that a relation may be symmetric as well as anti-symmetric at the same-time (not always). Here R_1 is both symmetric and anti-symmetric.

Transitive Relation

A relation R on a set A is called *transitive* if $x, y, z \in A$, and pair (x, y) & (y, z) both are in R then (x, z) must be in R. i.e. if (x, y) & $(y, z) \in R$ then $(x, z) \in R$

i.e. xRy & $yRz \Rightarrow xRz$

In the above examples R_1 is transitive by default, since there are no (x, y) & (y, z) in R so we need not look for (x, z) in R. Relation R_2 is also transitive. Verify that R_3 is also transitive. What about R_4 and R_5?

Relation R_6 is not transitive, because (1, 2) & (2, 1) are in R_6 but $(1, 1) \notin R$. What about R_7?

Intransitive Relation

A relation R defined on a set A is said to be *intransitive* if presence of pairs (x, y) & (y, z) together in R excludes the possibilities of $(x, z) \in R$. i.e.

If $(x, y) \in R$ and $(y, z) \in R$ then $(x, z) \notin R$

In the above examples R_1 is intransitive by default, since there are no (x, y) and (y, z) in R, so we need not check for $(x, z) \neq R$. Relation R_6 is intransitive relation on A. Verify about other relations given previously (e.g. R_2, R_3, R_4...).

Non-transitive Relation

A relation R defined on a set A is called *non-transitive*, if R is neither transitive nor in-transitive relation. Does a relation of this type exist?

Now we shall discuss a few examples to explain "how to show that a given relation is of a particular type".

Example 1: Let Z be a set of all integers. Let R be a relation on Z defined as "equal to". Discuss the nature of R on Z.

Solution: Given that R is a relation on Z, thus $R \subseteq Z \times Z$. Here, R contains only those pairs (x, y) of $Z \times Z$, in which $x = y$ (x is equal to y).

Thus, $R = \{(x, x) \mid x \in Z\}$. Here,

- R is not a universal relation on Z.
- R is not a NULL relation on Z.
- R is inverse of itself i.e. $R = R^{-1}$.
- Complement of R i.e. $R^c = \{(x, y) \mid x \neq y \text{ \& } (x, y)\ R\}$
- R is a reflexive relation, since every integer is equal to itself.
- R is not irreflexive, also not non-reflexive.

- R is symmetric, anti-symmetric & asymmetric by default.
- R is transitive and intransitive by default.
- R is not non-transitive.

Example 2: Let H be the set of all human beings on earth, and R be a relation defined on H as "is brother of". Discuss the nature of R on H.

Solution: Here again $R \subseteq H \times H$. R contains only those pairs (x, y) such that x & y are humans and x is brother of y. Note that there may be many human who are not brother of any humans (females).

- Since $R \neq H \times H$, R is not a universal relation.
- Male is a brother of himself, R is not empty and hence not a null relation.
- Next, if x is brother of y, then it is not necessary that y is also a brother of x [y may be female] thus, $R^{-1} = \{(x, y) \mid y$ is brother of $x\}$
- $R^c = H \times H - R$
  = collection of all pairs (x, y) of $H \times H$, in which x is not a brother of y.
- R is not reflexive, because $\exists\, x \in H$ such that $(x, x) \notin R$ i.e. for every human x, x is not brother of x, (x, may be female). Similarly R is not irreflexive, as $\exists\, x$ such that $(x, x) \in R$ [when x is a male]. Thus R is non-reflexive relation.
- R is not a symmetric relation, because whenever $(x, y) \in R$ i.e. x is brother of y, it is not necessary that y is also a brother of x (y may be sister of x) and hence (y, x) may not be in R. Similarly if $(x, y) \in R$ and $(y, x) \in R$ then x may not be equal to y always R is not an anti-symmetric relation. R is not asymmetric relation also, as x is brother of y does not rule out the possibility of y being brother of x. This relation is a perfect example of a non-symmetric relation.
- R is a transitive relation and hence its not non-transitive and not intransitive relation.

Example 3: Let Z_+ be the set of all positive integers and R be a relation defined on Z_+ as xRy for $x, y \in Z_+$ if $x = 2y$. Find what type of relation it is?

Solution: Here again $R \subseteq Z_+ \times Z_+$ and $R = \{ (x, y) \mid x = 2y \text{ and } x, y \in Z_+ \}$ a few examples of R are (1, 2), (2, 4), (3, 6), (4, 8) etc. It is obvious that R is neither universal nor null relation. $R^{-1} = \{(x, y) \mid 2x = y \text{ and } x, y \in Z_+\}$ i.e. elements of R^{-1} are (2, 1), (4, 2), (6, 3)........... $R^c = \{(x, y) \mid x \neq 2y \text{ and } x, y \in Z_+\}$ i.e. elements of R^c are (1, 1), (1, 3), (1, 4)...(2, 2), (2, 1), (2, 3).... Verify that R is not reflexive, it is irreflexive and not non-reflexive relation. [No integer is equal to twice of itself except zero, and zero $\notin Z_+$]

This relation R is not symmetric, it is asymmetric and anti-symmetric relation. Verify that, this relation is an intransition relation. It is neither transitive nor non-transitive.

1.2.2 Equivalence Relation

The concept of relation, specially equivalence relation and order relation (to be discussed in chapter 5), is very important so far as understanding of different topics of discrete mathematical structure are concerned. Therefore readers are advised to read this chapter well.

"Any relation R defined on a non-empty set A is said to be an equivalence relation if

- R is reflexive i.e. $\forall\ x \in A, (x, x) \in R$
- R is symmetric i.e. whenever $(x, y) \in R$, there is $(y, x) \in R$
- R is transitive i.e. if (x, y) & $(y, z) \in R$, then $(x, z) \in R$".

i.e. for any relation R to be an equivalence relation, it must be reflexive, symmetric and transitive. Read the following examples to know the way to prove that a given relation is an equivalence relation or not.

Example 1: Let A be the set of all triangles in a plane and let R be the relation on A defined as $R = \{(a, b) \mid a, b \in A$ and A is congruent to B prove that R is an equivalence relation on A.

Solution: To prove that whether a given relation R is an equivalence relation or not, we shall simply test that whether R is reflexive, symmetric and transitive or not.

- *Reflexivity:* Since every triangle $x \in A$ is congruent to itself, we can write $(x, x) \in R$, $\forall\ x \in A$. Thus R is reflexive.
- *Symmetry:* Let x and y be triangles of A and x is congruent to y so $(x, y) \in R$. Since x is congruent to y, y is also congruent

to x, so $(y, x) \in R$, i.e., whenever $(x, y) \in R$. Thus R is symmetric.

- *Transitivity*: Let x, y, z be any three triangles of A, and $(x, y) \in R$ and $(y, z) \in R$, i.e. x is congruent to y and y is congruent to z. Therefore x is congruent to z i.e., $(x, z) \in R$. Hence R is transitive.

Therefore, given relation R is an equivalence relation on set A.

Example 2: Let Z be the set of all integers and R be a relation defined on Z such that for $a, b \in Z$, $a\ R\ b$ iff $ab \geq 0$. Prove that R is an equivalence relation on Z.

Solution: (i) *Reflexivity:* Let $x \in Z$, then either or $x < 0$ or $x = 0$ or $x > 0$. In all these cases $x.x \geq 0$ i.e., $x\ R\ x\ \forall\ x \in Z$. R is reflexive.

(ii) *Symmetry:* Let x and y be any two integers. $x\ R\ y$ i.e. $x\ y$ then $yx \geq 0$, i.e. $y\ R\ x$ i.e. whenever $x\ R\ y$ we have $y\ R\ x$, i.e. if $(x, y) \in R$ then (y, x) is also in R. Thus R is symmetric.

(iii) *Transitivity:* Let x, y, z be any three elements of Z. s.t. $x\ R\ y$ and $y\ R\ z$ i.e. (x, y) and $(y, z) \in R$. Since (x, y) & $(y, z) \in R$, we have $xy \geq 0$ & $yz \geq 0$. This implies that

either $x = y = 0$ then $z = 0$ or < 0 or > 0
or $x < 0$ &$y < 0$ then $z = 0$ or $z < 0$ [$z > 0$]
or $x > 0$ &$y > 0$ then $z = 0$ or $z > 0$ [$z < 0$]

In all the tree cases listed above $xz \geq 0$, i.e., $x\ R\ z$. Therefore R is transitive. Since R is reflexive, symmetric and transitive. R is an equivalence relation.

Example 3: Let Z be set of integers and Let R be a relation defined on Z as: for $a, b, \in z$, $a\ R\ b$ iff a $\equiv_3$ b. Prove that R is an equivalence relation on Z.

Solution: Here $\equiv_3$ is read as congruence modulo 3. An a is said to be "congruence modulo 3" b, if a & b yield the same remainder when they are divided by 3. e.g. 1 and 4 are congruence modulo 3, since both yields the remainder of 1 when divided by 3. Similarly $2 \equiv_3 5$; $2 \equiv_3 8$; $2 \equiv_3 11$, $8 \equiv_3 11$ etc and $1 \equiv_3 4$, $1 \equiv_3 7$, $3 \equiv_3 6$ etc. We also know that zero is divisible by every number except by zero itself.

Now let us consider what remainder we will get, when a negative number is divided by a positive number. Readers may ask that why am I giving emphasis on this?

In most of the cases, I have noticed that students do not know the fact that remainder can never be negative i.e. if a number is divided by m, then remainder must be between 0 to $m - 1$. Thus when -5 is divided by 6, the remainder will be 1 and quotient will be -1. Similarly when -1 is divided by 3, the remainder will be 2 and quotient will be -1. Thus $-1 \equiv_3 2$. (and not $-1 \equiv_3 1$ as commonly perceived by many). Now let us solve the actual problem of proving that $\equiv_3$ is an equivalence relation on Z.

- *Reflexivity:* Let $x \in Z$, then $x \equiv_3 x$, since both yields the same remainder when divided by 3. Thus, $\forall\ x \in Z$, $x\ R\ x$ i.e $(x, x) \in R$. R is reflexive.
- *Symmetry:* Let x & $y \in Z$, and $x \equiv_3 y$, then x & y both yields same remainder when divided by 3. Then y and x also yields same remainder when divided by 3. i.e. $y \equiv_3 x$ i.e. $(y, x) \in R$. Whenever $(x, y) \in R$ we have $(y, x) \in R$, R is a symmetric relation.
- *Transitivity:* Let $x, y, z \in Z$, and $x \equiv_3 y$, $y \equiv_3 z$. Since x and y yields same remainder when divided by 3, then difference $x - y$ is divisible by 3. Similarly $y - z$ is divisible by 3. Whenever two numbers are divisible by a number, say m, their sum (as well as their difference) is divisible by that number m. Hence $(x - y) + (y - z)$ is divisible by 3, i.e $x - z$ is divisible by 3. Thus $x \equiv_3 z$ i.e. if (x, y) & $(y, z) \in R$ then $(x, z) \in R$. R is transitive also.
- Hence R $(\equiv_3)$ is an equivalence relation on Z.

In the remaining part of this section we will discuss "*partition of set*" and "*equivalence partition*". But, before discussing these topics, let us discuss some basic things regarding set arising from relations.

Let R be a relation from set $A = \{1, 2, 3, 4, 5, 6\}$ to set $B = \{x, y, z, w\}$ and is defined as

$$R = \{(1, x,), (5, y,), (3, z), (1, w), (2, x)\}$$

A relation has domain and a range set. In ordered pair $(x, y) \in R$, x comes from domain set of a R and y comes from range set of R. In the example of our consideration, *Domain* of relation R is $\{1, 2, 3, 5\}$ and Range of relation R is $\{x, y, z, w\}$.

Thus Domain of a relation R from set A to set B is defined as

Domain $(R) = \{x \mid (x, y) \in R$ for some $y \in B\} \times A$

Similarly range of a relation R from set A to set B is defined as

Ranges $(R) = \{y \mid (x, y) \in R$ for some $x \in A\} \times B$

Next to be discussed is R-relative set of an element x of A and R-relative set of a subset A_1 of A.

In the example of our consideration, the R-relative set if $1 \in A$ is $\{x, w\}$ i.e. $R(1) = \{x, w)$, similarly $R(2) = \{x\}$, $R(4) = \varphi$. The R-relative set of an element $x \in A$ is defined as:

$R(x) = \{y \mid y \in B$ and $xRy\}$

In the same way, if $A_1 = \{1, 2, 3, 4, 5\} \subseteq A$, then R-relative set of A_1 denoted by $R(A_1)$ is

$R = \{x, w, y\}$

i.e. $R(A_1) = \{y \mid (x, y) \in R$ for some $x \in A_1\}$

Example 1: Let $A = \{a, b, c, d)$ and let R be a relation defined on A in such a way that $R = \{(a, a), (b, b), (a, b), (b, c), (c, a), (d, c), (c, b)\}$, then find $R(a)$, $R(b)$. If $A_1 = \{c, d)$, find $R(A_1)$

Solution: Readers are advised to try as explained in the example above.

Now let us discuss some important theorems on R-relative sets.

Theorem 1: Let R and S be relations from A to B. If $R(a) = S(a)$ $\forall\ a \in A$. Then $R = S$.

Proof: Here, we have to prove that every ordered pairs of R are also in S, i.e. two sets R & S are equal.

Let (a, b) be any element of R then $a R\ b \Rightarrow b \in R(a)$.

$$aRb \Rightarrow b \in R(a)$$

$$\Rightarrow b \in S(a)$$

$$\Rightarrow aSb \qquad [\text{since } R(a) = S(a)]$$

$$\Rightarrow (a, b) \in s$$

$$\therefore R \subseteq S \text{ ———————— } (1)$$

Similarly, let $(a, b) \in S$ then $a\ S\ b$

$$\Rightarrow b \in S\ (a)$$

$$\Rightarrow b \in R\ (a) \qquad [\text{since } R(a) = S(a)]$$

$$\Rightarrow aRb$$
$$\Rightarrow (a, b) \in R$$
$$\therefore S \subseteq R \text{ ------------ } (2)$$

From 1 and 2, we have $S = R$

Proved.

Theorem 2: Let R be the relation from A to B and Let A_1 and A_2 be subsets of A then

- $A_1 \subseteq A_2 \Rightarrow R(A_1) \subseteq R(A_2)$
- $R(A_1 \cup A_2) = R(A_1) \cup R(A_2)$

Proof: (i) Let y be any element of $R(A_1)$, then $\exists\, x \in A_1$ st. xRy

$$\Rightarrow y \in R(A_1) \qquad [\text{Since } A_1 \subseteq A_2, \text{ so } x \in A_1 \Rightarrow x \in A_2]$$
$$\Rightarrow y \in R(A_2)$$
$$\therefore R(A_1) \subseteq R(A_2)$$

Proved.

(ii) From part 1 we can conclude that

$$R(A_1) \subseteq R(A_1 \cup A_2) \qquad [\text{Since } A_1 \subseteq A_2 \cup A_2]$$
$$\text{and } R(A_2) \subseteq R(A_1 \cup A_2) \qquad [\text{Since } A_2 \subseteq A_1 \cup A_2]$$

combining these two we get

$$R(A_1) \cup R(A_2) \subseteq R(A_1 \cup A_2) \text{ ------------ } (1)$$

$$[\text{Since } A \subseteq C \ \&\ B \subseteq C \Rightarrow A \cup B \subseteq C]$$

Now to show the equality, we have to prove that $R(A_1 \cup A_2) \subseteq R(A_1) \cup R(A_2)$

Let y be any element of $R(A_1 \cup A_2)$, then

$$\exists\, x \in A_1 \cup A_2 \text{ s.t. } xRy$$
$$\Rightarrow x \in A_1 \text{ or } x \in A_2 \text{ s.t. } xRy$$
$$\Rightarrow y \in R(A_1) \text{ or } y \in R(A_2)$$
$$\Rightarrow y \in R(A_1) \cup R(A_2)$$
$$\therefore R(A_1 \cup A_2) \subseteq R(A_1) \cup R(A_2) \text{ ------------ } (2)$$

Combining 1 and 2 we get $R(A_1 \cup A_2) = R(A_1) \cup R(A_2)$

Proved.

Theorem 3: Let R be a relation from A to B, then for all subsets A_1 and A_2 of A

$R(A_1 \cup A_2) \subseteq R(A_1) \cap R(A_2)$ iff $R(a) \cap R(b) = \phi$ for any $a, b \in A$ and $a \neq b$

Proof: We shall prove the theorem in two parts:

Part 1 when $A_1 \cap A_2 = \phi$

Part 2 when $A_1 \cap A_2 \neq \phi$

Part 1: Let a & b be any two elements of A and $a \neq b$. Since $A_1 \cap A_2 = \phi$, $R(A_1 \cap A_2) = \phi$ also, since R-relative set of null set is a null set. Now, to show the equality we have to prove

$$R(A_1) \cup R(A_2) = \phi$$

Without loss of generality, let us suppose that a be any element of A_1 & b be any element of A_2. Then $R(A_1) \cup R(A_2) = \phi$ if there is no common elements in R-relative set of a and R-relative set of b.

$$\Leftrightarrow R(a) \cap R(b) = \phi \ \forall \ a, b \in A \ \& \ a \neq b$$

$$\therefore R(A_1 \cap A_2) = R(A_1) \cap R(A_2) \text{ iff } R(a) \cap R(b) = \phi$$

Part 2: When $A_1 \cap A_2 \neq \phi$

Let $y \in R(A_1 \cap A_2)$,

$$\Rightarrow \exists\, x \in A_1 \cap A_2 \text{ s.t. } xRy$$

$$\Rightarrow \exists\, x \in A_1 \ \& \ x \in A_2 \text{ s.t. } xRy$$

$$\Rightarrow y \in R(A_1) \ \& \ y \in R(A_2)$$

$$\Rightarrow y \in R(A_1) \cap R(A_2)$$

$$\therefore R(A_1 \cap A_2) \subseteq R(A_1) \cap R(A_2) \text{ ———————— (1)}$$

Next Let $y \in R(A_1) \cap R(A_2)$

$$\Rightarrow y \in R(A_1) \ \& \ y \in R(A_2)$$

$$\Rightarrow \exists\, x \in A_1 \text{ s.t. } xRy \ \& \ \exists\, z \in A_2 \text{ s.t. } zRy$$

$\Rightarrow x$ must be equal to z [Since $R(x) \cap R(z) = \phi$ *for* $x \neq z$]

$$\Rightarrow \exists\, x \in A_1 \text{ s.t. } xRy \ \& \ \exists\, x \in A_2 \text{ s.t. } xRy$$

$$\Rightarrow \exists\, x \in A_1 \cap A_2 \text{ s.t. } xRy$$

$$\Rightarrow y \in R(A_1 \cap A_2)$$

$R(A_1) \cap R(A_2) \subseteq R(A_1 \cap A_2)$ ——————————— (2)

Combining (1) and (2) we have

$R(A_1 \cap A_2) = R(A_1) \cap R(A_2)$ iff $R(a) \cap R(b) = \phi$ for $a, b \in A$ & $a \neq b$ Proved

Special case: When $a, b \in A$ and $a \neq b$ $R(a) \cap R(b) \neq \phi$

$R(A_1 \cap A_2) \subseteq R(A_1) \cap R(A_2)$.

Example 1: Let $A = \{1, 2, 3\}$, $B = \{x, y, z, w, p, q\}$ and R from A to B is given as $R = \{(1, x), (1, z), (2, w), (2, p), (2, q), (3, y)\}$. For all possible subsets A_1 & A_2 of A prove or disprove that $R(A_1 \cap A_2) - R(A_1) \cap R(A_2)$

Solution: Readers are advised to try on their own.

Partitions of a Set

Let us see the following diagram of a set A and its divisions.

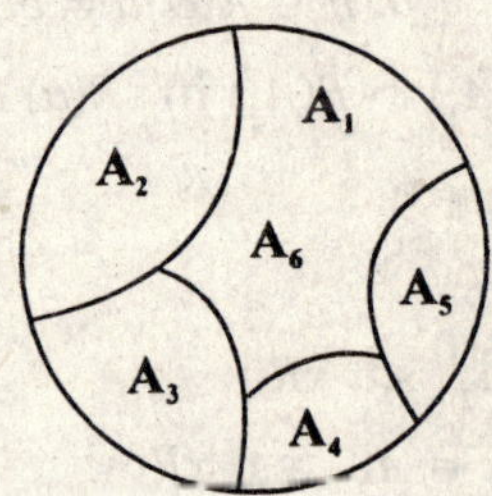

Let A be a set and A_1, A_2, A_3, A_4, A_5, and A_6 be subsets of A. From observation of the diagram, it is clear that

(i) $\bigcup_{i=1}^{6} = A$ and

(ii) $A_i \cap A_j = \phi$ for $i \neq j$

"A partition of a non empty set A is defined as a collection of non-empty subsets A_i of A such that

(i) Union of all such subsets A_i is equal to A i.e. $\bigcup_i A_i = A$, and
(ii) All such subsets A_i are mutually disjoint i.e. $A_i \cap A_j = \varphi$ for $i \neq j$

A *partition* is represented as P. Each subsets of A in P is called *blocks* or *cells*. A partition set is also called *Quotient set.*

Example 1: Let $A = \{1, 2, 3\}$, List all partitions of A.

Solution: Let us first list all possible subsets of A. These are: ϕ, $\{1\}$, $\{2\}$, $\{3\}$, $\{1, 2\}$, $\{1, 3\}$, $\{2, 3\}$, $\{1, 2, 3\}$. Since, we have to consider only non-empty subsets, we are not considering ϕ at all. Now let us construct various possible partitions of A.

1. $\{\{1\}, \{2\}, \{3\}\}$ 2. $\{\{1, 2, 3\}\}$ 3. $\{\{1\}, \{2, 3\}\}$
4. $\{\{2\}, \{1, 3\}\}$ 5. $\{\{3\}, \{1, 2\}\}$

Verify that, there exists no other partition for A.

Example 2: Let Z be a set of all integers, A_1 be the set of all positive integers and A_2 be the set of all negative integers including Zero, then

$\{A_1, A_2\}$ is a partition of Z.

Example 3: Let $A = \{a, b, c, d, e, f, g, h\}$, $A_1 = \{a, b, c, d\}$, $A_2 = \{b, d\}$, $A_3 = \{a, c, e, f, g, h\}$, $A_4 = \{a, c, e, g\}$, $A_5 = \{f, h\}$, then $\{A_2, A_4, A_5\}$ is a partition of A. $\{A_1, A_3\}$ is not a partition as $A_1 \cap A_3 \neq \phi$.

Equivalence Partition

This is another basic concept required to understand many discrete mathematical structures which are to follow in subsequent chapters. Thus, let us understand this with the help of an example.

Let $A = \{1, 2, 3, 4\}$ and R be a relation defined on A as $R = \{(1, 1), (2, 2), (2, 3), (4, 4), (1, 2), (2, 1), (3, 4), (4, 3)\}$.

By simple observation we can find that R is an equivalence relation on A. Also $R(1) = \{1, 2\} = R(2)$, $R(3) = \{3, 4) = R(4)$ and $\{\{1, 2\}, \{3, 4\}\}$ is a partition of A.

This partition $\{\{1,2\}, \{3,4\}\}$ of A has been obtained by grouping related elements of A, w.r.t R, in one subset. The subset $\{1, 2\}$ & $\{3, 4\}$ are equivalence classes of A. thus, $\{1, 2\}$ is equivalence class of elements 1 and 2 of A. Similarly $\{3, 4\}$ is equivalences class of elements 3 and 4 of A. Now, an equivalence class can be defined as:

"If A be a non-empty set and R be an *equivalence relation* on A, then *equivalence class* of any element $a \in A$–denoted on $[a]$ or $R(a)$ is collection of all elements $x \in A$ such that $a\,R\,x$ i.e. $[a] = \{x \mid x \in A$ and $a\,R\,x\}$

Having discussed these, we can now define an equivalence partition if a non-empty set as:

"Let A be a non-empty set, and R be an equivalences relation defined on A, then equivalence partition of A by R denoted as A/R, is defined as collection of equivalence classes $A_1, A_2 ..., A_n$ such that

1. $\bigcup_{i=1}^{n} A_i = A,$
2. $A_i \cap A_j = \phi$ for $i \neq j$,
3. If $x, y \in A_i$ then $x \, R \, y$,
4. If $x \in A_i$, & $y \in A_j$, $i \neq j$ then $x \sim R \, y$ and
5. $|A_i| \neq 0$ $1 \leq i \leq n$ i.e. All A_i's are non-empty.

Example 1: Find the equivalence partition of Z- the set of all integers by the equivalence relation $\equiv_m$ on Z.

Solution: We know that $\equiv_m$ stands for "*congruence modulo m'*. Two integers a and b are congruence modulo m if they yield same remainder when divided by m. In other words a & b are congruence modulo m if $a - b$ is divisible by m. Now total possible remainders when any numbers are divided by m are 0, 1, 2, ...m – 1. Let us group all the elements which yield remainder 0 in [0], which yield remainder 1 in [1] and so on up to $[m - 1]$

$[0] = \{....-3m, -2m, -m, 0, m, 2m, 3m...\}$

$[1] = \{...-3m + 1, -2m + 1, -m + 1, 1, m + 1, 2m + 1, 3m + 1...\}$

$[2] = \{.....-3m + 2, -2m + 2, -m + 2, 2, m + 2, 2m + 2, ...\}$

$[10] = \{...-3m + 10, -2m + 10, -m + 10, 10, m + 10, 2m + 10, ...\}$

$[m - 1] = \{...-3m + m - 1, -2m + m - 1, -m + m - 1, m - 1, m + m - 1, ...\}$

$= \{.....-2m - 1, -m - 1, -1, m - 1, 2m - 1, ...\}$

Here, {[0], [1], ..., $\{m - 1]\}$ is an equivalence partition of Z by $\equiv_m$, because

- $[0] \cup [1] \cup ... \cup [m - 1] = Z$
- $[i] \cap [j] = \phi$ for $i \neq j$ and $0 \leq i \leq m-1$ & $0 \leq i \leq m - 1$
- If $x, y \in [i]$ then $x \equiv_m y$
- If $x \in [i]$ and $y \in [j]$ then $x \equiv_m y$, and
- $|\,[i]\,| \neq 0$ for $0 \leq i \leq m - 1$

Example 2: Show that aRb iff $R(a) = R(b)$, where R is an equivalence relation on A and a, b are any two elements of A.

Solution: Let $R(a) = R(b)$

Since R is a reflexive, $b \in R(b)$

$\Rightarrow b \in R(b)$ [Since $R(a) = R(b)$]

$\Rightarrow aRb$

i.e., $a R b$ if $R(a) = R(b)$ —————— (1)

Conversely suppose that aRb, then

$b \in R(a)$, since R is an equivalence relation so both a & b belongs to same equivalence class which is equal to $R(a)$ or $R(b)$

$\therefore R(a) = R(b)$

i.e $R(a) = R(b)$ if $a R b$ —————— (2)

Combining 1 and 2 we have

$$a R b \text{ iff } R(a) = R(b)$$

Proved.

Theorem: (i) Let A be a non-empty set and R be an equivalence relation defined on A, Then show that R always induces an equivalence partition of

(ii) Let A be a non-empty set and P be a partition of A. Let R be a relation defined on A as xRy if x & y belongs to same blocks of A, then show that R is an equivalence relation.

Proof: Let us make blocks of elements of A as follows:

Step 1: Choose any element a of A and find $R(a)$

Step 2: If $R(a) = A$, its over, other-wise, choose another element $b \in A$ such that $b \notin R(a)$ and find $R(b)$.

Step 3: If A is not the union of previously obtained equivalence classes, then choose $a\ x \in A$ such that x is left out and not included in any of the previously obtained equivalence classes, find $R(x)$.

Step 4: Repeat steps 3 untill every elements of A has been included in some classes.

Now collection of these equivalence classes is a partition of A induced by equivalence relation R.

This proves part (i)

(ii) *Reflexivity:* Let A be any element of A, then A belongs to same block as itself so $a\ R\ a\ \forall\ a \in A$. R is a reflexive relation.

Symmetry: Let a & b be two elements of A such that $a\ R\ b$, then a and b both belongs to same block, by definition of R i.e., $b\ R\ a$. Hence R is a symmetric relation.

Transitivity: Let a, b, c be any three elements of A such that $a\ R\ b$, $b\ R\ c$ then all a, b, c belong to same block, thus $a\ R\ c$. Hence R is a transitive relation.

$\therefore$ R is an equivalence relation.

This proves part (ii)

1.2.3 Diagraph and Matrix Representation of Relation

In computer science, we always deal with finite elements, because there is no way to deal with infinite amount of information. Since a set may be infinite or finite, a relation can also be finite or infinite according to the number of ordered pairs it contains. If a Relation is defined on a finite set, it can be represented in *diagraph form* or in *matrix form.*

First we shall be discussing the diagraph representation of a relation. We shall be discussing about graph in detail in *chapter 7.* For now only a brief description of what a graph is all about. A graph contains a set of points (also called nodes or vertices) and a set of arcs (also called edges) connecting a pair of points. Now let us consider an example to know how to draw a graph for a given relation.

Example 1: Let $A = \{1, 2, 3, 4\}$ and R be a relation on A given as $R = \{ (1, 1), (1, 2), (2, 1), (2, 2), (2, 3), (2, 4), (3, 4), (4, 1) \}$. Consider A as set of vertices, and R as set of directed arcs. This means $(1, 2) \in R \Rightarrow$ there is an arc from vertex 1 to 2. Thus diagraph for R can be drawn as

$$M_R = \begin{matrix} 1 \\ 2 \\ 3 \\ 4 \end{matrix}\begin{bmatrix} 1 & 1 & 0 & 0 \\ 1 & 1 & 1 & 1 \\ 0 & 0 & 0 & 1 \\ 1 & 0 & 0 & 0 \end{bmatrix}_{4\times 4}$$

Fig. 1.2.3.1

Example 2: Let $A = \{a, b, c, d\}$, $B = \{1, 2, 3\}$ $R = \{(a, 1), (a, 2), (b, 1), (c, 2), (d, 1)\}$. Draw a diagraph for R.

Solution: Here, we have two sets of vertices and R is a set of arcs from one set of vertices to other. We draw a graph as below:

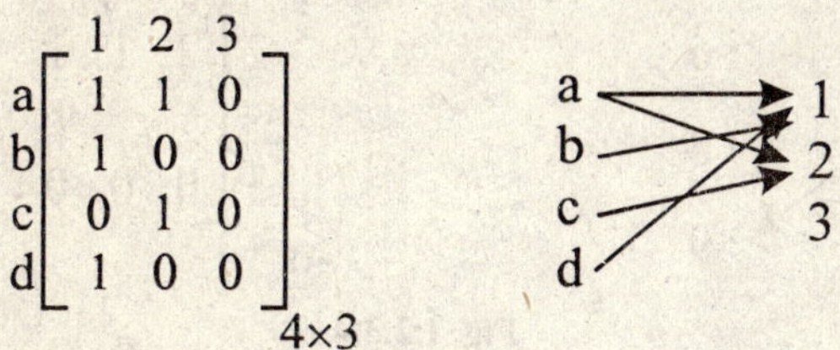

Fig. 1.2.3.2

This completes the answer.

Example 3: Find the set A and relation R defined on A from the following diagraph.

Solution: Here A is set of vertices i.e. $\{1, 2, 3, 4\}$, and R is set of arcs i.e.

$R = \{(1, 1), (1, 3), (2, 3), (3, 3), (3, 2), (4, 3)\}$

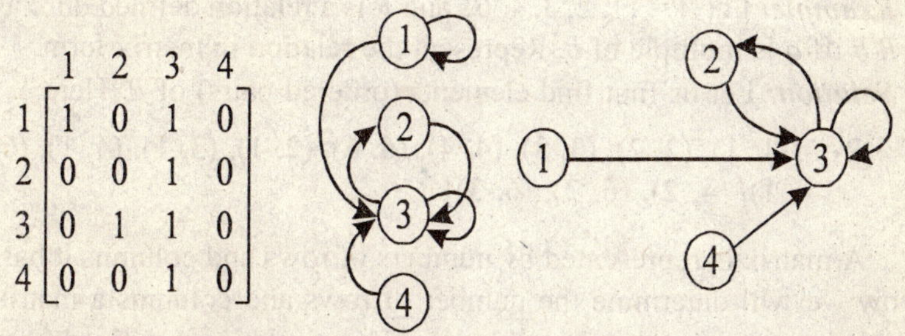

Fig. 1.2.3.3

This completes the answer.

Example 4: Let $A = \{1, 2, 3, 4\}$ and $B = \{1, 4, 6, 8, 9\}$ and R be a relation from A to be defined as $a\ R\ b$ iff $b = a^2$

Solution: The relation $R = \{(1, 1), (2, 4), (3, 9)\}$, Thus diagraph can be drawn as

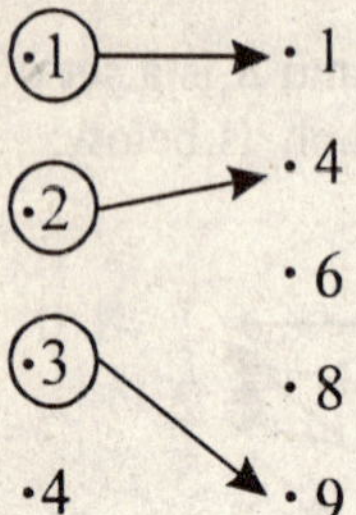

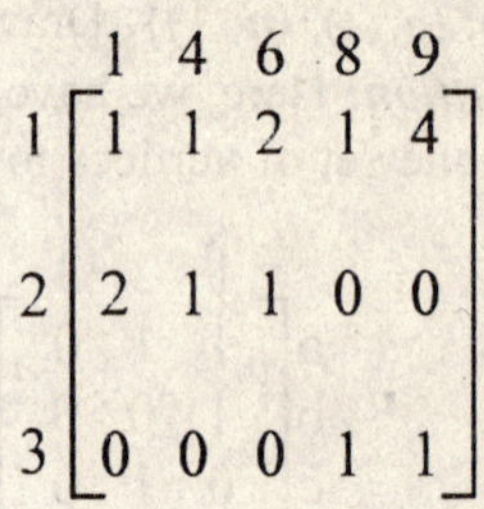

Fig. 1.2.3.4

This Completes the answer.

Let us discuss now, the *matrix representation* of a given relation. A matrix is a rectangular arrangement of set of elements in columns & rows form. Each element is referred as a_{ij} of ith row and jth column. Readers are advised to go through atleast once the basic matrix operations and different types of matrices. Teachers are suggested to spend one class to review the basics of matrices. Here, again we shall begin with an example.

Example: Let $A = \{1, 2, 3, 4, 6\}$ and R is a relation defined on A as $a\ R\ b$ iff a is multiple of b. Represent the relation in matrix form.

Solution: Let us first find elements (ordered pairs) of R. Here

$R = \{(1, 1), (2, 2), (3, 3), (4, 4), (6, 6), (2, 1), (3, 1), (4, 1), (6, 1), (4, 2), (6, 2), (6, 3)\}$

A matrix is represented by numbers of rows and columns it has. Now we will determine the number of rows and columns a matrix needed to represent the given relation. Since a relation is defined from a sets A to another set B. i.e. $R: A \rightarrow B$, $|A|$ determines number of rows and $|B|$ determines number of columns of the matrix needed to represent the R. If M is the matrix then its order will be $|A| \times |B|$. Having determined number of rows & columns, we have to find elements of the matrix. Only thing required to be represented at is the presence of association between $x \in A$ and $y \in B$, i.e. whether pair (x, y) is in R or not. Thus two symbols 0 and 1 will suffice. If a_{ij} is the element of M then

$$a_{ij} = \begin{cases} 0 \; if\,(i,j) \notin \text{R} \\ 1 \; if\,(i,j) \notin \text{R} \end{cases}$$

Here, elements of A has been assumed to the numbered form 1 to $|A|$ and is referred as i, and that of B from 1 to $|B|$ and is referred as j. Now let us represent R.

$$M_R = \begin{array}{c} \\ 1 \\ 2 \\ 3 \\ 4 \\ 6 \end{array} \begin{array}{c} \begin{array}{ccccc} 1 & 2 & 3 & 4 & 6 \end{array} \\ \begin{bmatrix} 1 & 0 & 0 & 0 & 0 \\ 1 & 1 & 0 & 0 & 0 \\ 1 & 0 & 1 & 0 & 0 \\ 1 & 1 & 0 & 1 & 0 \\ 1 & 1 & 1 & 0 & 1 \end{bmatrix}_{5\times 5} \end{array}$$

This completes the answer.

Example 2: Let A be a set $\{a_1, a_2, a_3\}$ and $B = \{b_1, b_2, b_3, b_4\}$ and R be a relation from set A to set B defined as $R = \{(a_1, b_1), (a_1, b_4), (a_2, b_2), (a_2, b_3), (a_3, b_1), (a_3, b_3)\}$. Find the matrix representation of R i.e. M_R.

Solution: Here, number of rows of $M_R = |A| = 3$, and number of columns of $M_R = |B| = 4$

$$\therefore M_R = \begin{array}{c} \\ a_1 \\ a_2 \\ a_3 \end{array} \begin{array}{c} \begin{array}{cccc} b_1 & b_2 & b_3 & b_4 \end{array} \\ \begin{bmatrix} 1 & 0 & 0 & 1 \\ 0 & 1 & 1 & 0 \\ 1 & 0 & 1 & 0 \end{bmatrix}_{3\times 4} \end{array}$$

This completes the answer.

Example 3: Let $A = \{1, 2, 3\}$ and R is a relation defined on A as $R = \{(1, 1), (2, 2), (3, 3), (1, 2), (2, 1)\}$. Find the matrix representation of R i.e. M_R.

Solution: Here, number of rows of $M_R = |A| = 3$, and number of columns of $M_R = |A| = 3$

$$\therefore M_R = \begin{array}{c c} & \begin{array}{ccc} 1 & 2 & 3 \end{array} \\ \begin{array}{c} 1 \\ 2 \\ 3 \end{array} & \begin{bmatrix} 1 & 1 & 0 \\ 1 & 1 & 0 \\ 0 & 0 & 1 \end{bmatrix} \end{array}$$

This completes the answer.

Note: If R is relation on a non-empty set A, and M_R is matrix for R, then

1. R is reflexive iff diagonal elements of M_R are all 1.
2. R is symmetric iff $M_R = M_R^T$
3. R is transitive iff $M_R = M_R^2$, (Boolean multiplication)
4. A relation R which is reflexive & symmetric is called compatible relation.

1.2.4 Closures of Relation

In this section, we will discuss the way to let a relation, R, defined on a non-empty set A, to attain a particular type by adding a few ordered pairs to R (if it lacks it). For example, suppose a relation R can be made reflexive by adding a few ordered pairs (x, x), then a relation R_1, so obtained is the smallest relation containing R which is reflexive. Such R_1 will be called reflexive closures of R. We shall discuss reflexive, symmetric and transitive closure only of a relation.

Reflexive Closure

Let $\triangle$ be a diagonal relation i.e. a relation defined on a non-empty set A such that

$\triangle = \{(x, x) \mid x \in A\}$. Now, let R be any relation on a non-empty set A. If R is reflexive then $R \cup \triangle = R$. If R is not reflexive then $\exists$ some element $x \in A$ such that pair $(x, x) \notin R$, i.e. some pairs (x, x) needed for R to be a reflexive, are not in R. Let R_1 $R \cup \triangle$, then R_1 is a smallest reflexive relation containing R. A reflexive closure of R is $R \cup \triangle$ and is defined as "Reflexive Closure of a relation R is the smallest reflexive relation containing R."

Example: Let $A = \{x, y, z\}$ and R be a relation defined on A as $R = \{(x, x), (x, y), (y, z), (x, z)\}$. Find reflexive closure of R.

Solution: Here $\triangle = \{(x, x), (y, y), (z, z)\}$, the given R is not reflexive. R does not contain (y, y) & (z, z), which are required for R to be reflexive. Thus $R \cup \triangle = \{(x, x), (y, y), (z, z), (x, y), (y, z), (x, z)\}$ is reflexive closure of R

This completes the solution.

Symmetric Closure

We can define symmetric closure of a given relation R as "Symmetric closure of a relation R is the smallest symmetric relation containing *R*".

Symmetric closure of a relation $R = R \cup R^{-1}$

Example: In the previous example, R on A is not symmetric, Find Symmetric closure of R

Solution: Since $(x, z), (x, y), (y, z) \in R$, but $(z, x), (y, x)$, and (z, y) do not belong to R, R is not symmetric. If we add these three elements in R, then R becomes Symmetric. Also

$R^{-1} = \{ (x, x), (z, x), (y, x), (z, y) \}$

$\therefore\ R \cup R^{-1} = \{(x, x), (x, z), (x, y), (y, z), (z, x), (y, x), (z, y)\}$ which smallest symmetric relation containing R.

Transitive Closure

Let R be a relation defined on a non-empty set A. "Transitive closure of R is the smallest transitive relation containing R".

The method to find transitive closure of a given relation R is not as straight forward as we have seen in case of reflexive and symmetric closure. In this case we shall have to use boolean matrix multiplication. We shall begin with an example.

Example: Let $A = \{1, 2, 3, 4\}$ and R be a relation on A given as $R = \{(1, 2), (2, 3), (3, 4), (2, 1)\}$. Find the transitive closure of R.

Solution: *Step 1:* First we convert the given relation in matrix form and call it M_R. So

$$M_R = \begin{bmatrix} 0 & 1 & 0 & 0 \\ 1 & 0 & 1 & 0 \\ 0 & 0 & 0 & 1 \\ 0 & 0 & 0 & 0 \end{bmatrix}$$

Step 2: Compute different powers of M_R using boolean matrix multiplication. Stop the computation when M_R^n is equal to any of M_R, $M_R^2 \ldots M_R^{n-1}$. Here M_R^n is the nth power of M_R.

$$M_R^2 = \begin{bmatrix} 0 & 1 & 0 & 0 \\ 1 & 0 & 1 & 0 \\ 0 & 0 & 0 & 1 \\ 0 & 0 & 0 & 0 \end{bmatrix} \begin{bmatrix} 0 & 1 & 0 & 0 \\ 1 & 0 & 1 & 0 \\ 0 & 0 & 0 & 1 \\ 0 & 0 & 0 & 0 \end{bmatrix} = \begin{bmatrix} 1 & 0 & 1 & 0 \\ 0 & 1 & 0 & 1 \\ 0 & 0 & 0 & 0 \\ 0 & 0 & 0 & 0 \end{bmatrix}$$

Here readers should note that in Boolean multiplication 0.0 = 1.0 = 0; 1.1 = 1, 1+1 = 1, 0+1 = 1+0 = 1 and 0+0 = 0.

Now

$$M_R^3 = M_R^2 . M_R = \begin{bmatrix} 1 & 0 & 1 & 0 \\ 0 & 1 & 0 & 1 \\ 0 & 0 & 0 & 0 \\ 0 & 0 & 0 & 0 \end{bmatrix} \begin{bmatrix} 0 & 1 & 0 & 0 \\ 1 & 0 & 1 & 0 \\ 0 & 0 & 0 & 1 \\ 0 & 0 & 0 & 0 \end{bmatrix} = \begin{bmatrix} 0 & 1 & 0 & 1 \\ 1 & 0 & 1 & 0 \\ 0 & 0 & 0 & 0 \\ 0 & 0 & 0 & 0 \end{bmatrix}$$

Similarly

$$M_R^4 = M_R^3 \, M_R = \begin{bmatrix} 0 & 1 & 0 & 1 \\ 1 & 0 & 1 & 0 \\ 0 & 0 & 0 & 0 \\ 0 & 0 & 0 & 0 \end{bmatrix} \begin{bmatrix} 0 & 1 & 0 & 0 \\ 1 & 0 & 1 & 0 \\ 0 & 0 & 0 & 1 \\ 0 & 0 & 0 & 0 \end{bmatrix} = \begin{bmatrix} 1 & 0 & 1 & 0 \\ 0 & 1 & 0 & 1 \\ 0 & 0 & 0 & 0 \\ 0 & 0 & 0 & 0 \end{bmatrix}$$

Similarly $M_R^5 = \begin{bmatrix} 0 & 1 & 0 & 1 \\ 1 & 0 & 1 & 0 \\ 0 & 0 & 0 & 0 \\ 0 & 0 & 0 & 0 \end{bmatrix}$ [Verify it]

Here, $M_R^5 = M_R^3$ & $M_R^4 = M_R^2$. This pattern will be repeated and $M_R^n = M_R^2$ when n is even and $M_R^n = M_R^3$ when n is odd.

Step 3: Select distinct M_R^n from the computed list of matrices M_R, M_R^2, M_R^3 as distinct matrices.

Step 4: Join the matrices so selected in step 3 using boolean OR on the elements of matrices while adding these matrices.

$\therefore$ Matrix for transitive closure of R denoted as M_R^∞, (R^∞ is transitive closure) is given as

$$M_R^\infty = M_R \vee M_R^2 \vee M_R^3$$

$$= \begin{bmatrix} 0 & 1 & 0 & 0 \\ 1 & 0 & 1 & 0 \\ 0 & 0 & 0 & 1 \\ 0 & 0 & 0 & 0 \end{bmatrix} + \begin{bmatrix} 1 & 0 & 1 & 0 \\ 0 & 1 & 0 & 1 \\ 0 & 0 & 0 & 0 \\ 0 & 0 & 0 & 0 \end{bmatrix} + \begin{bmatrix} 0 & 1 & 0 & 1 \\ 1 & 0 & 1 & 0 \\ 0 & 0 & 0 & 0 \\ 0 & 0 & 0 & 0 \end{bmatrix}$$

$$= \begin{bmatrix} 1 & 1 & 1 & 1 \\ 1 & 1 & 1 & 1 \\ 0 & 0 & 0 & 1 \\ 0 & 0 & 0 & 0 \end{bmatrix}$$

Step 5: Convert the M_R^∞ into R^∞ i.e. transitive closure

$\therefore$ $R^\infty = \{(1, 1), (1, 2), (1, 3), (1, 4), (2, 1), (2, 2), (2, 3), (2, 4), (3, 4)\}$

This completes the answer.

The five steps algorithm explained in the previous example is used to compute reachability matrix in a graph. The same algorithm is also used to find whether a graph is connected or not. All these and more about the application of transitive closure in graph, we will learn in Chapter 7. There we shall define what is connected relation. Till then, we shall wait and learn here a move efficient algorithm due to Warshall to find transitive closure of a given relation. The same algorithm is used to find all pair shortest path in a given weighted graph, thus readers are advised to understand the technique to make themselves comfortable while using this in Chapter 7.

Warshall's Algorithm

We shall use the same relation as given in previous example to explain this algorithm.

Step 1: Convert the given relation into matrix form M_R Call it initial Warshall matrix W_0

$$\therefore W_0 = \begin{bmatrix} 0 & 1 & 0 & 0 \\ 1 & 0 & 1 & 0 \\ 0 & 0 & 0 & 1 \\ 0 & 0 & 0 & 1 \end{bmatrix}$$

Step 2: For $k = 1$ to $|A|$, A is the set on which R is defined, compute W_k called Warshall matrix at kth state. To compute W_k from W_{k-1}, we shall proceed as below:

Step 2.1: Transfer all 1's from W_{k-1} to W_k;

Step 2.2: List the rows in column K of W_{k-1} where the entry in W_{k-1} is 1, say it $p_1, p_2, p_3, \ldots\ldots$;
Similarly, list columns in row K of W_{k-1} where entry in W_{k-1} is 1, say it $q_1, q_2, q_3, \ldots\ldots$;

Step 2.3: Place 1 at all locations (p_i, q_j) in w_k, if 1 not already there.

Let us now use this algorithm to find transitive closure of the given relation R. Here we have to find w_1, w_2, w_3 and w_4 since $|A| = 4$

$$\therefore W_1 = \begin{bmatrix} 0 & 1 & 0 & 0 \\ 1 & 1 & 1 & 0 \\ 0 & 0 & 0 & 1 \\ 0 & 0 & 0 & 0 \end{bmatrix}$$

Step 2.1: Transfer all 1's from W_0 to W_1.

Step 2.2: $k = 1$, in column 1, W_0 has 1 in row = 2, and in row 1, W_0 has 1 in column 2. Therefore in W_1 we have additional 1 at (2, 2).

$$\therefore W_2 = \begin{bmatrix} 1 & 1 & 1 & 0 \\ 1 & 1 & 1 & 0 \\ 0 & 0 & 0 & 1 \\ 0 & 0 & 0 & 0 \end{bmatrix}$$

Step 2.1: Transfer all 1's from W_1 to W_2.

Step 2.2: $k = 2$ in column 2, W_1 has 1 at rows 1 & 2, and in row 2, W_1 has 1 in columns 1, 2, 3, thus in W_2, we shall have additional 1's at location (1, 1), (1, 2), (1, 3), (2, 1), (2, 2), (2, 3) if 1 is not already there.

$$\therefore W_3 = \begin{bmatrix} 1 & 1 & 1 & 1 \\ 1 & 1 & 1 & 1 \\ 0 & 0 & 0 & 1 \\ 0 & 0 & 0 & 0 \end{bmatrix}$$

Step 2.1: Transfer all 1's from W_2 to W_3.

Step 2.2: $k = 3$, in column 3, W_2 has 1 at rows 1 and 2 and in rows 3 W_2 has 1 at column 4, so we shall have additional 1's at (1, 4) & (2, 4).

$$\therefore W_4 = \begin{bmatrix} 1 & 1 & 1 & 1 \\ 1 & 1 & 1 & 1 \\ 0 & 0 & 0 & 1 \\ 0 & 0 & 0 & 0 \end{bmatrix}$$

Step 2.1: Transfer all 1's from W_3 to W_4.

Step 2.2: $k = 4$, in column 4, W_3 has 1 at rows 4 W_3 has 1 at no place. So no additional 1 is added to W_3.

Therefore W_3 is the matrix for transitive closures of the given relation R. Hence

R^∞ = {(1, 1), (1, 2), (1, 3), (1, 4), (2, 1), (2, 2), (2, 3), (2, 4), (3, 4)}

By this time reader may be thinking to ask why to use this complicated procedure once we have a simple method of matrix multiplication to do the same job. The answer lies in the efficiency of the Warshall's algorithm.

1. In Warshall's algorithm it is not required to store past results as in previous method.
2. Once we have computed $W_{|A|}$, we have the result available. No boolean addition is required on previously obtained matrices as in previous method.

Note: Sometimes in Warshall's algorithm, say on a relation R on a set A having $|A| = n$, we may have $W_k = W_{K+1}$ for $k, K+1 < n$, it does not mean that we should stop procedure there. In Warshall's algorithm the computation must proceed upto $|A|$.

Example: Let $A = \{1, 2, 3, 4, 5\}$ and R is a relation defined on A as R = {(1, 1), (1, 2), (2, 1), (2, 2), (3, 3), (3, 4), (4, 3), (4, 4), (4, 5), (5, 4), (5, 5)}. Find transitive closure of R using Warshall's algorithm.

Solution: Readers are advised to do that and verify the fact mentioned in note.

Exercise 1.2

1. In each part, find x or y so that the statement is true.
   (a) $(x, 3) = (4, 3)$ (b) $(a, 3y) = (a, 9)$
   (c) $(3x + 1, 2) = (7, 2)$ (d) (C++, PASCAL) = (y, x)
2. In each part, find x and y so that the statement is true.
   (a) $(4x, 6) = (16, y)$ (b) $(2x - 3, 3y - 1) = (5, 5)$
   (c) $(x2 + 25) = (49, y)$ (d) $(x, y) = (x^2, y^2)$
3. Let $A = \{a, b\}$ and $B = \{4, 5, 6\}$. List the elements in
   (a) $A \times B$ (b) $B \times A$
   (c) $A \times A$ (d) $B \times B$
4. Let A = {Fine, Yang} and B = {president, vice-president, secretary, treasurer}. Give each of the following:
   (a) $A \times B$ (b) $B \times A$
   (c) $A \times A$
5. A genetics experiment classifies fruit flies according to the following two criteria:
   Gender: male(m), female (f)
   Wing span: short(s), medium(m), long(l)
   (a) How many categories are there in this classification scheme?
   (b) List all the categories in this classification scheme.
6. A car manufacturer makes three different types of car frames and two types of engines.
   Frame type: sedan (s), coupe (c) van (v)
   Engine type: gas (g), diesel (d)
   List all possible model of cars.
7. If $A = \{a, b, c\}$, $B = \{1, 2\}$, and $C = \{\#, *\}$, list all the elements of $A \times B \times C$.
8. If A has three elements and B has $n \geq 1$ elements, use mathematical induction to prove that $|A \times B| = 3n$.

9. If $A = \{a \mid a$ is real number$\}$ and $B = \{1, 2, 3\}$, sketch each of the following in the Cartesian plane.

   (a) $A \times B$ (b) $B \times A$

10. If $A = \{a \mid a$ is real number and $-2 \le a \le 3\}$ and $B = \{b \mid b$ is a real number and $1 \le b \le 5\}$, sketch each of the following in the Cartesian plane.

    (a) $A \times B$ (b) $B \times A$

11. Show that if A_1 has n_1 elements, A_2 has n_2 elements, and A_3 has n_3 elements, then $A_1 \times A_2 \times A_3$ has $n_1 . n_2 . n_3$ elements.
12. If $A \subseteq C$ and $B \subseteq D$, prove that $A \times B \subseteq C \times D$.
13. Let $A = \{1, 2, 3, 4, 5, 6, 7, 8, 9, 10\}$ and

    $A_1 = \{1, 2, 3, 4\}$, $A_2 = \{5, 6, 7\}$

    $A_3 = \{4, 5, 7, 9\}$, $A_4 = \{4, 8, 10\}$,

    $A_5 = \{8, 9, 10\}$, $A_6 = \{1, 2, 3, 6, 8, 10\}$.

    Which of the following are partitions of A?

    (a) $\{A_1, A_2, A_5\}$ (b) $\{A_1, A_3, A_5\}$

    (c) $\{A_3, A_6\}$ (d) $\{A_2, A_3, A_4\}$

14. If A_1 is the set of positive integers and A_2 is the set of all negative integers, is $\{A_1, A_2\}$ a partition of Z? Explain your conclusion.
15. If $B = \{0, 3, 6, 9, \ldots.\}$, give a partition of B containing

    (a) two infinite subsets (b) three infinite subsets.

16. List all partitions of $A = \{1, 2, 3\}$
17. List all partitions of $B = \{a, b, c, d\}$
18. Let A, B and C be subsets of U. Prove that $A \times (B \cup C) = (A \times B) \cup (A \times C)$
19. Use the sets $A = \{1, 2, 4\}$, $B = \{2, 5, 7\}$, and $C = \{1, 3, 7\}$ to investigate whether $A \times (B \cap C) = (A \times B) \cap (A \times C)$.
20. Let A denote the set of shirts and B denote the set of slacks a man owns. What possible interpretation can be given to the cartesian product $A \times B$? To a binary relation from A to B?
21. Let $A = \{1, 2\}$. Construct the set $P(A) \times A$.
22. (a) Given that $A \subseteq C$ and $B \subseteq D$, show that $A \times B \subseteq C \times D$.

    (b) Given that $A \times B \subseteq C \times D$, does it necessarily follow that $A \subseteq C$ and $B \subseteq D$?

23. (a) Let A be an arbitrary set. Is the set $A \times \phi$ well defined?
    (b) Given that $A \times B = \phi$, what can one say about the set A and B?
    (c) Is it possible that $A \ \phi \ A \times A$ for some set A?

24. Let A, B, C, D be arbitrary sets.
    (a) Show that

$$(A \cap B) \times (C \cap D) = (A \times C) \cap (B \times D)$$

    (b) Confirm or disprove the following identities:

$$(A \cup B) \times (C \cup D) = (A \times C) \cup (B \times D)$$
$$(A - B) \times (C - D) = (A \times C) - (B \times D)$$
$$(A \oplus B) \times (C \oplus D) = (A \times C) \oplus (B \times D)$$

25. Let A, B, C be arbitrary sets.
    Confirm or disprove the following identities:

$$(A \cup B) \times C = (A \times C) \cup (B \times C)$$
$$(A - B) \times C = (A \times C) - (B \times C)$$
$$(A \oplus B) \times C = (A \times C) \oplus (B \times C)$$

26. Let $A = \{6 : 00, 6 : 30, 7 : 00, ..., 9 : 30, 10 : 00\}$ denote the set of nine half-hour periods in the evening. Let $B = \{3, 12, 15, 17\}$ denote the set of the four local television channels, Let $R_1$and R_2 be two binary relations from A to B. What possible interpretations can be given to the binary relations R_1, R_2, $R_1 \cup R_2$, $R_1 \cap R_2$, $R_1 \oplus R_2$, and $R_1 - R_2$?

27. Let 1 be the set of all integers
    (a) Is there a natural way to interpret the ordered pairs in $I \times I$ as geometric points in the plane?
    (b) Let R_1 be a binary relation on $I \times I$ such that the ordered pair (of rendered pair) $((a, b), (c, d))$ is in R_1 if and only if $a - c = b - d$. What is a geometric interpretation of the binary relation R_1?
    (c) Let R_2 be a binary relation on $I \times I$ such that $((a, b), (c, d))$ is in R_2 if and only if $\sqrt{(a-c)^2 + (b-d)^2} \leq 10$. What is a geometric interpretation of R_2? of $R_1 \cup R_2$? of $R_1 \cap R_2$? of $R_1 - R_2$? of $R_1 \oplus R_2$?

28. Let A be a set of workers and B be a set of jobs. Let R_1 be a binary relation from A to B such that (a, b) is in R_1 if worker a

is assigned to job b. (We assume that a worker might be assigned to more than one job and more than one worker might be assigned to the same job). Let R_2 be a binary relation on A such that (a_1, a_2) is in R_2 if a_1 and a_2 can get along with each other if they were assigned to the same job. State a condition in terms of R_1, R_2 and (possibly) binary relations derived from R_1 and R_2 such that an assignment of the workers to the jobs according to R_1 will not put workers that cannot get along with one another on the same job.

29. Let A be a set of books.

    (a) Let R_1 be a binary relation on A such that (a, b) is in R_1 if book a costs more and contains fewer pages than book b. In general, is R_1 reflexive? Symmetric? Antisymmetric? Transitive?

    (b) Let R_2 be a binary relation on A such that (a, b) is in R_2 if book a costs more or contains fewer pages than book b. In general, is R_2 reflexive? Symmetric? Antisymmetric? Transitive?

30. (a) Let R be a binary relation on the set of all positive integers such that

    $$R = \{(a, b) \mid a - b \text{ is an odd positive integer}\}$$

    Is R reflexive? Symmetric? Antisymmetric? Transitive? An equivalence relation? A partial ordering retaliation?

    (b) Repeat part (*a*) if $R = \{(a, b) \mid a = b^2\}$

31. Let P be the set of all people. Let R be a binary relation on P such that (a, b) is in R if a is a brother of b. (Disregard half-brothers and fraternity brothers). Is R reflexive? Symmetric? Antisymmetric? Transitive? An equivalence relation? A partial ordering relation?

32. Let R be a binary relation on the set of all strings of 0s and 1s such that $R = \{(a, b) \mid a$ and b are strings that have the same number of 0s$\}$. Is R reflexive? Symmetric? Antisymmetric? Transitive? An equivalence relation?

33. Use the fact that "better than" is a transitive binary relation to prove the claim "A ham sandwich is better than eternal happiness?

    Hint: Nothing is better eternal happiness.

34. Let A be a set with 10 distinct elements.
   (a) How many different binary relations on A are there?
   (b) How may of them are reflexive?
   (c) How may of them are symmetric?
35. $A = \{a, b, c, d\}$, $B = \{1, 2, 3\}$, $R = \{(a, 1), (a, 2), (b, 1), (c, 2), (d, 1)\}$
36. A = {IBM, COMPAQ, Dell, Gateway, Zenith},
   B = {750, PS60, 450V, 4/33S, 525SX, 466V, 486SL}
   R = {(IBMm 750C), (Dell, 466V),
   (COMPAQ, 450SV), (Gateway, PS60) }
37. $A = \{1, 2, 3, 4\}$, $B = \{1, 4, 5, 8, 9\}$: $a\ R\ B$ if and only if $b = a^2$.
38. $A = \{1, 2, 3, 4, 8\}$, $B = \{1, 4, 6, 9\}$, $a\ R\ B$ if and only if $a \backslash b$.
39. $A = \{1, 2, 3, 4, 8\} = B$; $a\ R\ b$ if and only if $a + b \le 9$.
40. Let $A = Z^+$, the positive integers, and R be the relation defined by $a\ R\ b$ if and only if there exists a k in Z^+ so that $a = b^k$ (k depends on a and b). Which of the following belong to R?
   (a) (4, 16) (b) (1, 7)
   (c) (8, 2) (d) (3, 3)
   (e) (2, 8) (f) (2, 32)
41. Let $A = R$. Consider the following relation R on A, $a\ R\ b$ if and only if $2a + 3b = 6$. Find Dom (R) and Ran (R).
42. Let $A = R$. Consider the following relation R on A; $a\ R\ b$ if and only if $a^2 + b^2 = 25$. Find Dom (R) and Ran (R).
43. Let $A = R$. Give a description of the relation R specified by the shaded region in Figure 2.8.

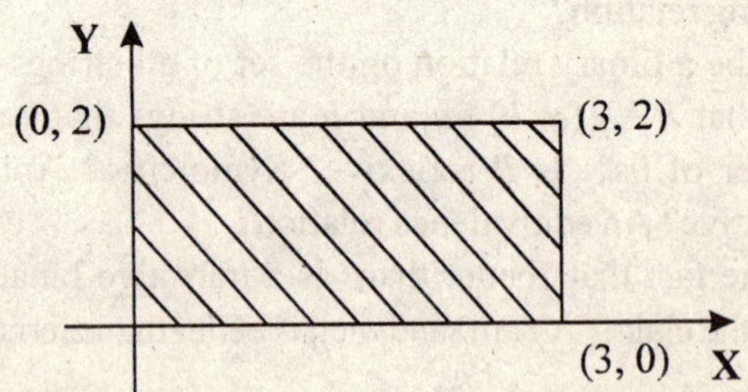

44. If A has n elements and B has m elements, how many different relations are there from A to B? In Exercises 45, and 46 give the relation R defined on A and its digraph.

45. Let $A = \{1, 2, 3, 4\}$ and $M_R = \begin{bmatrix} 1 & 1 & 0 & 1 \\ 0 & 1 & 1 & 0 \\ 0 & 0 & 1 & 1 \\ 1 & 0 & 0 & 0 \end{bmatrix}$

46. Let $A = \{a, b, c, d, e\}$ and $M_R = \begin{bmatrix} 1 & 1 & 0 & 0 & 0 \\ 0 & 0 & 1 & 1 & 0 \\ 0 & 0 & 0 & 1 & 1 \\ 0 & 1 & 1 & 0 & 0 \\ 1 & 0 & 0 & 0 & 0 \end{bmatrix}$

In Exercises 47 and 48 find the relation determined by the digraph and give its matrix.

47. Diagram

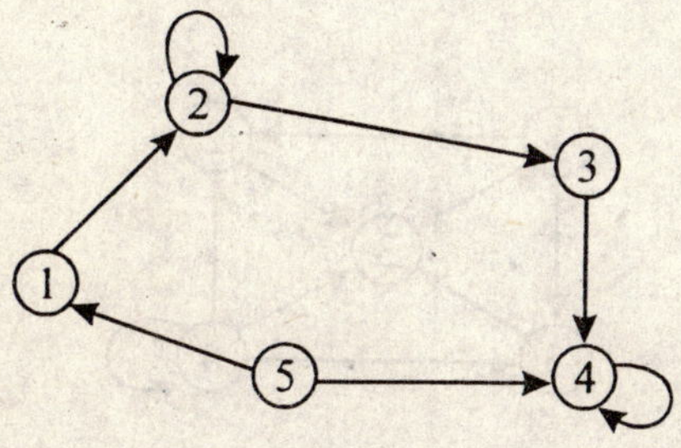

48. Diagram

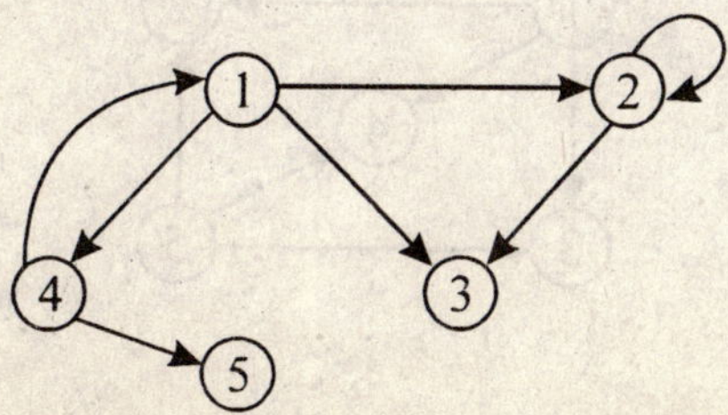

49. Let $A = \{1, 2, 3, 4, 5, 6, 7\}$ and $R = \{(1, 2), (1, 4), (2, 3), (2, 5), (3, 6), (4, 7)\}$. Compute the restriction of R to B for the subset of A given.

 (a) $A = \{1, 2, 4, 5\}$ (b) $B = \{2, 3, 4, 6\}$

In Exercise 50 through 57, Let $A = \{1, 2, 3, 4\}$. Determine whether the relation is reflexive, irreflexive, symmetric, asymmetric, antisymmetric or transitive.

50. $R = \{(1, 1), (1, 2), (2, 1), (2, 2), (3, 3), (3, 4), (4, 3), (4, 4)\}$
51. $R = \{(1, 2), (1, 3), (1, 4), (2, 3), (2, 4), (3, 4)\}$
52. $R = \{(1, 3), (1, 1), (3, 1), (1, 2), (3, 3), (4, 4)\}$
53. $R = \{(1, 1), (2, 2), (3, 3)\}$
54. $R = \phi$
55. $R = A \times A$
56. $R = \{(1, 2), (1, 3), (3, 1), (1, 1), (3, 3), (3, 2), (1, 4), (4, 2), (3, 4)\}$
57. $R = \{(1, 3), (4, 2), (2, 4), (3, 1), (2, 2)\}$

In Exercises 58 and 59, let $A = \{1, 2, 3, 4, 5\}$. Determine whether the relation R whose digraph is given is reflexive, irreflexive, symmetric, asymmetric, antisymmetric, or transitive.

58. Diagram

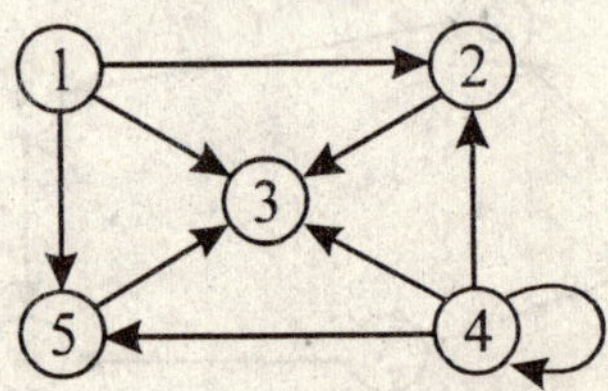

59. Diagram

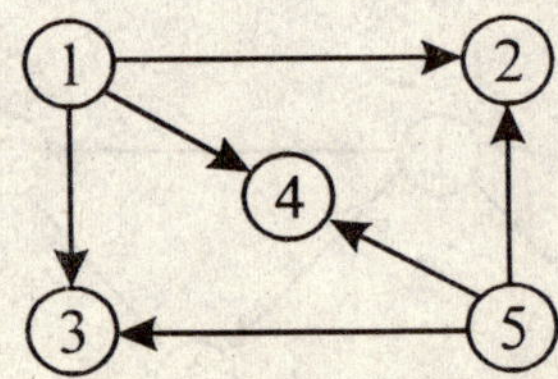

In Exercises 60 through 64 determine whether the relation R on the set A is reflexive, irreflexive, symmetric, asymmetric, antisymmetric, or transitive.

60. $A = Z$; $a \; R \; b$ if and only if $a \le b + 1$
61. $A = Z^+$; $a \; R \; b$ if and only if $|a-b| \le 2$
62. $A = Z^+$; $a \; R \; b$ if and only if $a = b^k$ for some $k \in Z^+$
63. $A = Z$; $a \; R \; b$ if and only if $a + b$ is even
64. $A = Z^+$; $a \; R \; b$ if and only if GCD $(a, b) = 1$. In this case, we say that a and b are relatively prime.
65. Let R be the following symmetric relation on the set $A = \{1, 2, 3, 4, 5\}$.

    $R = \{(1, 2), (2, 1), (3, 4), (4, 3), (3, 5), (5, 3), (4, 5), (5, 4), (5, 5)\}$

    Draw the graph of R.
66. Let $A = \{a, b, c, d\}$ and let R be the symmetric relation

    $R = \{(a, b), (b, a), (a, c), (c, a), (a, d), (d, a)\}$

    Draw the graph of R.
67. Show that if a relation on a set A is transitive and irreflexive, then it is asymmetric.
68. Prove that if a relation R on a set A is symmetric, then the relation R^2 is also symmetric.
69. Let R be a nonempty relation on a set A. Suppose that R is symmetric and transitive. Show that R is not irreflexive.
70. $A = B = \{1, 2, 3\}$; $R = \{ (1, 1), (1, 2), (2, 3), (3, 1) \}$; $S = \{ (2, 1), (3, 1), (3, 2), (3, 3) \}$
71. $A = \{a, b, c\}$; $B = \{1, 2, 3\}$; $R = \{(a, 1), (b, 1), (c, 2), (c, 3)\}$; $S = \{(a, 1), (a, 2), (b, 1), (b, 2)\}$

In Exercises 72 and 73, let R and S be two relations whose corresponding digraphs are shown there. Compute

(a) $\overline{R}$ (b) $R \cap S$

(c) $R \cup S$ (d) S^{-1}

72. Diagram

$R =$

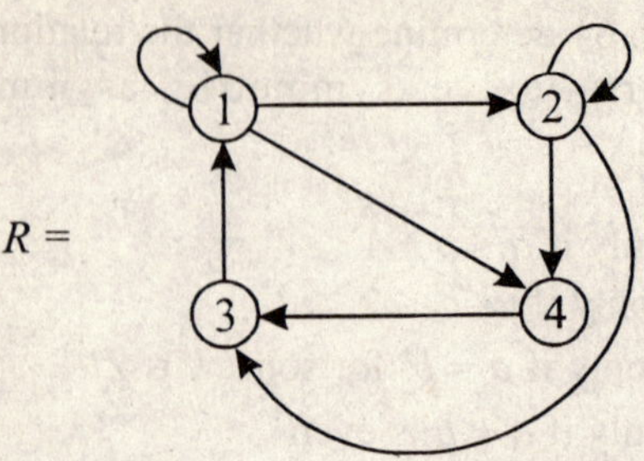

$S =$

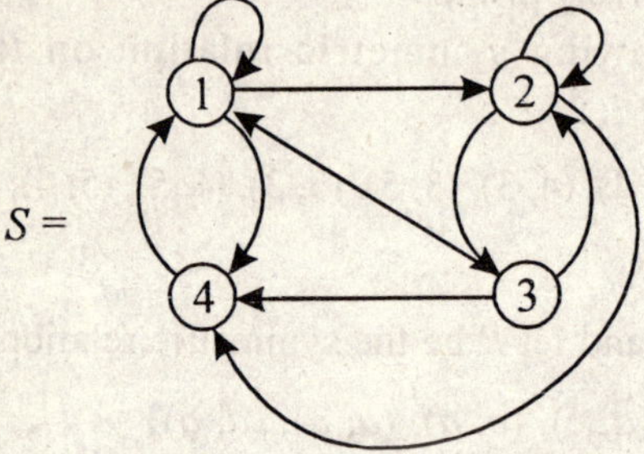

73. Diagram

$R =$

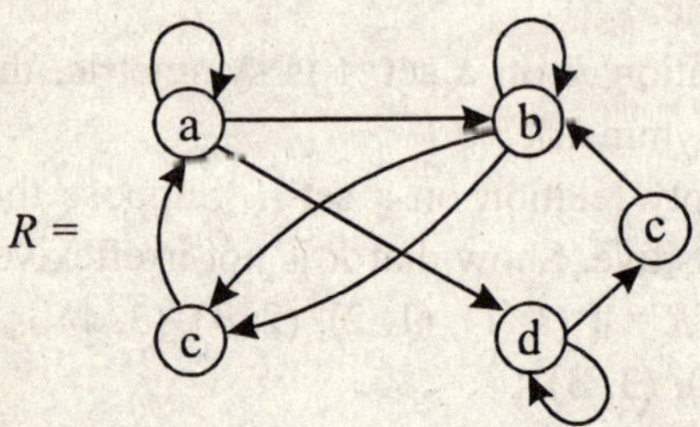

$S =$

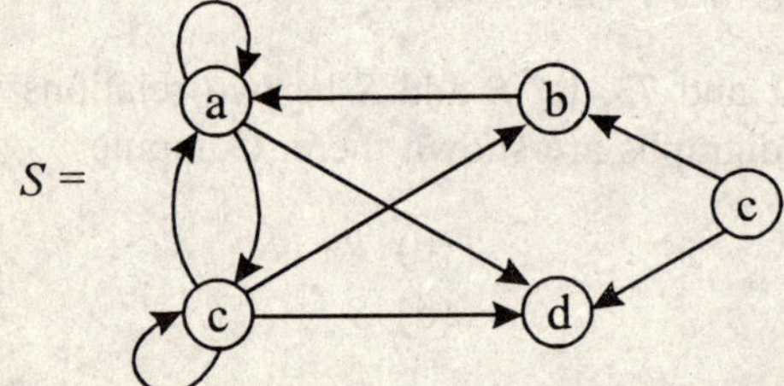

In Exercises 74 and 75, let $A = \{1, 2, 3\}$ and $B = \{1, 2, 3, 4\}$. Let R and S be the relations from A to B whose matrices are given. Compute

(a) S (b) $R \cap S$

(c) $R \cup S$ (d) R^{-1}

74. $M_R = \begin{bmatrix} 1 & 1 & 0 & 1 \\ 0 & 0 & 0 & 1 \\ 1 & 1 & 1 & 0 \end{bmatrix}$, $M_S = \begin{bmatrix} 0 & 1 & 1 & 0 \\ 1 & 0 & 0 & 1 \\ 1 & 1 & 0 & 0 \end{bmatrix}$

75. $M_R = \begin{bmatrix} 1 & 0 & 1 & 0 \\ 0 & 0 & 0 & 1 \\ 1 & 1 & 1 & 0 \end{bmatrix}$, $M_S = \begin{bmatrix} 1 & 1 & 1 & 1 \\ 0 & 0 & 0 & 1 \\ 0 & 1 & 0 & 1 \end{bmatrix}$

76. A relation R on set A is called circular if $a\,R\,b$ and $b\,R\,c$ imply $c\,R\,a$. Show that R is reflexive and circular if and only if it is an equivalence relation.
77. Show that if R_1 and R_2 are equivalence relations on A, then $R_1 \cap R_2$ is an equivalence relation on A.
78. Define an equivalence relation R and Z, the set of integers, whose corresponding partition contains exactly three infinite sets.
79. Let R be a symmetric and transitive relation on a set A. Show that if for every a in A there exists b in A such that (a, b) is in R, then R is an equivalence relation.
80. Let R be a transitive and reflexive relation on A. Let T be a relation on A such that (a, b) is in T if and only if both (a, b) and (b, a) are in R. Show that T is an equivalence relation.
81. Let R be a binary relation. Let $S = \{ (a, b) \backslash (a, c) \in R$ and $(c, b) \in R$ for some $c\}$. Show that if R is an equivalence relation, then S is also an equivalence relation.
82. Let R be reflexive relation on a set A. Show that R is an equivalences relation if and only if (a, b) and (a, c) are in R implies that (b, c) is in R.
83. A binary relation on a set that is reflexive and symmetric is called a *compatible relation.*

(a) Let A be a set of people and R be a binary relation on A such that (a, b) is in R if a is a friend of b. Show that R is a compatible relation.

(b) Let A be a set of English words and R be a binary relation on A such that two words in A are related if they have one or more letters in common. Show that R is a compatible relation.

(c) Give more examples of compatible relations.

(d) Let R_1 and R_2 be two compatible relations on A. Is $R_1 \cap R_2$ a compatible relation? Is $R_1 \cup R_2$ a compatible relation?

(e) Let A be a set. A cover of A is a set of nonempty subsets of A, $\{A_1, A_2, \ldots A_k\}$, such that the union of the A_i's is equal to A. Suggest a way to define a compatible relation on A from a cover of A. Give an interpretation of the notion of a cover in terms of the example in part (*a*).

(f) Suggest a way to define a cover of A from a compatible relation on A. Does your suggested way define *uniquely* a cover of A?

84. (a) Show that the transitive closure of a symmetric relation is symmetric.

(b) Is the transitive closure of an antisymmetric relation always antisymmetric?

(c) Show that the transitive closures of compatible relation (see Prob. 4.20) is an equivalence relation.

85. Let R be a binary relation from A to B. The converse of R, denoted R^{-1}, is a binary relation from B to A such that

$R^{-1} = \{ (b, a) \mid (a, b) \in R \}$

(a) Let R_1 and R_2 be binary relations from A to B. Is it true that $(R_1 \cup R_2)^{-1} = R_1^{-1} \cup R_2^{-1}$?

(b) Let R be a binary relation on A. If R is reflexive, is R^{-1} necessarily reflexive? If R is symmetric, is R^{-1} necessarily symmetric? If R is transitive, is R^{-1} necessarily transitive?

86. (a) Let $A = \{1, 2, 3\}$ and let $R = \{(1, 1), (1, 2), (2, 3), (1, 3), (3, 1), (3, 2)\}$. Compute the matrix M_{R^∞} of the transitive closure R by using the formula

$$M_{R^\infty} = M_R \vee (M_R)^2 \vee (M_R)^3$$

(b) List the relation R^∞ whose matrix was computed in part (a).

87. For the relation R of Exercise 1, compute the transitive closure R^∞ by using Warshall's algorithm.

88. Let $A = \{a_1, a_2, a_3, a_4, a_5\}$ and let R be a relation on A whose matrix is

$$M_R = \begin{bmatrix} 1 & 0 & 0 & 1 & 0 \\ 0 & 1 & 0 & 0 & 0 \\ 0 & 0 & 0 & 1 & 1 \\ 1 & 0 & 0 & 0 & 0 \\ 0 & 1 & 0 & 0 & 1 \end{bmatrix} = W_0$$

Compute W_1, W_2, and W_3 as in Warshall's algorithm.

89. Find R^∞ for the relation in Exercise 3.

90. Prove that if R is reflexive and transitive, then $R^n = R$ for all n.

In Exercises 91 and 94, let $A = \{1, 2, 3, 4\}$, For the relation R whose matrix is given, find the matrix of the transitive closure by using Warshall's algorithm.

91. $M_R = \begin{bmatrix} 1 & 0 & 0 & 1 \\ 1 & 1 & 0 & 0 \\ 0 & 0 & 1 & 0 \\ 0 & 0 & 0 & 1 \end{bmatrix}$

92. $M_R = \begin{bmatrix} 1 & 1 & 0 & 0 \\ 1 & 0 & 0 & 0 \\ 0 & 0 & 0 & 0 \\ 0 & 0 & 1 & 0 \end{bmatrix}$

93. $M_R = \begin{bmatrix} 1 & 0 & 0 & 1 \\ 0 & 1 & 1 & 0 \\ 0 & 1 & 1 & 0 \\ 1 & 0 & 0 & 1 \end{bmatrix}$

94. $M_R = \begin{bmatrix} 0 & 0 & 0 & 1 \\ 1 & 0 & 0 & 1 \\ 0 & 1 & 0 & 1 \\ 0 & 0 & 1 & 0 \end{bmatrix}$

In Exercises 95 and 96 let $A = \{a, b, c\}$. Determine whether the relation R whose matrix M_R is given is an equivalence relation.

95. $M_R = \begin{bmatrix} 1 & 0 & 0 \\ 0 & 1 & 1 \\ 0 & 1 & 1 \end{bmatrix}$

96. $M_R = \begin{bmatrix} 1 & 0 & 1 \\ 0 & 1 & 0 \\ 0 & 0 & 1 \end{bmatrix}$

In Exercises 97 and 98 (Figures 1 and 2), determine whether the relation R whose digraph is given is an equivalence relation.

97. Diagram 1

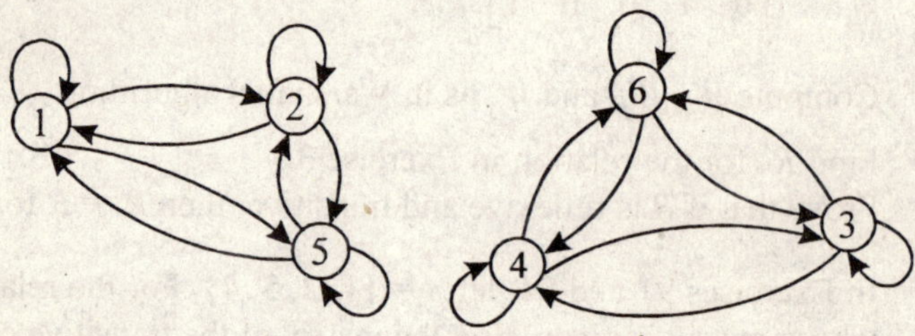

98. Diagram 2

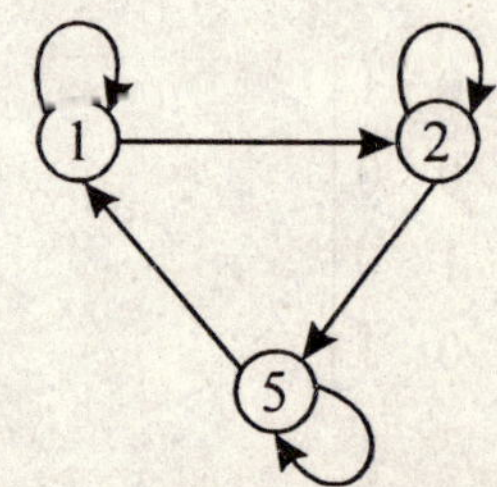

In Exercises 99 through 106, determine whether the relation R on the set A is an equivalence relation.

99. $A = \{a, b, c, d\}$, $R = \{(a, a), (b, a), (b, b), (c, c), (d, d), (d, c)\}$

100. $A = \{1, 2, 3, 4, 5\}$, $R = \{(1, 1), (1, 2), (1, 3), (2, 1), (2, 2), (3, 1), (2, 3), (3, 3), (4, 4), (3, 2), (5, 5)\}$

101. $A = \{1, 2, 3, 4, \}$, $R = \{(1, 1), (1, 2), (2, 1), (2, 2), (3, 1), (3, 3), (1, 3), (4, 1), (4, 4)\}$

102. A = the set of all members of the Software-of-the Month Club; $a \, R \, b$ if and only if a and b buy the same programs.
103. A = the set of all members of the Software-of-the Month Club; $a \, R \, b$ if and only if a and b buy the same number of programs.
104. A = the set of all people in the Social Security database; $a \, R \, b$ if and only if a and b have the same last name.
105. A = the set of all triangles in the plane; $a \, R \, b$ if and only if a is similar to b.
106. $A = Z^+ \times Z^+$; $(a, b) \, R \, (c, d)$ if and only if $b = d$
107. If $\{\{a, c, e\}, \{b, d, f\}\}$ is a partition of the set $A = \{a, b, c, d, e, f\}$, determine the corresponding equivalence relation. R.
108. Let $S = \{1, 2, 3, 4, 5\}$ and let $A = S \times S$. Define the following relation R on A: $(a, b) \, R \, (a', b')$ if and only if $ab' = a'b$.

    (a) Show that R is an equivalence relation.

    (b) Compute A/R..

2

Function and Generating Function

A function is defined in terms of some variables. If x is a variable then $f(x)$ is said to be a function of one variable. Similarly, if x and y are two variables then $f(x, y)$ is called a function of two variables. A function of n variables can be denoted by $f(x_1, x_2, x_3, x_4..., x_n)$. Here, $x_1, x_2, x_3, x_4, ..., x_n$ are called arguments or parameters of function f. Each of these parameters has a defined type say, integer, float etc. The value returned by the function is again of some type. A function, simply, maps a set of values to another set of values. In this chapter, we shall learn this and other related concepts. Sequence is another important topics in the discrete mathematical structure. We shall learn this and about a function, called Generating Function that generates sequences.

Function

A function is defined from a set A to a set B. Let $A = \{a, b, c, d, e\}$ and $B = \{1, 2, 3, 4, 5, 6\}$, and $f: A \rightarrow B$ be a function defined from set to set as shown below.

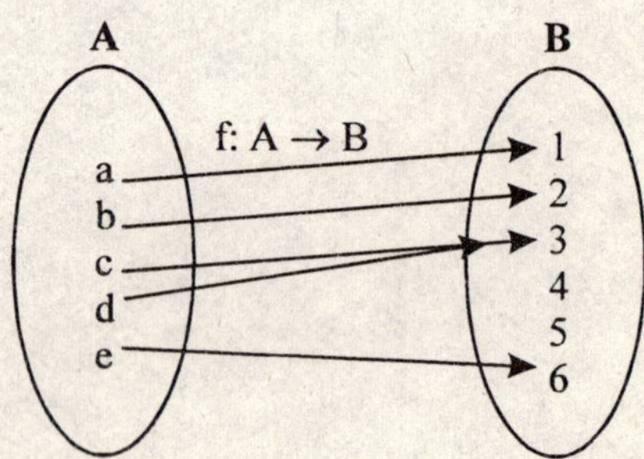

Fig. 2.1.

The diagram shows that element ***a*** of A is associated to element **1** of B, ***b*** of A to **2** of B, ***c*** of A to **3** of B, ***d*** of A to **3** of B and ***e*** of A to **6** of B. We say that, **1** is a ***f*-image** of ***a*** (or **image** of ***a*** under *f*), ***a*** is called **pre-image** of **1 under *f*.** If we arrange a and its image 1 as ordered pair $(a, 1)$, then f can be represented as $f = \{(\boldsymbol{a}, \mathbf{1}), (\boldsymbol{b}, \mathbf{2}), (\boldsymbol{c}, \mathbf{3}), (\boldsymbol{d}, \mathbf{3}), (\boldsymbol{e}, \mathbf{6})\}$, a set of ordered pairs of all the associations shown in the above diagram. Here, $f \subseteq A \times B$. We know that any subset of $A \times B$ is a relation from set A to set B, f is also a relation from A to B. Every function is a relation but the reverse is not true. A function $f: A \to B$ said to be defined from set A to set B if and only if f is **everywhere defined** on set A, i.e.

1. Every element of A is associated with some element of B, and
2. There exists no element in set A such that it is associated with more that one element of B i.e. for any two elements ***a*** and ***b*** of A if $a = b$ then $f(\boldsymbol{a}) = f(\boldsymbol{b})$.

The **figure 2.1**, is an example of a valid function from a nonempty set A to a nonempty set B. Whereas, the diagram shown in **figure 2.2** does not correspond to a valid function $f: A \to B$, since $-2 \in A$ and $2 \in A$ have no image in B under f.

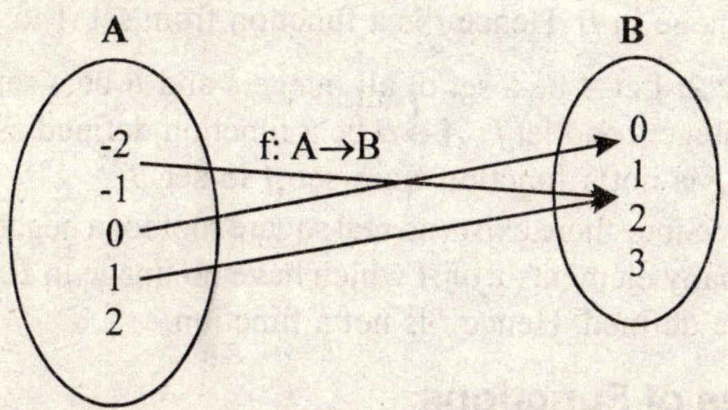

Fig. 2.2.

Another example of an invalid function can be given by diagram shown in **figure 2.3.** Where an element $4 \in A$ has two images -2 and 2 in B under function f. The function $f: A \to B$ is defined as $f(x) = \sqrt{x}$. This function is not defined on the set of integers, on the set of rational numbers, on the set of real numbers and many more such sets containing at least one negative number.

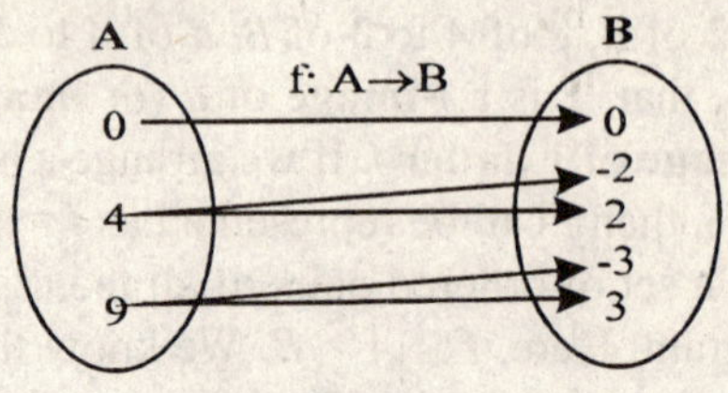

Fig 2.3.

In function f: $A \to B$ set A is called **Domain** set of f and set B is called **Range** set of f. Thus any relation R: $A \to B$ is a function if **Domain(*R*)** = A and ***R*-relative** set of every element of A is a **Singleton** set. A function is also called **Mapping** or **Transformation.**

Example 1: Let A be a set of all even integers and B be a set of all odd integers and let f: $A \to B$ be a function defined as $f(x) = x + 1$. Show that f is a function from set **A** to set B.

Solution: Let x be any element of set A. Since x is even, $x + 1$ is odd. Therefore $f(x) \in B$. For every even integer $x \in$ A, $\exists\, x + 1 = y$, an odd integer in B such that $f(x) = y$. And for any two elements x_1 and x_2 of A if $x_1 = x_2$ then $f(x_1) = f(x_2)$ i.e. no element of A has more than one image in B. Hence f is a function from set A to set B.

Example 2: Let A be a set of all integers and B be a set of all non negative integers and let f: $A \to B$ be a function defined as $f(x) = \sqrt{x}$. Show that f is not a function from set A to set B.

Solution: Since there exists no real square root for a negative integer, there are many elements x of A which have no image in B, i.e. f is not everywhere defined. Hence f is not a function.

2.1 Types of Functions

In a function f: $A \to$B one element of A cannot be associated with more than one elements of B, but more than one elements of A may get associated with one element of B. This gives two types of functions. One is **One to One** function and other is **Many to One** function. Further, there may be some elements left in set B which is (are) not associated with any of the element of A. Accordingly, we can have **Into** and **Onto** function. Let A be a set {**'One to One'**, **'Many to One'**} and B be another set {**Onto, Into**}. Then all possible functions can be given by the elements of $A \times B$. These are **One to One Onto, One to One Into, Many to One Onto, and Many to One Into** functions. Before discussing these functions, we shall discuss **Identify** function because it is required for understanding some important properties of a function. Let $\boldsymbol{I}$ be a function defined from set A to itself, called $\boldsymbol{I}_A$, such that $\boldsymbol{I}_A(x) = x$, $\forall\ x \in$ A. This function is called **Identity** function on set A.

Let f: $\boldsymbol{A \to B}$ and g: $\boldsymbol{B \to C}$ be two functions. The $\boldsymbol{g}$ 0 $\boldsymbol{f}$, called **composition** of functions f and g, is a function from set A to set C. The diagram shown in **figure 2.1.1** demonstrates a composition of functions f and g.

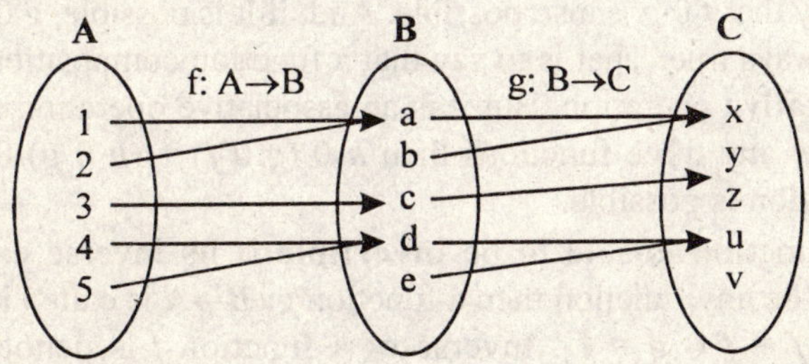

Fig. 2.1.1.

In the above diagram, $f = \{(1, a), (2, a), (3, c), (4, d), (5, d)\}$ and $g = \{(a, x), (b, x), (c, y), (d, u), (e, u)\}$. The composition of these two functions $\boldsymbol{g}$ 0 $\boldsymbol{f}$ is given by the set of ordered pairs $\{(1, x), (2, x), (3, y), (4, u), (5, u)\}$. Clearly, $\boldsymbol{g}$ 0 $\boldsymbol{f}$ is a function from set A to set C. The method to find $\boldsymbol{g}$ 0 $\boldsymbol{f}$ can be illustrated as below.

The elements of f can be arranged in two lines: the first line contains elements from set A and second line contains images of corresponding elements under f. Similarly, we can have arrangement for the elements of g. The arrangement is shown in **figure 2.1.2** on the next page.

The $\boldsymbol{g}$ 0 $\boldsymbol{f}(x)$ is also written as $g(f(x))$. Now, $g(f(1)) = g(a) = x$, $g(f(2)) = g(a) = x$, $g(f(3)) = g(\text{c}) = y$, $g(f(4)) = g(d) = u$, and $g(f(5)) = g(d) = u$. This is the way to find the composition of two functions. In general, if $f_1, f_2, f_3, \ldots \text{f}_n$ are n functions then theirs composition is given as $f_n \, 0 \, f_{n-1} \, 0 \ldots 0 \, f_3 \, 0 \, f_2 \, 0 \, f_1$.

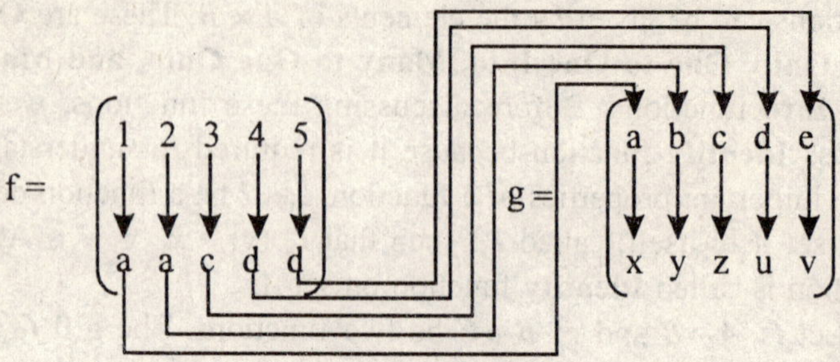

Fig. 2.1.2.

It is important to note that whenever g 0 f is possible, it is not necessary that f 0 g is also possible. And, if it is possible, g 0 f = f 0 g is not always true. That is to say that a function composition is not a **commutative** operation. But it is an associative operation i.e. if f, g and h are any three functions then h 0 (g 0 f) = (h 0 g) 0 f, if the composition is possible.

A function is said to be **invertible** if its inverse exists.. Let f: A→B be any function then a function g: B→A is called inverse of f if g 0 f = f 0 g = I_A. Inverse of a function f is denoted as f^1. Therefore, f^1 0 f = f^1 0 f = I_A.

Example 3: Let A is set of all negative integers and B a set of all positive integers. Let f be a function f: A→B defined as $f(x) = -x$. Let g be another function g: B→A defined as $g(y) = -y$. Show that g is an inverse function of f.

Solution: Here g 0 f $(x) = g(f(x)) = g(-x) = -(-x)) = x$, by the definition of functions f and g. Similarly, f 0 $g(y)) = f(g(y)) = f(-y) = -(-y) = y$, again by the definition of functions f and g. Also we know that if I_A is an identity function on A then I_A $(x) = x$ for every element x of A. Thus, we have g 0 f = f 0 g = I_A. Therefore f and g are inverse of each other i.e. $f^{-1} = g$ and $g^{-1} = f$.

Note:

1. Inverse of a function is the function itself i.e. $(f^{-1})^{-1} = f$.
2. $(g \, 0 \, f)^{-1} = f^{-1} \, 0 \, g^{-1}$ and $(f \, 0 \, g)^{-1} = g^{-1} 0 f^{-1}$.
3. In general, if $f_1, f_2, f_3, \ldots, f_n$ are n functions then $(f_n \, 0 \, f_{n-1} \, 0 \ldots 0 \, f_3 \, 0 \, f_2 \, 0 \, f_1)^{-1} = f_1^{-1} \, 0 \, f_2^{-1} \, 0 \, f_3^{-1} \, 0 \ldots 0 \, f_{n-1}^{-1} \, 0 \, f_n^{-1}$.

One to One Onto function

A function *f*: **A→*B*** is said to be One to One and Onto from a set *A* to a *B* if and only if

1. ***F*** is One to One i.e. for any two elements x_1 and x_2 of *A* if $x_1 \neq x_2$ then $f(x_1) \neq f(x_2)$, and
2. ***F*** is **onto,** i.e. for every element *y* in *B* ∃ an element *x* in *A* such that $f(x) = y$.

The diagram shown in **Figure 2.1.3** is an example of an One to One Onto function from set *A* to set *B* defined as $f(x) = x+3$.

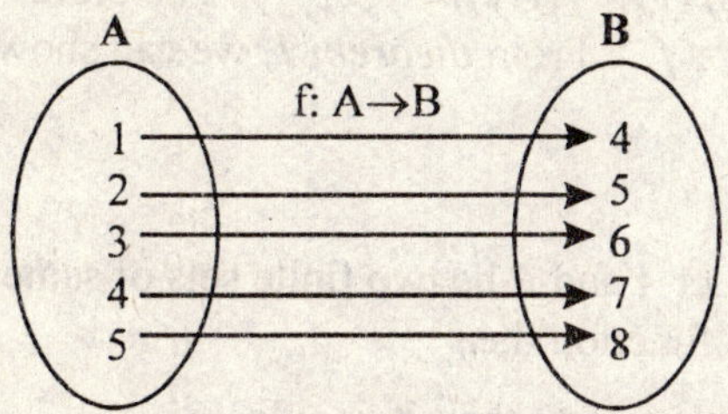

Fig. 2.1.3.

When a function is One to One it is called **injection** and if it is One to One Onto it is called **bijection.** The following theorem shows that inverse of a function *f* exists if and only if *f* is One to One and Onto.

Theorem 1: Let *f* be a function *f*: ***A→B*** such that *f* is **bijection**, then show that f^{-1} is also a function from *B* to *A*, and it is a **bijection**.

Proof: Let *f*: ***A→B*** be a one to one and onto function from *A* to *B*. Then Dom (*f*) = A and Range (*f*) = *B*. Let f^{-1} be a function from *B* to *A*. Then we have to show that f^{-1}: ***B→A*** is a function and it is one to one and onto.

Since Range (*f*) = *B*, f^{-1} is everywhere defined on *B*. Also *f* is one to one, so no two or more elements of *A* have same image in *B*. Therefore, no two or more elements of *B* can have same pre-image in *A*. Hence, f^{-1} is a function from *B* to *A*.

Let y_1 and y_2 be any two elements of *B* then ∃ x_1 and x_2 in A such that $f(x_1) = y_1$ and $f(x_2) = y_2$. Now, $f(x_1) = y_1$ and $f(x_2) = y_2 \Rightarrow x_1 = f^{-1}(y_1)$ and $x_2 = f^{-1}(y_2)$. Since *f* is one to one, we have $y_1 \neq y_2$ for

$x_1 \neq x_2$, i.e., for $y_1 \neq y_2$ we have $f^{-1}(y_1) \neq f^{-1}(y_2)$. Therefore f^{-1} is one to one function. Since f is a function from A to B, every element of A is pre-image of some element of B, i.e. every element of A is image of some element of B under f^{-1}. Hence f^{-1} is onto.

Proved.

Theorem 2: Let f be a function f: $A \rightarrow B$, defined as $f(x) = y$ and that f is a **bijection.** Show that a function **g:** $B \rightarrow A$ defined as $g(y) = x$ is inverse of f and it is also a **bijection.**

Proof: Let x be any element of A. Then there exists a y in B such that $f(x) = y$ by definition of f. Now, $(g \, 0 \, f)(x) = g(f(x)) = g(y) = x$. Similarly, $(f \, 0 \, g)(y) = f(g(y)) = f(x) = y$. Therefore, we have $g \, 0 \, f = f \, 0 \, g = I_A$, i.e. $g = f^{-1}$. From ***theorem 1,*** we can show that g is one to one and onto.

Proved.

Theorem 3: Let A and B be two finite sets of same cardinality, and let f: $A \rightarrow B$ is a function then

(i) If f is one to one then it is onto.
(ii) If f is onto then it is one to one.

Proof: (i) Let $|A|$ be n and since f is one to one, B must have at least n elements for f to be defined from A to B. Since $|B| = |A|$, $\exists$ no element in B which is not an image of any element of A. This proves that function f is onto.

(ii) Let f is not one to one then f is many to one. Since $|A| = |B|$, there must exist at least one element in B which is not associated with any element of A. This leads to conclusion that f is not onto. This is contrary to that fact that f is onto. Thus our assumption that f is not one to one is wrong. Hence f is one to one.

Proved.

Example 4: Let A is a set of all positive real numbers and B is a set of all real numbers. Let f be a function f: $A \rightarrow B$ defined as $f(x) = \log_e x$. Show that f is one to one and onto function.

Solution: Let x_1 and x_2 be any two elements of A such that

$$\log_e x_1 \neq \log_e x_2$$

Now,

$$\log_e x_1 \neq \log_e x_2 \Leftrightarrow e^{x_1} \neq e^{x_2} \Leftrightarrow x_1 \neq x_2$$

Therefore, *f* is one to one. Now, let *y* be any element of *B*, then e^x is a positive real number in A, such that $\log_e e^y = y$, i.e. for every element *y* in $B \exists e^y$ in *A*, such that $\log_e e^y = y$. Hence *f* is onto also.

One to One Into function

A function $f: A \rightarrow B$ is said to be One to One and Into from a set *A* to a set *B* if and only if

1. *f* is **One to One** i.e. for any two elements x_1 and x_2 of *A* if $x_1 \neq x_2$ then $f(x_1) \neq f(x_2)$, and
2. *f* is **into** i.e. $\exists$ at least one element *y* in *B* such that it is **not** an image of any element *x* of *A*.

Example 5: Let *A* is the set of positive integers and *B* is a set of even integers. Let *f* be a function $f: A \rightarrow B$ defined as $f(x) = 2x$. Show that *f* is one to one and into function.

Solution: Let x_1 and x_2 be any two distinct elements of *A* then $2x_1 \neq 2x_2$. i.e. *f* is one to one. Since there are many (**infinite**) negative even integers in *B* which are not associated with any of the element of *A* (twice of a positive integer is positive integer only). *f* is an into function.

A diagrammatic example of an One to One Into function can be given as in **Figure 2.1.4.** The function *f* from set *A* to set *B* is defined as $f(x) = x + 3$.

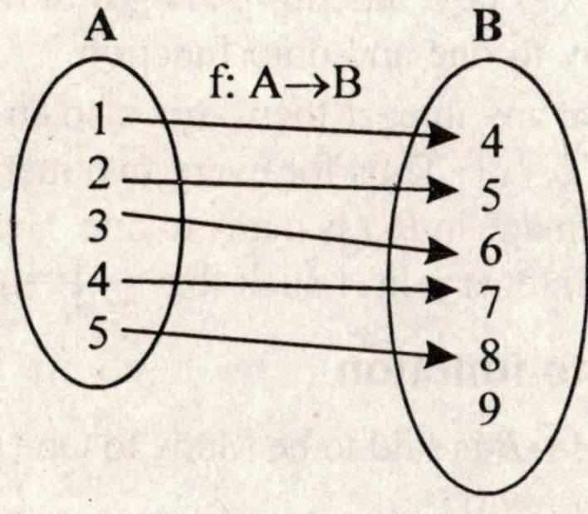

Fig. 2.1.4.

Many to One Onto function

A function $f: A \to B$ is said to be Many to One and Onto from a set A to a set B if and only if

1. f is **Many to One** i.e. $\exists$ at least one pair of distinct elements x_1 and x_2 in A such that $f(x_1) = f(x_2)$, and
2. f **is onto**, i.e. for every element y in B $\exists$ an element x in A such that $f(x) = y$.

In the following examples, shown in **Figure 2.1.5**, f is a function from A to B. Distinct elements pair 1 and 2 have same image 4 in B. F is many to one function. Further, $\exists$ no element in B that is not an image of any element of A. Thus f is onto.

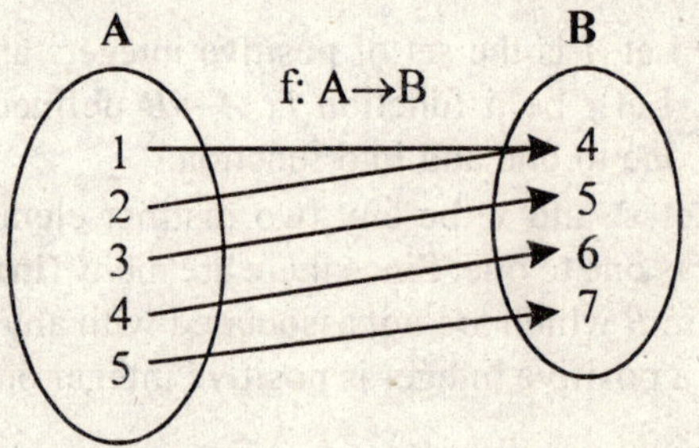

Fig. 2.1.5.

Example 6: Let A is the set of integers and B is the set of all non-negative integers. Let f be a function $f: A \to B$ defined as $f(x) = |x|$. Show that f is many to one and onto function.

Solution: Let x be any integer then $-x$ is also an integer and $|\pm x|$ is magnitude of x i.e. $|x|$. Thus for every two members of A (except zero) we have one image in B. f is many to one. Next, for any element y in B we have an integer y in A such that $|y| = y$. f is onto.

Many to One Into function

A function $f: A \to B$ is said to be Many to One and Into from a set A to a set B if and only if

1. f is **Many to One** i.e. $\exists$ at least one pair of distinct elements x_1 and x_2 in A such that $f(x_1) = f(x_2)$, and

2. f is **into** i.e. ∃ at least one element y in B such that it is **not** an image of any element x of A.

Example 7: Let A and B be set of real numbers. Let f be a function f: $A \to B$ defined as $f(x) = x^2$. Show that f is many to one and into function.

Solution: Let x be any real number then x is also a real number and x^2 is a positive real number. Thus for every two members of A (except zero) we have one image in B. f is many to one. Next, ∃ real numbers in B (all negative), which are not square of any real numbers. Therefore, f is into function.

In **Figure 2.1.6**, f is a function from A to B. Distinct elements pair 1 and 2 have same image 4 in B. f is many to one function. Further, ∃ an element 8 in B, which is not an image of any element of A, thus f is into.

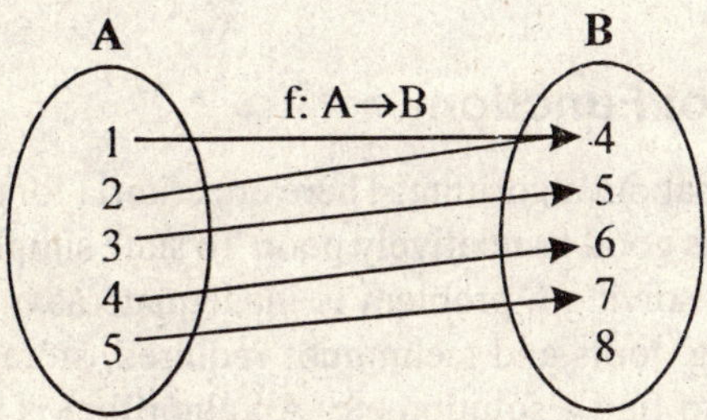

Fig. 2.1.6.

Example 8: Let $A = \{-\pi, \pi\}$ and B be set of real numbers. Let f be a function f: $A \to B$ defined as $f(x) = \sin(x)$. Show that f is many to one and into function.

Solution: Since $\sin(-\pi) = \sin(\pi) = 0$ i.e. for two distinct elements of A, we have same image in B, f is many to one function. Further there are many elements (infact, all except zero) in B which are not image of either $-\pi$ or π. Therefore, f is into function.

It is important to mention, here, two more types of functions, which are frequently used. One is **Constant function** and other is **real valued function.** A function is said to be real valued if its range set is the set of real numbers. Examples 4, 7 and 8 mentioned in this section are examples of a **real valued function.** A function f: $A \to B$

is called a **Constant function** if for every $x \in A$, there is only one image say k in B. For example $f(x) = 5$ is a constant function from R to R. A diagrammatic example is given in **figure 2.1.7.**

In this example $f(x) = 5\ x \in A$. Thus f is a constant function. **Can we say that f is many to one into also in this example?**

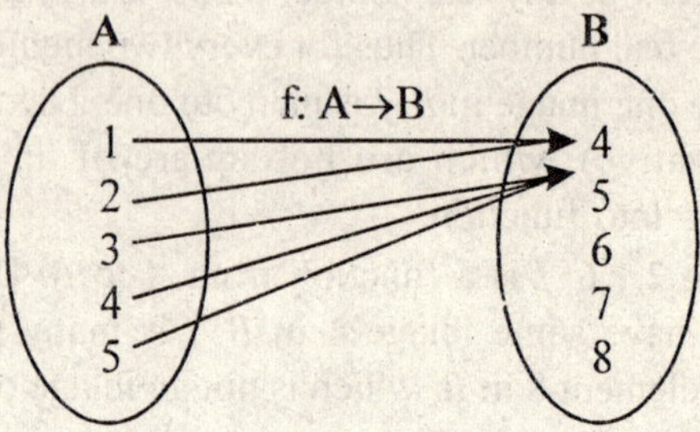

Fig. 2.1.7.

2.2 Order of Function

We know about algorithm. There are criteria for deciding whether an algorithm is good or relatively poor. To state simply an algorithm is a method for solving a problem is inadequate as many assumptions are hidden e.g. tools and techniques required, storage requirements, time needed to find a solution etc. An algorithm is then defined as a discrete, deterministic method for solving a problem that terminates in a finite number of steps, whatever be the input data.

This definition of an algorithm does not provide criteria for the goodness of an algorithm. Several concepts are involved in such a judgment. Some of these are: correctness, tolerant to erroneous data, number of steps needed to find solution, main memory requirement, elegance, specificity, generality etc. Some of these are measurable criteria whereas others are subjective in nature. **Time,** number of steps needed to find the solution, and **storage,** main memory requirement can be measured but criteria like elegance, specificity etc, are not measurable. Order of a function is used to measure time and storage requirement (complexity) of an algorithm. *Time complexity of a problem is defined as time complexity of a best possible algorithm for solving that problem.* Let us consider the following algorithm of sorting an array of n items in ascending order.

```
BUB_SORT (List, no_of_item)
{
    for (int i=0; i< no_of_item; i++)
    for (int j = i; j< no of item -1; j++)
        if (list | j | > list | j + 1 |)
            {
                int tmp = list | j |;
                list | j | = list | j + 1 |;
                list | j + 1 | = tmp;
            }
    }
```

Algorithm 2.2.1.

Let number of items is n. Size of every item in the list of this algorithm, is fixed. Thus storage complexity is a function of n, and it is directly proportional to n, i.e. as n increases the storage requirement increases. Let f and g are functions whose domains are subsets of Z^+. We say f is O (g) if there exists constants c and k such that

If $(n) | \leq c. |g(n)| \ \forall \ n \geq k$.

If f is O (g), then f increases no faster than g. In the above example, let f be storage complexity function then $f(n) = n + 3$; n is number of items, two counter variable i & j and one tmp variable used for swapping.

$f(n) = n + 3 \leq - n + n \leq 2\text{n} = 2.g(n)$

Where, $g(n) = n$. thus $f(n)$ is $O(g)$ i.e. $f(n)$ is $O(n)$. Let us consider, now, the time complexity of the same algorithm. To find the required time, we take into account the number of comparisons needed to perform the above task. There are $n - 1$ comparisons in first pass, $n - 2$ in second pass, $n - 3$ in third pass & so on, and 1 comparison in last pass. Therefore, there are in total

$$1 + 2 + 3 + \cdots + n - 1 = \frac{n(n-1)}{2} \ \textit{comparisons.}$$

Let $f(n)$ be the function of time complexity of the above algorithm then

$$f(n) = \frac{1}{2}(n^2 - n)$$

$$\text{or} \leq \frac{1}{2}n^2 + n$$

$$\text{or} \leq \frac{1}{2}n^2 + n^2$$

$$\text{or} \leq n^2$$

∵ time complexity is of $O(n^2)$

Now, let us consider another algorithm of finding whether a relation R defined on a set A is transitive or not. Let $|A| = n$ and $|R| = P$. Let M_R be the matrix of R and $A = \{1, 2, 3, \ldots N)$. R is stored in a two dimensional array, say MAT. If ordered pair $(i, f) \in R$, we have MAT $(i, f) \leftarrow 1$

```
TRANS (MAT, n)
{
   RESULT ¬ T
   for (int i = 0; i < n; i++)
      for (int j = 0; j < n; j++)
         if (MAT (i, j) == 1)
         {
            for (int k = 0; k < n; k++)
               if (MAT (j, k) == 1) and (MAT (j, k) == 0))
               RESULT ← F
         }
      }
```

Algorithm 2.2.2.

Amount of storage required, for matrix MAT, is n^2. Thus if f is the measure of storage complexity of this algorithm then $f(n)$ is $O(n^2)$. We shall now find time complexity of the algorithm. Observe that outer two loops corresponding to i & j run n times each. If (i, j) is not in R, the only test "if (MAT(i, j) == 1)" is performed. If the test is false, the rest of statements are not executed. Since only p ordered pairs are in R, "if (MAT (i, j) == 1)" will be true for p times only.

That means $n^2 - p$ times, only up to "if (MAT (i, j) == 1)" is executed and the rest of the instructions are not executed. The case when condition "if (MAT (i, j) == 1)" is true, an additional loop corresponding to k is executed n times. This means the test "if ((MAT (j, k) == 1) and (MAT (j, k) == 0))" is executed np times. Thus, total number of execution required by the algorithm 2.2.2 is given as $n^2 - p + np$. If f be the time complexity function of n, then $f(n)$ is given as

$$f(n) = n^2 - p + np$$

Since p must be between 0 and n^2, let $p = k\,n^2$, where $0 \leq k \leq 1$. Therefore,

$$f(n) = n^2 - k\,n^2 + k\,n^3 = k\,n^3 + (1 - k)\,n^2$$

So $f(n)$ is $O(n^3)$. Having discussed this, we shall now discuss a few examples before taking up the rules to determine order of a function.

Example 1: Show that

$$f(n) = \frac{n^3}{2} + \frac{n^2}{2}$$

is $O(g)$ for $g(n) = n^3$

Solution: To see this we proceed as below:

$$f(n) = \frac{n^3}{2} + \frac{n^2}{2} \leq \frac{n^3}{2} + \frac{n^3}{2} = 1 * n^3$$

or, $f(n) \leq 1 * n^3$

If $c = 1$ and $k = 1$, we have $|\,f(n)\,| \leq c.\,|\,g(n)\,|\ \forall\ k \geq n$ and so f is $O(g)$.

Example 2: Show that $f(n) = 3n^4 - 5n^2$ and $g(n) = n^4$ are of same order.

Solution: $f(n) = 3n^4 - 5n^2$

$$\begin{aligned} &\leq 3n^4 - 5n^2 \\ &\leq 3n^4 + 5n^2 \\ &\leq 3n^4 + 5n^4 \\ &= 8n^4 \end{aligned}$$

Now, let $c = 8$ and $k = 1$, then $|\,f(n)\,| \leq c.\,|\,g(n)\,|$ for all $n \geq k$.

Thus f is $O(g)$. And

$$\begin{aligned} g(n) &= n^4 \\ &= 3n^4 - 2n^4 \\ &\leq 3n^4 - 5n^4 \text{ for } n \geq 2 \end{aligned}$$

If we put $c = 1$ and $k = 2$. then $| g(n) | \leq$ c. If $| (n) |$ for all $n \geq$ k. Therefore g is $O(f)$.

Example 3: Show that every logarithmic function $f(n) = \log_b n$ has the same order as $g(n) = \log_2 n$.

Solution: We know the logarithmic change of base identity $\mathbf{\log_a x = \log_a b * \log_b x}$. Let $a = 2$, then for any positive integer b ≥ 3, $\mathbf{\log_a b} \geq 1$ and is a constant. Also $1/\mathbf{\log_a b}$ is a constant. So for any base b ≥ 3, we have,

$$\log_2 n = \log_2 b * \log_b n$$

$$\text{or, } | \log_2 n | = \log_2 b * | \log_b n |$$

$$\text{or } | g(n) | = \log_2 b * | f(n) |$$

i.e., g is $O(f)$. Similarly, we can write

$$\log_b n = \frac{1}{\log_2 b} * \log_2 n$$

$$\text{or, } |\log_b n| = \frac{1}{\log_2 b} * |\log_2 n|$$

$$\text{or, } |f(n)| = \frac{1}{\log_2 b} * |g(n)|$$

Or, f is $O(g)$. This shows that every logarithmic function is of same order as that of $\log_2 n$. That means while mentioning the complexity of an algorithm and if it is of logarithmic order, base of log is not important.

Let F be the set of all functions whose domains are subsets of set of positive integer. Clearly an element of F can be used to determine complexity, of an algorithm. Let Θ (big theta) be a relation on F defined as $f \Theta g$ if and only if both f and g are of the same order for any function f and g belonging to F.

Theorem 1: Relation Θ defined as above on F is an equivalence relation.

Proof: Since every function f of F is of order itself, the relation Θ is reflexive. Let f and g be any two elements of F and if $f \Theta g$, then both f and g are of same order, so $g \Theta f$, i.e., Θ is symmetric. Next, let f, g and h be any three elements of F such that $f \Theta g$ and $g \Theta h$. This implies that f and g are of same order & g and h are of same order. This shows that f and h are also of same order, i.e., $f \Theta h$ and so Θ is transitive. Therefore Θ is an equivalence relation.

Proved.

We know that every equivalence relation on a set induces an equivalence partition. The relation Θ partitions F into equivalence classes. Each equivalence class contains functions that are of same order. As usual, we use any simple function of an equivalence class to represent that class and hence to represent the order of all functions in that class. One Θ class $|f|$ is said to be **lower** than another Θ class $|g|$ if a representative function f from $|f|$ is of lower order than that of any g from $|g|$. This implies that a function from class $|f|$ grows more slowly than a function from class $|g|$ as magnitude of n, value from domain of function of number of items to be processed by the algorithm, increases. It is the Θ class of a function that gives the information we need for analysis of algorithm.

Rules for Determining the Θ Class of a Function

1. A constant function has zero growth as n increases. This class of function is represented as Θ (1) class. In the hierarchy of time complexity, this class of function corresponds to the lowest class. All **hash functions** belongs to this class.
2. Any logarithmic function grows more slowly than any power function with positive exponent, i.e., Θ ($\log n$) is lower than Θ (n^k) for $k > 0$.
3. Any exponential function with base greater than 1 grows more rapidly than any power function, i.e., Θ (n^k) is lower than Θ (a^n) for any k and $a > 1$.
4. Θ (n^a) is lower than Θ (n^b) if and only if $a < b$, i.e. in case of two power functions the hierarchy is decided on the basis of theirs exponents.
5. Among the exponential functions the hierarchy is determined on the basis of the base of the function, i.e., Θ (a^n) is lower than Θ (b^n) if and only if $a < b$.

6. If a function is multiplied by a non zero constant, its order remains unchanged, i.e., for any non zero k $\Theta(kf) = \Theta(f)$ for any f.
7. While determining the hierarchy of order of composition of functions, the hierarchy of constituent functions is the determining factor. If h is a non zero function and $\Theta(f)$ is lower than $\Theta(g)$, then $\Theta(fh)$ is lower than $\Theta(gh)$.
8. Finally, if $\Theta(f)$ is lower than $\Theta(g)$, then $\Theta(f+g) = \Theta(g)$.

In computer science while analyzing algorithm, most frequently used Θ–classes of functions are constant $\Theta(1)$, logarithmic, power $\Theta(n^k)$, exponential $\Theta(a^n)$ and composition of these functions only. The above set of rules is sufficient to determine the complexity of any algorithm and of any problem.

Example 4: Determine the Θ class of the following functions.

(i) $f(n) = 4n^4 - 6n^7 + 25n^3$ (ii) $g(n) = \log n - 3n$

(iii) $h(n) = (1.001)n + n^{15}$

Solution:

1. Here, $f(n) = 4n^4 - 6n^7 + 25n^3$. From rule of determining order of power function, $\Theta(f) = \Theta(n^7)$.
2. From rule 2, we know that any logarithmic function grows more slowly than any power function, so ($\log n - 3n$) grows more slowly than n, thus $\Theta(g) = \Theta(n)$.
3. Rule 3 above states that any exponential function with base greater than 1 grows more rapidly than any power function, so n^{15} grows more slowly than $(1.001)n$, thus $\Theta(h) = \Theta((1.001)n)$.

2.3 Sequences and Series

Let us consider the following collection of numbers.

(i) 28, -2.25, 27, 32, 7, ...

(ii) 2, 7, 11, 19, 31, 51, ...

(iii) 1, 2, 3, 4, 5, ...

(iv) 20, 18, 16, 14, 12, ...

(v) 4, 8, 16, 32, 64, ...

In the collection (i), it is not possible to know which number will follow 7 as no definite rule is followed by the elements in this collection.

In the **second** collection no rule is noticed except that all are positive and are in ascending order. In collection **three,** if one asks "what is the nth element?" Prompt reply comes from some corner that it is n. This is possible because every element is arranged according to some rule. The rule here is, succeeding element is one greater than its predecessor. Similarly, we can predict that nth term in (iv) is **$20 - 2n$ (for $n = 0, 1, 2, \ldots, n$)** and that in (v) it is **$4x(n + 1)$ ($n = 0, 1, 2, \ldots, n$).** Therefore collections in (iii), (iv) and (v) are **sequence** but that in (i) and (ii) are not. We can, then, define a sequence as:

"An ordered collection of numbers $a_0, a_1, a_2, a_3, \ldots a_n, \ldots$; is a sequence if according to some definite rule or law, there is a definite value of an called the term or elements of the sequence corresponding to any value of the natural number n."

Clearly, t_0 is the initial term of the sequence t_1 is the first term, t_2 is the second term..., t_n is the nth term. In the nth term, t_n, by successively substituting 0, 1, 2, 3, ..., for n we get t_0, t_1, t_2, t_3, and so on. It is obvious now that nth term of a sequence is a function of natural number n. A sequence is denoted by $\{t_n\}$ or by (t_n) or simply by t, in which t_n is the nth term of sequence. The nth term t_n is called **general term** of the sequence.

Example 1: Let t be a sequence in which general term t_n is given by $= 2n + 3$, then by substituting $n = 0, 1, 2, \ldots$ we get t_0, t_1, t_2 and so respectively as given below.

When $n = 0, t_0 = 2 * 0 + 3 = 3$
$n = 1, t_1 = 2 * 1 + 3 = 5$
$n = 2, t_2 = 2 * 2 + 3 = 7$
$n = 3, t_3 = 3 * 3 + 3 = 9$

Similarly, when $n = 10, t_{10} = 2 * 10 + 3 = 23$ and so on.

The above example shows that there is correspondence of one to one between the set of natural numbers and the elements of a sequence. If f is function from N, the set of natural number, to R, the set of real number then depending upon the $f(N) \subset R$, we have different sequences. A function f: $N \rightarrow \boldsymbol{R}$ gives a sequence of real numbers. There could be infinite many such functions, and so are infinite many possible sequences of real numbers. We call these functions as **discrete numeric functions** or simply **numeric functions.** We can say $f(0)$

is a_0, $f(1)$ is a_1, $f(2)$ is a_2 and so on. The sequence a, is called **numeric function.** We shall discuss how to find a **generating function** for a given sequence i.e. for a given **numeric function** and how to find a **numeric function** from a given **generating function.** Before that, let us learn about different types of sequences and way to represent a sequence.

Depending upon the number of terms in a sequence, we may have a finite or **infinite** sequence. A sequence is said to be infinite if number of terms is finite i.e. there is a last term in the sequence. For example, the sequence of dates in the month of August 2001, which are Mondays, i.e.

6, 13, 20, 27

This is, obviously, a finite sequence. A sequence is said to be **infinite** if there are infinite many terms in that sequence i.e., each term is followed by another term. In such a sequence there is no last term. For examples.

(i) Sequence of positive odd integers

1, 3, 5, 7, ...

(ii) Sequence of positive even integers

2, 4, 6, 8, ...

(iii) Sequence of prime numbers

2, 3, 5, 7, 11, 13, 17, 19, 23, ...

Further, if the absolute value of any terms of a sequence do not exceed a definite value, then the sequence is called a **bounded sequence.** Thus a sequence t is said to be bounded if there exists a positive number K such that

$$|t| < K \ \forall \ n.$$

If there exists no such numbers K satisfying the condition mentioned above, the sequence is said to be **unbounded.** All finite sequences are bounded sequence. The following sequences are infinite and bounded.

(i) $1, \frac{1}{2}, \frac{1}{3}, \frac{1}{4}, \ldots\ldots, \frac{1}{n}, \ldots\ldots$

(ii) $1, \frac{1}{2}, \frac{2}{3}, \frac{3}{4}, \ldots\ldots, \frac{n}{n+1}, \ldots\ldots$

(iii) 1, –1, 1 –1, 1, ... $(-1)^n$, ...

However, the following sequences are unbounded

(i) 1, 2, 3, 4, ... $(n + 1)$, ...

(ii) 1, 2, 4, 8, 16, ..., 2^n , ...

(iii) 1, 3, 5, 7, ..., $2n + 1$, ...

A sequence t in which a term t_{n+1} is never less than any of its previous terms i.e. $t_{n+1} \geq t_n$ for every value of n, then t is called **monotonic increasing** sequence. Similarly, a sequence t in which a term t_{n+1} is never greater than any of its previous terms i.e. $t_{n+1} \leq t_n$ for every value of n, then t is called **monotonic decreasing** sequence. A monotonic increasing or monotonic decreasing sequence is called **monotonic sequence.** The following sequences are monotonic increasing.

(i) 1, 2, 3, 4, ... $(n + 1)$, ...

(ii) 1, 2, 4, 8, 16, ..., 2^n, ...

(iii) 1, 3, 5, 7, ..., $2n + 1$, ...

However, the sequences given below are monotonic decreasing.

$$1, \frac{1}{2}, \frac{1}{3}, \frac{1}{4}, \ldots\ldots$$

The following is an example of non-monotonic sequence.

$$1 - \frac{1}{3!}, 1 - \frac{1}{3!} + \frac{1}{5!}, 1 - \frac{1}{3!} + \frac{1}{5!} - \frac{1}{7!}, \ldots\ldots$$

In principle, a sequence can be specified by exhaustively listing the values of terms (0th, 1st, 2nd etc. elements) of that sequence. But, in practice, it is not possible to enlist all the elements of an infinitely long sequence. Thus we need a representation of sequence which is not infinitely long. A sequence t may be expressed by an algebraic formula describing its general term t_n. This formula for t_n may be in terms of only its position number n in the sequence or in terms of previous sequence terms $t_{n-1}, t_{n-2}, t_{n-3}\ldots$etc. The former is called **explicit formula** and the latter is called **recurrence formula** or **recurrence equation.**

Example 2: The sequence 5, 10, 20, 40, ..., can be represented by an explicit formula

$a_n = 5 * 2^n$ *for* $n \geq 0$

And, by an recurrence formula as

$a_n = 2a_{n-1}$ *for* $n \geq 1$ *with initial condition* $a_0 = 5$

Example 3: The sequence 3, 7, 11, 15, ..., can be represented by an explicit formula

$a_n = 4n + 3$ *for* $n \geq 0$

And, by an recurrence formula as

$a_n = a_n + 4$ *for* $n \geq 1$ *with initial condition* $a_0 = 3$

Example 4: The sequence 87, 82, 77, 72, 67, ..., can be represented by an explicit formula

$a_n = 87 - 5n$ *for* $n \geq 0$

And, by an recurrence formula as

$a_n = a_n - 1 - 5$ *for* $n \geq 1$ *with initial condition* $a_0 = 87$

Example 5: The sequence –4, 16, –64, 256, 1024, ..., can be represented by an explicit formula

$a_n = (-4)^{n+1}$ for $n \geq 0$

And, by an recurrence formula as

$a_n = (-4) * a_{n-1}$ *for* $n \geq 1$ *with initial condition* $a_0 = -4$

We have noticed that in recurrence formula an **initial condition,** also called the **terminating condition,** is used. As in a recursive function, a recursive formula must have a terminating condition. The number of terminating conditions depends upon the order of a recurrence relation. We shall learn about this in the next chapter.

Example 6: Find the different terms of numeric function

$$a = a_n = \begin{cases} 2r & 0 \leq r \leq 11 \\ 3^r - 1 & r > 11 \end{cases}$$

Solution: The sequence terms for 0 to 11 is given by formula $2r$ and then it is given by the formula 3^r-1. Putting $r = 0, 1, 2, .., 11, 12, 13, \ldots$, we get the sequence as below.

$\boldsymbol{a} = 1, 2, 4, 6, \ldots, 22, 531440, 1594322, \ldots$

A set contains only distinct element, as we have learn in the first chapter. The order of appearance of an element in a set is not important. Whereas, in a sequence order of term is important but an element may be repeated any number of times. **A set corresponding to a sequence** is collection of all **distinct elements** forming that sequence. For example, set corresponding to sequence 1, 2, 3, 1, 2, 3, 1, 2, 3, ... is {1, 2, 3}. Similarly, a set corresponding to sequence 1, 2, 3, 4, ..., is {1, 2, 3, 4, ...}, the set of positive integers. On the other hand, if a set is given, we can form a sequence from the elements of that set. For example, let $A = \{0, 1\}$. Then we can have sequences like

1, 0, 1, 0, ...
1, 1, 1, ...
0, 0, 0, ...
And so on.

All such sequences arising from set A. In fact, we can have infinite many sequences arising from the set A. Some of the sequences will be finite and some infinite.

The indicated sum $\boldsymbol{t_0 + t_1 + t_2 + t_3 + \ldots + t_n + \ldots}$, of the terms of a sequence $\boldsymbol{t} = (t_0, t_1, t_2, t_3, \ldots)$ is called **series**. For example,

(i) $1 + 2 + 3 + 4 + \ldots$
(ii) $1 - 2 + 4 - 8 + 16 - \ldots + (-2)n + \ldots$

are series. A series may be finite or infinite according to the sequence it corresponds to. If $\boldsymbol{S_n = t_0 + t_1 + t_2 + t_3 + \ldots + t_n}$, then S_n is called the sum to n terms of $\boldsymbol{t}$, or **nth partial sum** of the series

$$\sum_{i=0}^{n} t_n$$

The sequence in example 6 has been represented using **characteristic function** notation. Many sequences are represented in that way. Particularly, when nature of algebraic formula differs from one subset of natural numbers to another subset of natural number for a sequence. This gives the concept of selection of definition (**block of statements**) accordinr to situation. If there are two definitions and one is to be selecteu the selection is called **Binary Branching,** c ::nply **Branching** (If Then Else). If one selection is to be made . of many (but finite), the selection is called **Multiple Branching** (Switch case). This concept has come from the definition of the **characteristic function.** A function f**:** $A \to \{0, 1\}$ from a nonempty set A to set $\{0, 1\}$ defined as

$$f(x) = \begin{cases} 0 & if \quad x \notin A \\ 1 & if \quad x \in A \end{cases}$$

is called characteristic function of set.

Theorem 1: Characteristic functions of subsets satisfy the following properties:

(i) $f_{A \cap B} = f_A f_B$

(ii) $f_{A \cup B} = f_A + f_B - f_A f_B$

(iii) $f_{A \oplus B} = f_A + f_B - 2 f_A f_B$

Proof: (i) Let x is any element of A∩B. Then, $x \in$A or $x \in B$. Therefore,

$$f_{A \cap B}(x) = 1 \iff f_A(x) = 1 \textit{ and } f_B(x) = 1$$
$$\iff f_A(x)\ f_B(x) = 1$$

Similarly, if $x \notin A \cap B$ *then,*

$$f_{A \cap B}(x) = 0 \iff f_A(x) = 0 \text{ or } f_B(x) = 0$$
$$\iff f_A(x)\ f_B(x) = 0$$

Therefore, $f_{A \cap B} = f_A f_B$

Proved.

(ii) Let x is any element of $A \cup B$. Then $x \in B$. There are three possibilities only. (a) $x \in A$ and $x \in B$, (b) $x \notin A$ and $x \in B$ and (c) $x \in A$ and $x \notin B$. Therefore,

$$f_{A \cup B}(x) = 1 \iff f_A(x) = 1 \textit{ and } f_B(x) = 1$$

$$or,\ f_A(x) = 1\ and\ f_B(x) = 0$$
$$or,\ f_A(x) = 0\ and\ f_B(x) = 1$$

Similarly, if $x \notin A \cap B$ then,

$$f_{A\cup B}(x) = 0 \Leftrightarrow f_A(x) = 0\ and\ f_B(x) = 0$$
$$and\ f_A(x) f_B(x) = 0$$

Therefore, $f_{A\cup B} = f_A + f_B - f_A f_B$

Proved.

(iii) Let x is any element of $A \oplus B$. Then, there are two possibilities only. (a) $x \in A$ and $x \notin B$ and (b) $x \notin A$ and $x \in B$. Further, if x is not a member of $A \oplus B$, then either x is in both A and B or not in A and B. Therefore,

$$f_{A\oplus B}(x) = 1 \Leftrightarrow f_A(x) = 1\ and\ f_B(x) = 0$$
$$or,\ f_A(x) = 0\ and\ f_B(x) = 1$$

Similarly, if $x \notin A \cap B$ then,

$$f_{A\oplus B}(x) = 0 \Leftrightarrow f_A(x) = 0\ and\ f_B(x) = 0$$
$$or,\ f_A(x) = 1\ and\ f_B(x) = 1$$

Therefore, $f_{A\oplus B} = f_A + f_B - 2 f_A f_B$

Proved.

This completes the proof.

2.4 Generating Function

An important idea in mathematics is to establish connection between two fields in order to apply knowledge in one field to the other field, or at least take a problem of one field and transform it to a problem in the other field. This idea motivates the idea of a **generating function,** which establishes a connection between functions of a real variable and sequences of numbers.

A **generating function** can be defined as "Given a numeric sequence $a = a_0, a_1, a_2, \ldots, a_n, \ldots.$, the series

$$f(x) = a_0 + a_1 x + a_2 x^2 + \ldots\ldots + a_n x^n + \ldots\ldots$$

is called the **generating function** of the sequence."

Suppose $h = h_0, h_1, h_2, \ldots, h_n, \ldots,$ is a sequence of numbers. We write it as an infinite sequence, but we mean to include finite sequences also as well. **If**

$$g(x) = h_0 + h_1x + h_2x^2 + \ldots + h_nx^n + \ldots = \sum_{n=0}^{\infty} h_n x^n$$

$h_0, h_1, h_2, \ldots, h_n$ is finite sequence, we make it of infinite length, simply, by setting $h_r = 0$ for $r > n$. A generating function $g(x)$ for the sequence h is an infinite series such that

Thus, the function generates the sequence as its sequence of coefficients. If the sequence is finite then there is an m for which **$h_r = 0$ for $r > m$.** In this case $g(x)$ is an ordinary polynomial in x of degree m. The inspiration for this idea is the **Binomial Theorem.** The function $g(x) = (1 + x)^m$ generates the binomial coefficient **$h_r = C$ (m, r)**. Therefore a generating function in which coefficients of x^n are sequence terms of the sequence **h**, is called **Binomial Generating function** of the sequence **h**. If we recall our basic mathematical knowledge of school days, the binomial coefficient **$C\,(m, r)$** gives us the total number of **combinations** of **r** selections from **m** objects. Thus we can infer from here that the concept of binomial generating function can be used to solve problems of combinations. How to do it, we will learn in the **chapter 4.**

Permutation is a sister term (one may use brother to avoid gender bias) while discussing combination. One comes to mind whenever other is referred. We know that **$P(m, r) = r! * C(m, r)$.** This means if **$C(m, r)$** is coefficient of x^r then **$P(m, r)$** is coefficient of **$x^r/r!$**. Now, let **$h(x)$** is a **generating function** for a sequence a given in series form as below.

$$h(x) = a_0 + a_1x + a_2\frac{x^2}{2!} + a_2\frac{x^3}{3!} + \cdots = \sum_{n=0}^{\infty} a_n \frac{x^n}{n!}$$

This series is exponential. The generating function $h(x)$ defined as above for a sequence $a = a_0, a_1, a_2, \ldots, a_n, \ldots,$ is called **Exponential Generating** function. Clearly coefficients of exponential generating function gives the value of permutation when r selections are made from n objects. This is only tip of the iceberg. The answers to many-

many counting problems are just coefficient of certain polynomial generating functions. Actually, **De Moivre** introduced the notion of generating function in 1730 AD to solve recurrence equations. It look almost 530 years to find a closed form formula for a recurrence equation defined by **Fabonacci** in 1200 AD. In this book we shall discuss **Binomial** and **Exponential** generating function only.

To find the generating function for a sequence means to find a closed form formula for $f(x)$, one that has no ellipses(...). In essence, a generating function is that function whose coefficients are really the values that we seek. That is, we do not really care about evaluating the function but only about examining its coefficients.

Example 1: Find the generating function for the sequence a = 1, 1, 1, 1, ...

Solution: In this sequence the general term $a_n = 1$. Let $f(x)$ be the binomial generating function for this sequence. Then, $f(x)$ can be written as

$$f(x) = \sum_{n=0}^{\infty} a_n x^n = \sum_{n=0}^{\infty} 1 * x^n \qquad [\because a_n = 1]$$

$$= 1 + x + x^2 + x^3 + \ldots\ldots\ldots$$

$$= \frac{1}{1-x}$$ [∵ *The seires is geometric with x as common ratio*]

Therefore, $f(x) = \dfrac{1}{1-x}$ *is the binomial generating function for the sequence* ***a*** – 1, 1, 1, ...

Next, let $g(x)$ be the exponential generating function for the same sequence then $g(x)$ can be written as

$$g(x) = \sum_{n=0}^{\infty} a_n \frac{x^n}{n!} = \sum_{n=0}^{\infty} 1 * \frac{x^n}{n!} \qquad [\because a_n = 1]$$

$$= 1 + x + \frac{x^2}{2!} + \frac{x^3}{3!} + \ldots\ldots\ldots$$

$= e^x$ [∵ *The series is exponential with x as exponent*]

Therefore, $f(x) = e^x$ *is the exponential generating function for the sequence* $\boldsymbol{a} = 1, 1, 1, \ldots$

Example 2: Find generating function for the sequence $b = 1, 3, 9, \ldots, 3^n, \ldots$

Solution: In this sequence the general term $b_n = 3^n$. Let $f(x)$ be the binomial generating function for this sequence. Then, $f(x)$ can be written as

$$f(x) = \sum_{n=0}^{\infty} b_n x^n = \sum_{n=0}^{\infty} 3^n * x^n \qquad [\because b_n = 3^n]$$

$$= 1 + 3x + (3x)^2 + (3x)^3 + \ldots\ldots.$$

$$= \frac{1}{1-3x}$$ [∵ *The series is geometric with* 3x *as common ratio*]

Therefore, $f(x) = \dfrac{1}{1-3x}$ *is the binomial generating function for the sequence* $\boldsymbol{b} = \mathbf{1, 3, 3^2, 3^3} \ldots.$

Let $g(x)$ be the exponential generating function for the same sequence. Then, $g(x)$ can be written as

$$g(x) = \sum_{n=0}^{\infty} b_n \frac{x^n}{n!} = \sum_{n=0}^{\infty} 3^n \frac{x^n}{n!} \qquad [\because b_n = 3^n]$$

$$= 1 + 3x + \frac{(3x)^2}{2!} + \frac{(3x)^3}{3!} + \ldots\ldots$$

$= e^{3x}$ [∵ *The series is exponential with* 3x *as exponent*]

Therefore, $f(x) = e^{3x}$ is the exponential generating function for the sequence $\boldsymbol{b} = 1, 3, 3^2, 3^3, \ldots\ldots.$

Example 3: Find exponential generating function for sequence

$t = {}^nP_0, {}^nP_1, {}^nP_2, \ldots, {}^nP_n$

Solution: The given sequence is finite. All terms in this sequence for $m > n$ are zero. Let $h(x)$ be the exponential generating function for the given sequence then $h(x)$ can be written as

$$h(x) = \sum_{i=0}^{\infty} a_i \frac{x^i}{i!} = \sum_{i=0}^{\infty} {}^nP_1 \frac{x^i}{i!} \qquad [\because a_i = {}^nP_i]$$

$$= {}^nP_0 + {}^nP_1 x + {}^nP_2 \frac{x^2}{2!} + \ldots\ldots\ldots\ldots + {}^nP_n \frac{x^n}{n!}$$

$$= {}^nC_0 + {}^nC_1 + {}^nC_2 x^2 + \ldots\ldots\ldots + {}^nC_n x^n \quad [\because {}^nP_r = r! * {}^nC_r]$$

$$= (1 + x)^n$$

Therefore, the exponential generating function for the sequence

t $= {}^nP_0, {}^nP_1, {}^nP_2, \ldots\ldots., {}^nP_n$ *is given by*

$$h(x) = (1 + x)^n$$

Ans.

Example 4: Find the binomial generating function for the sequence

$a = 1, 2, 3, \ldots, r, \ldots$

Solution: In this sequence the general term $a_r = r$. Let $f(x)$ be the binomial generating function of the given sequence, then it can be written as

$$f(x) = \sum_{r=0}^{\infty} a_r x^r = \sum_{r=0}^{\infty} r * x^r \qquad [\because a_r = r]$$

$$or,\ f(x) = 1 + x + 2x^2 + 3x^3 + \ldots \qquad \text{———} (1)$$

The equation (1) is arithmetic-geometric in which first factor is following arithmetic progression and second following geometric. We find the sum of the series by multiplying equation (1) with x, and then by subtracting the result from equation (1). Thus, we get

$$(1 - x) f(x) = 1 + x + x^2 + x^3 + \ldots\ldots\ldots$$

$$or;\quad f(x) = \frac{1}{(1-x)^2}.$$ *This is the required binomial ge nerating function.*

We have seen some examples to find a generating function for a given sequence (discrete numeric function). Similarly, if a generating function $f(x)$ is given, we can find the sequence as coefficient of different powers of x. If $f(x)$ is a binomial generating function we take coefficient of x^n for all $n \geq 0$. If, on the other hand, $f(x)$ is an exponential generating function we collect the coefficient of $x^n/n!$ for all $n \geq 0$.

Example 5: Find the discrete numeric function (sequence) whose exponential generating function is given by $\mathbf{2e^x}$.

Solution: The given generating function can be written in series form as

$$2e^x = 2\left[1 + x + \frac{x^2}{2!} + \frac{x^3}{3!} + \frac{x^4}{4!} + \ldots\ldots\ldots\right]$$

Clearly, the coefficient $x^n/n!$ for $n \geq 0$, are 2, 2, 2, …. Therefore, if the discrete numeric function is $\boldsymbol{a}$, then $\boldsymbol{a} = 2, 2, 2, 2, \ldots..$

Ans.

Sometimes two or more sequences are considered together to know the combined behavior of these sequences. For example, if $\boldsymbol{a}$ is a sequence for the monthly income of husband and $\boldsymbol{b}$ is a sequence corresponding to monthly income of the wife, then sequence $\boldsymbol{a} + \boldsymbol{b}$ is the sequence representing joint monthly income of the couple. Similarly, if e is the sequence representing monthly balance in a savings account and d is the sequence corresponding to the monthly interest rate, which fluctuates from month to month, then the terms of sequence ***cd*** represents interest earned in each month.

Example 6: For the following two sequences $\boldsymbol{a}$ and $\boldsymbol{b}$, whose general terms are given find $\boldsymbol{a} + \boldsymbol{b}$ and $\boldsymbol{ab}$.

$$a_r = \begin{cases} 0^r & \textit{for} \quad 0 \leq r \leq 2 \\ 2^{-r} + 5 & \textit{for} \quad r \geq 3 \end{cases}$$

and

$$b_r = \begin{cases} 3 - 2^r & \textit{for} \quad 0 \leq r \leq 1 \\ r + 2 & \textit{for} \quad r \geq 2 \end{cases}$$

Solution: Let ***c*** be the sequence for ***a*** + ***b***. At $r = 2$, the general definitions of ***a*** and ***b*** cannot be simply added. For $r \geq 3$ and $0 \leq r \leq 1$, bottom and top definitions will be added, but at $r = 2$ the value will be determined and will be placed in the definition. Next let ***d*** = ***ab***. In this case, we shall find the sequence formula for d by multiplying the definitions of a and b and setting the range for sequence position ***r*** accordingly. The result is shown below.

$$c_r = a_r + b_r = \begin{cases} 3 - 2^r + 0 & for\, 0 \leq r \leq 1 \\ 4 + 0 & for \;\; r = 2^r \\ 2^{-r} + r + 7 & for \;\; r \geq 3 \end{cases}$$

and

$$c_r = a_r * b_r = \begin{cases} 0 & for\, 0 \leq r \leq 2 \\ r2^{-r} + 2^{-r+1} + 5r + 10 & for \;\; r \geq 3 \end{cases}$$

Ans.

Theorem 1: Let $f(x)$ is the generating function for ***a*** and $g(x)$ is the generating function for ***b***. Show that $f(x)+g(x)$ is the generating function for ***a*** + ***b***.

Proof: Let

$$f(x) = \sum_{n=0}^{\infty} a_n x^n \;\; and \;\; g(x) = \sum_{n=0}^{\infty} b_n x^n$$

be the generating functions for ***a*** and ***b***, then

$$f(x) + g(x) = \sum_{n=0}^{\infty} a_n x^n + \sum_{n=0}^{\infty} b_n x^n = \sum_{n=0}^{\infty} (a_n + b_n) x^n$$

Therefore, the sequence generated by $f(x) + g(x)$ *is*

$$(a_0 + b_0), (a_1 + b_1), (a_2 + b_2), \ldots\ldots.$$

These terms are precisely the sum of the corresponding terms from ***a*** and ***b***. Thus, $f(x)+g(x)$ is generating function for the sum of the sequence ***a*** + ***b***.

Proved.

In general, the above theorem can be applied to any finite number of sequences. The same composition cannot be applied for the product of two or more sequences, i.e., if $f(x)$ and $g(x)$ are generating functions for sequences a and b respectively then $f(x) * g(x)$ is not the generating function for $a * b$.

Theorem 2: Let $f(x)$ and $g(x)$ are generating functions for finite sequences ***a*** and ***b*** respectively. The **convolution** of two sequences ***a*** and ***b*** is the sequence ***e*** defined as

$$c_n = \sum_{r=0}^{n} a_r b_{n-r}$$

Show that $f(x)\, g(x)$ is the generating function for **convolution** of ***a*** and ***b***.

Proof: Let as an exercise to the readers.

The generating function method can be used to solve many of the counting problems. We shall be using this method to solve problems related to permutation and combination in the chapter 4. In the chapter 3, we shall see how to use this method to solve some recurrence equations as well. But, it is important to discuss one problem, here, before leaving this topic. For $k > 0$ the function $g(x) = 1/(1-x)^k$ generates the sequence $a = \{C(n + k - 1, n) \mid n = 0, 1, 2, 3, \ldots\}$. Thus the nth coefficient is the number of ways to select ***n*** **objects** of ***k*** **types**. Obviously,

$$\frac{1}{(1-x)^k} = \left[\frac{1}{(1-x)}\right]^k = (1 + x + x^2 + x^3 + \ldots\ldots)^k$$

If we carry out this multiplication k times, then x^n appear as many times as there are nonnegative integer solutions to the equation

$$x_1 + x_2 + x_3 + \ldots\ldots + x_k = n$$

And, we know that this number is $C(n + k - 1, n)$. We shall use this concept in the next example.

Example 7: Find the number of positive integral solution to the equation

$$x + y + z = 10$$

Solution: Here x, y and z are positive integers satisfying the given equations. Our task is to find the number of possible solutions to the above equation. Here x belongs to $\{1, 2, 3, 4, \ldots\}$ and so are y and z. If different possible positive integers for x is represented by the powers of x then series for x can be written as

$$x + x^2 + x^3 + \ldots.$$

This problem is reduced to selection of 10 objects of 3 kinds. The generating function $f(x)$ for this problem is then written as

$$f(x) = (x + x^2 + x^3 + \ldots.)^3$$

$$= \frac{x^3}{(1-x)^3}$$

Then, the required result is given by the coefficient of x^{10} in $f(x)$. This is equal to the coefficient of x^7 in $(1-x)^3$. Which is

$$^{7+3-1}C_7 \qquad [\because n = 7, k = 3 \textit{ in the above formula}]$$

$$= {}^9C_7 = \frac{9*8}{2} = 36$$

Ans.

The same approach can be used to find number of solutions to various such equations in which values allowed for variables may be different. For example, a variable may take only digits as its value whereas, in the same equation another variable may be allowed to assume any positive integers. The generating function is found accordingly before calculating the coefficient. Readers are suggested to try problems given in the exercises at the end of this chapter.

Exercise 2

1. Let $A = \{a, b, c, d\}$ and $B = \{1, 2, 3\}$. Determine whether the relation R from A to B given below is a function. If it is a function, give its range.

 (a) $R = \{(a, 1), (b, 2), (c, 1), (d, 2)\}$ (b) $R = \{(a, 1), (b, 2), (a, 2), (c, 1), (d, 2)\}$

(c) $R = \{(a, 3), (b, 2), (c, 1)\}$ (d) $R = \{(a, 1), (b, 1), (c, 1), (d, 1)\}$

2. Determine whether the relation R from A to B is a function.

   (a) A = the set of all holders of motor driving license in India, $B = \{x \mid x \text{ is a seven-digit number}\}$, $a\text{R}b$ if b is a's driving license number.

   (b) A = the set of citizen of India, $B = \{x \mid x \text{ is a seven-digit number}\}$, $a\text{R}b$ if b is a's passport number.

3. Let $A = B = C = \text{R}$, the set of real number, and let $f: A \to B$ and $g: B \to C$ be defined by $f(a) = a^{-1}$ and $g(b) = b^2$. Find

   (a) (fog) (2) (b) (gof) (2)

   (c) (fog) (x) (d) (gof) (x)

   (e) (fof) (y) (f) (gog) (y)

4. Let A = $B = C = R$, the set of real number, and let $f: A \to B$ and $g: B \to C$ be defined by $f(a) = a + 1$ and $g(b) = b^2 + 2$. Find

   (a) (fog) (–2) (b) (gof) (–2)

   (c) (fog) (x) (d) (gof) (x)

   (e) (fof) (y) (f) (gog) (y)

5. If A has n elements then how many functions are there from A to A? How many bijections are there from A to A?

6. If A has m elements and B has n elements, how may functions are there from A to B?

7. Let $f: A \to B$ be a function with finite domain and range. Suppose that $|\text{Dom}(f)| = n$ and $|\text{Ran}(f)| = m$. Prove that:

   (a) If f is one to one, then $m = n$, and

   (b) If f is not one to one, then $m < n$.

8. Let $|A| = |B| = n$ and let $f: A \to B$ is function. Prove that the following three statements are equivalent.

   (a) f is one to one,

   (b) f is onto, and

   (c) f is a bijection.

9. Let $A = R$, set of real number and $B = Z$, set of integers, Let $f: A \to B$ be a function defined as $f(a)$ = the greatest integer less than equal to a. Verify that f is a function.
10. Prove that if $f: A \to B$ and $g: B \to C$ are bijection then gof is a bijection.
11. Let $f: A \to B$ and $g: B \to C$ are function. Show that if gof is one to one then f is one to one and when gof is onto then g is onto.
12. A bijection is also called a set bijection or set isomorphism. Show that a function from set of integer to the set of even integer defined as $f(x) = 2x$ is a set isomorphism.
13. Let C denote the set of complex numbers and let R denote the set of real numbers. Prove that the mapping $f: C \to R$ given by $f(x + iy) = x$, where x and y are real, is an onto mapping.
14. Let C denote the set of complex numbers and let R denote the set of real numbers. Prove that the mapping $f: C \to R$ given by $f(x + iy) = |x + iy|$, where x and y are real, is neither one to one nor onto.
15. Let R be the set of real numbers. Using the fact that every cubic equation with real coefficients has real root, show that $f(x) = x^3 - x$ defines a function from R to R.

In exercise 16 through 20 analyze the operation performed by the given pseudo code and give a function describing the number of steps required to carry out the job. Give the order of the function.

16.
```
A ← 1
B ← 1
UNTIL (B > 100)
a. B ← 2a – 2
b. A ← A + 3
```

17.
```
A ← 1
B ← 100
WHILE (A < B)
a. A ← A + 2
b. B ← B / 2
```

18.
```
X ← 1
Y ← 0
READ N
WHILE (X ≤ N)
a. Y ← Y + 1
b. X ← X + 1
```

19.
```
SUM ← 0
B ← 1
READ N
FOR (I = 0; I< 2N – 1; I+= 2)
SUM ← SUM +
```

20. Let A be an array of size N and X be an item being searched in the array A. The pseudo code is given below.

```
FUNCTION (A, X)
    FOUND ← FALSE
    K ← 1
    WHILE (NOT FOUND) AND (K < N))
    IF (A | K | = = X) FOUND ← FALSE
    ELSE K ← K + 1
RETURN
```

21. Show that $g(n) = n!$ is of $O(n^n)$.
22. Show that $g(n) = 1 + 2 + 3 + \ldots + n$ is of $O(n^2)$.
23. Show that $f(n) = n^2(7n - 2)$ is of $O(n^3)$.
24. Show that f and g have the same order for
    (a) $f(n) = 5n^2 + 4n + 3$ and $g(n) = n^2 + 100n$.
    (b) $f(n) = \log(n^3)$ and $g(n) = \log_5(6n)$.
25. Show that $f(n) = n\log(n)$ is of $O(g)$ for $g(n) = n^2$, but that g is not of $O(f)$.
26. Show that $f(n) = n^{100}$ is of $O(g)$ for $g(n) = 2^n$, but that g is not of $O(f)$.
27. We study the problem of computing x^n for given x and n, assuming that all intermediate results of computation are available to us. For given integer, let $b_k b_{k-1} b_{k-2} \ldots b_2\ b_1\ b_0$ denote its binary number representation. For each b_i, replace b_i by SX if $b_i = 1$ and by S if $b_i = 0$. For example the binary number representation for 29 is 11101, we get the sequence ***SXSXSXSSX***. Removing leading SX, we are left with ***SXSXSSX***. Now, if we interpret each S by 'square' and each X by 'multiply by x', we get the sequence of steps as {square multiply by x, square, multiply by x, square, square, multiply by x}, where square means to compute the square of current result and multiply by x means to multiply the current result by x. Show that starting with x as the current result, for $n = 29$, this sequence of steps indeed will complex x^{29}.
    (a) Given x, can you compute x^{16} in four multiplication?
    (b) Given x, in how many multiplication can you compute x^{15}?

(c) Show how we can compute x^{73} using the above algorithm.

(d) Find the time complexity of the algorithm.

28. Given an integer n, an addition chain for n is a sequence of integers $b_0 b_1 b_2 \ldots b_m$ such that $b_0 = 1$, $b_m = n$, and $b_i = b_j + b_k$ for $k \le j \le i$. For example 1, 2, 3, 6, 7, 14, 28, 29 is an addition chains for 29. What is the relationship between addition chains and evaluation of powers of x? Determine addition chains for 19, 33, 46, 79 and 87. In how many multiplication can you evaluate x^{19}, x^{33}, x^{46}, x^{79} and x^{87}?

29. Give the set corresponding to the following sequence.
    (i) 2, 1, 2, 1, 2, 1, 2, 1
    (ii) 0, 2, 4, 6, 8, 10, ...
    (iii) *aabbccddee…zz*

30. Give three different sequences that can arise from the following set
    (i) $\{x, y, z\}$ (ii) $\{1, 2, 3, \ldots\}$

31.. Write down the first ten terms of the sequence whose general term is given below.
    (i) $a_n = 5^n$ (ii) $b_n = 3n^2 + 2n - 6$
    (iii) $c_n = c_{n-1} + 1.5$ and $c_1 = 2.5$ (iv) $d_n = -2_{dn-1}$ and $d_1 = -3$
    (v) $f_{n+2} = 2f_n + f_{n+1}$ and $f_0 = 0, f_1 = 1$ (vi) $g_{n+2} = g_n^{\,2} + g_{n+1}$ and $f_0 = 1, f_1 = 2$

32. Write the formula for the nth term of the sequence given below. Identify your formula as recursive (difference) or explicit (closed).
    (i) 1, 3, 5, 7, … (ii) 0, 3, 8, 15, 24, 35, …
    (iii) 1, –1, 1, –1, 1, –1, … (iv) 0, 2, 0, 2, 0, …
    (v) 1, 4, 7, 10, 13, … (vi) 2, 5, 8, 11, 14, 17, …
    (vii) 2, 5, 7, 12, 19, ...

33. Let $A = \{x \mid x \text{ is real and } 0 < x < 1\}$,
    $B = \{x \mid x \text{ is real and } x^2 + 1 = 0\}$,
    $C = \{x \mid x = 4m, m \in Z\}$,
    $D = \{(x, 3) \mid x \text{ is an English word of length } 3\}$, and
    $E = \{x \mid x \in Z \text{ and } x^2 \le 100\}$.

Identify each set as finite, countable, or uncountable.

34. Using characteristic functions, prove that $(A \oplus B) \oplus C = A \oplus (B \oplus C)$.

35. We define T-numbers recursively as follows:

    a. 0 is a T-number

    b. If X is a T-number, $X+3$ is a T-number.

    Write a description of the set of T-numbers.

36. We define S-numbers recursively as follows:

    a. 8 is a S-number

    b. If X is a S-number and Y is a multiple of X then Y is a S-number,

    c. If X is a S-number and X is a multiple of Y then Y is a S-number.

    Write a description of the set of S-numbers.

37. Find the sum to infinite number of terms of the following series.

    (i) $1 + 2r + 3r^2 + 4r^3 + \ldots. (r < 1)$

    (ii) $1.1 + 2.3x + 4.5x^2 + \ldots (x < 1)$

    (iii) $1 + 3x + 5x^2 + 7x^3 + \ldots(x < 1)$

    (iv) $1.2 + 2.3x + 3.4\,x^2 + \ldots.(x < 1)$

    (v) $1 + \frac{2}{3} + \frac{3}{9} + \frac{4}{27} + \ldots\ldots$

    (vi) $1 + \left(1 + \frac{1}{2}\right)\frac{1}{3} + \left(1 + \frac{1}{2} + \frac{1}{2^2}\right)\frac{1}{3^2} + \ldots\ldots.$

38. Sum up to n terms the following series

    (i) $7 + 77 + 777 + \ldots$

    (ii) $.4 + .44 + .444 + \ldots$

    (iii) $.3 + .33 + .333 + \ldots$

    (iv) $1.2.3 + 2.3.5 + 3.4.7 + \ldots$

    (vi) $(x^2 + \frac{1}{x^2} + 2) + (x^4 + \frac{1}{x^4} + 5) + (x^6 + \frac{1}{x^6} + 8) + \ldots..$

39. The natural numbers have been grouped as (1), (2, 3), (4, 5, 6), ... Prove that the sum of the nth such group is $n(n^2 + 1)/2$. Using this approach prove that the sum of numbers in any

of the group given as (1), (1, 3, 5), (1, 3, 5, 7, 9), ...is the square of an odd number.

40. A ping pong ball is dropped to the floor from a height of 20 *m*. Suppose that the ball always rebounds to reach half of the height from which it falls. If a_r be height of a ball, that it reaches in the *r*th rebound, find the numeric function *a*.

41. Let **a** be a numeric function such that a_r is equal to the remainder when the integer *r* is divided by 17. Let ***b*** be a numeric function such that b_r is equal to 0 if the integer *r* is divisible by 3 and is equal to 1 otherwise.

    (a) Let $c_r = a_r + b_r$. For what value of *r* is $c_r = 0$? For what values of *r* is $c_r = 1$?

    (b) Let $d_r = a_r + b_r$. For what value of *r* is $d_r = 0$? For what values of *r* is $d_r = 1$?

42. Every particle inside a nuclear reactor splits into two particles in each second. Suppose one particle is injected into the reactor every second beginning at $t = 0$. How many particles are there in the reactor at the nth second? Find the numeric function and then a generating function for the numeric function so found.

43. After a careful analysis, a student concluded that a certain numeric function ***a*** is O (*n*log *n*). Upon hearing that her mate reminded her that she had forgotten to specify the base of the logarithm. Had she forgotten something important?

44. Find generating function for each of the following discrete numeric functions:

    (i) 1, –2, 3, –4, 5, -6, ... (ii) 1, 2/3, 3/9, 4/27, ...

    (iii) 1, 1, 2, 2, 3, 3, 4, 4, ... (iv) 0 * 1, 1 * 2, 2 * 3, 3 * 4, ...

    (vi) $0 * 5^0, 1 * 5^1, 2 * 5^2, 3 * 5^3$, ...

45. If *x*, *y* and *z* are digits, then find the number of possible solutions to the following equations:

    (i) $x + y + z = 10$ (ii) $x - y + z = 17$

    (iii) $x + 2y + 3z = 8$ (iv) $x - 2y + 2z = 37$

46. If x, y and z are non-negative integers, then find the number of possible solution to the following equations:

    (i) $x + y + z = 100$ (ii) $x - y + z = 17$

    (iii) $x + 2y + 3z = 10$ (iv) $x - 2y + 2z = 71$

3

Solving Recurrence Relation

While learning programming languages, we have come across the term ‘recursive function’. In our early school days we have learnt about ‘recurring number’. Now we are facing the term ‘recurrence relation’ or ‘**recurrence equation**’. All these contains a common term ‘recurrence’, i.e. there is something that occurs again and again and repeats itself until it meets some terminating conditions.

In previous chapter, we have studied that a sequence of term $a_0, a_1, a_2, a_3, ..., a_n, ...$, can be expressed in explicit way (also called closed form) or in recurrence form. In recurrence form, we express sequence term a_n in terms of previous sequence terms $a_{n-1}, a_{n-2}, a_{n-3}$ etc. and terminating (also called initial) conditions. For example,

$a_n = a_{n-1} + a_{n-2}$; with $a_0 = 0$ and $a_1 = 1$

In this chapter, we shall learn how to solve a recurrence equation i.e. how to convert a recurrence form of presentation of a sequence into its closed form of representation. There are only few types of recurrence relations, which can be solved, in closed form. It means any term in the sequence can be evaluated by plugging numbers into an equation instead of having to calculate the entire sequence. Classifying the recurrence relation helps to decide which, if any, techniques can be used to solve it.

3.1 Classification of Recurrence Relation

Here is a generic recurrence equation consisting of a sequence of terms a_n, $a_n + 3a_{n-1} - 2\sqrt{n}\, a_{n-2} + 2\sqrt{n}\, a_{n-2}a_{n-3} = f(n)$ ———— (1)

In the above equation, the right side, $f(n)$ is some function of n- the place at which an is in the sequence. In the standard form, the right side of a recurrence equation can contain any function of n but must not contain any of the a_n. There are four characteristics of any recurrence equation which are needed to be understood to classify a given recurrence equation.

Homogeneous or Non-homogeneous

A recurrence equation is homogeneous if the right side of the standard form of the recurrence equation is zero, i.e. $f(n) = 0$. The above recurrence equation (1) is non-homogeneous but the equations (2) and (3) —given below, are homogeneous.

$$a_n + 3_{n-1} - 2\sqrt{n}\, a_{n-2} + 2\sqrt{n}\, a_{n-2}a_{n-3} = 0 \quad \text{——— (2)}$$

$$a_n + 3a_{n-1} - 2\sqrt{n}\, a_{n-2} + 2\sqrt{n}\, a_{n-3} = 0 \quad \text{——— (3)}$$

Linear or Non-linear

A recurrence equation is linear if there are no products or powers of the sequence terms. The equation (1) and (2) above are non-linear whereas, equation (3) is a linear recurrence equation. The following equations are linear recurrence equation.

$$a_n + 3a_{n-1} - 2a_{n-2} = 0 \quad \text{——— (4)}$$

$$b_n + 4b_{n-1} = \sqrt{n} + 17 \quad \text{——— (5)}$$

$$c_n + \sqrt{n}\, c_{n-1} + n^n \quad \text{——— (6)}$$

$$d_n - 2d_{n/2} - 7c_n + \sqrt{n}\, c_{n-1} = n^n \quad \text{——— (7)}$$

$$e_n - \sqrt{n}\, c_{\sqrt{n}} = n^n \quad \text{——— (8)}$$

The recurrence equation listed below are examples of non-linear recurrence equations.

$$a_n^2 + 2a_n a_{n-1} + 2a_{n-1}^2 = 0 \quad \text{——— (9)}$$

$$b_n \times 2b_{n-1} \times 3b_{n-2} \times \ldots \times nb_1 = \sqrt{n} + 17 \quad \text{——— (10)}$$

$$c_n + c_{n-1}c_{n-1} = n^n \quad \text{——— (11)}$$

Order of the Recurrence Equation

The order of a recurrence equation is the number of steps along the sequence from the first to the last member of the sequence in the equation. Some examples are given below to help you understand the order of a recurrence equation.

Table 3.1

Sl. No.	*Recurrence Equation*	*Order*
1.	$a_n + 4a_{n-1} = n^2$	First order
2.	$b_n - 2b_{n-1} + b_{n-2} = 0$	Second order
3.	$c_n - c_{n-2} = 2^n$	Second order
4.	$d_n - 3d_{n-2} + d_{n-4} = 1$	Fourth order
5.	$c_n - c_{n/2} = 2^n$	Huh?

The last recurrence equation does not have a constant order. That complicates the solution. The number of initial conditions required to express a sequence in recurrence form is equal to order of the recurrence equation if the equation is constant order.

Constant verses Non-constant Coefficients

We make a distinction between recurrence equation in which the coefficient which multiply the sequence terms are constant, such as these:

$$a_n + 4a_{n-1} + a_{n-2} = n^2 \quad (12)$$

$$b_n - b_{n-1} - b_{n-2} = n \quad (13)$$

and those recurrence equations in which the coefficients are non constant –they vary with n, such as these:

$$a_n + 4na_{n-1} + a_{n-2} = n^2 \quad (14)$$

$$nb_n - (n-1)b_{n-1} - (n-2)b_{n-2} = n \quad (15)$$

If we form three sets, H, L and C of characteristics of a recurrence equations and let H = {Homogeneous, Non-constant}, L = {Linear, Non-linear} and C = {Constant-coefficient, Non-constant-coefficient}. Then a recurrence equation can be of any type of ordered triplet of $H \times L \times C$, and each equation can be of either constant or variable order. So what? There are some classes of recurrence equations, which are always solvable. So it is important to recognize them.

First-order homogeneous or non-homogeneous, linear recurrence equations in which the coefficients are never zero; are always solvable.

Any constant order linear recurrence equations with constant coefficients which are homogeneous or whose right sides can be expressed as the product of polynomials in n and constants to the nth power; are also always solvable.

The techniques for solving these two classes of recurrence equations are discussed in detail in this chapter. Other recurrence equations are generally harder to solve although a few special cases have been solved.

3.2 Backtracking and Forward Chaining Method

This is a very useful method to solve a first order linear recurrence equation with constant coefficient. The core of this technique is to identify a pattern emerging out from the arrangements of terms. **Once you have got the pattern, you have got the solution.** The solution so obtained is the closed form formula for the given recurrence equation. We shall learn this method through the following examples.

Example 1: Solve the recurrence equation $a_n = a_{n-1} + 3$ with $a_1 = 2$.

Solution: We shall use the backtracking method to solve this problem. In this method we shall start from a_n and move backward towards a_1 to find a pattern, if any, to solve the problem. To backtrack, we keep on substituting the definition of a_n, a_{n-1}, a_{n-2} and so on until a recognizable pattern appears. The method is illustrated below.

$$
\begin{aligned}
a_n &= a_{n-1} + 3 \\
&= a_{n-2} + 3 + 3 && [\text{since } a_{n-1} = a_{n-2} + 3] \\
&= a_{n-2} + 2 \times 3 \\
&= a_{n-3} + 3 + 2 \times 3 && [\text{since } a_{n-2} = a_{n-3} + 3\ ] \\
&= a_{n-3} + 3 \times 3 \\
&= a_{n-4} + 3 + 3 \times 3 && [\text{since } a_{n-3} = a_{n-4} + 3\ ] \\
&= a_{n-4} + 4 \times 3 \\
&= a_{n-(n-1)} + (n-1) \times 3 && [\textbf{note the pattern}] \\
&= a_1 + 3\,(n-1) \\
&= 2 + (n-1) \quad [\text{since } a_1 = 2 \text{ is the terminating condition}]
\end{aligned}
$$

Therefore, $a_n = 2 + 3(n-1)$.

Ans.

Now, to illustrate the forward chaining method, we shall continue with the same example. In this method, we begin from initial

(terminating) condition and keep on moving to-wards the nth term until we get a clear pattern. The method is illustrated below.

$$
\begin{aligned}
a_1 &= 2 \\
a_2 &= a_1 + 3 \\
a_3 &= a_2 + 3 \\
&= a_1 + 2 \times 3 \\
a_4 &= a_3 + 3 \\
&= a_1 + 2 \times 3 + 3 \\
&= a_1 + 3 \times 3 \\
&= a_1 + (4 - 1) \times 3 \\
a_5 &= a_1 + (5 - 1) \times 3 \\
a_0 &= c_1 + (n - 1) \times 3
\end{aligned}
$$

Therefore, $a_n = 2 + 3(n - 1)$.

Ans.

The next and very useful method is **summation method** to solve a first order linear recurrence relation with constant coefficient. This is a form of backtracking method only, but is more convenient to use. In this method we arrange the given equation in the following form:

$$a_n - ka_{n-1} = f(n)$$

and then backtrack till terminating condition. In the process, we get *a* number of equations. Add these equations in such *a* way that all intermediate terms get cancelled. Finally we get the required solution. We illustrate the method on the same example under consideration. The given equation can be rearranges as

$$
\begin{aligned}
a_n - a_{n-1} &= 3 \\
a_{n-1} - a_{n-2} &= 3 \\
a_{n-2} - a_{n-3} &= 3 \\
&\cdot \\
&\cdot \\
a_3 - a_2 &= 3 \\
a_2 - a_1 &= 3 \qquad \text{[we stop here, since } a_1 = 2 \text{ is given]}
\end{aligned}
$$

Adding all, we get $a_n - a_1 = 3 + 3 + 3 + \ldots + (n - 1)$ times

or, $a_n - a_1 = 3\,(n - 1)$

Therefore, $a_n = 2 + 3\,(n - 1)$.

Ans.

Example 2: Solve the recurrence equation $t_n = 2t_{n/2} + n$ with $t_1 = 1$.

Solution: We shall solve this recurrence equation first by **backtracking method** and then by **summation method.** Readers should try to solve it by **forward chaining method.**

Backtracking method: We have $t_n = 2t_{n/2} + n$ ——— (1)

By repeated substitution of the definition of $t_{n/2}$, $t_{n/4}$, $t_{n/8}$, etc., we get the following results:

$$t_n = 2[2t_{n/4} + n/2] + n$$
$$= 2^2 t_{n/4} + 2 * n/2 + n$$
$$= 2^2 t_{n/4} + n + n \text{ ——— (2)}$$

$$t_n = 2^2[2t_{n/8} + n/4] + n + n$$
$$= 2^3 t_{n/8} + 2^2 * n/4 + n + n$$
$$= 2^3 t_{n/8} + n + n + n \text{ ——— (3)}$$

By closely observing equations (1), (2) and (3) we can write them as:

$$t_n = 2^{\log_2 2}\, t_{n/2} + n \log_2 2 \text{ ——— from (1)}$$
$$t_n = 2^{\log_2 4}\, t_{n/4} + n \log_2 4 \text{ ——— from (2)}$$
$$t_n = 2^{\log_2 8}\, t_{n/8} + n \log_2 8 \text{ ——— from (3)}$$

Similarly we can write,

$$t_n = 2^{\log_2 n}\, t_{n/n} + n \log_2 n$$
$$t_n = 2^{\log_2 n}\, t_1 + n \log_2 n$$
$$t_n = 2^{\log_2 n} + n \log_2 n, \text{ ——— } [\because t_1 = 1]$$

Therefore, $t_n = 2^{\log_2 2} + n \log_2 n$

Ans.

Summation method: Arranging the given recurrence equation we get

$$t_n - 2t_{n/2} = n$$

$$t_{n/2} - 2t_{n/4} = \frac{n}{2}$$

$$t_{n/4} - 2t_{n/8} = \frac{n}{4}$$

...

$$t_4 - 2t_2 = 4$$
$$t_2 - 2t_1 = 2$$

The number of equations in this set of equations is $\log_2 n$. Now if we multiply the above equations by

$$1, 2, 3, 4, \ldots\ldots.., 2^{\log_2 n}$$

respectively from top to bottom, and then by adding them all together we get

$$t_n - 2^{\log_2 n} = n + n + n + \ldots \log_2 n \ \ times$$
$$or,\ t_n - \ 2^{\log_2 n} = n \log_2 n$$
$$\therefore t_n = 2^{\log_2 n} + n \log_2 n$$

Ans.

Example 3: Solve the recurrence equation

$t(n) = t(\sqrt{n}) + c \log_2 n$ with $t(1) = 1$.

Solution: Sometimes a sequence term t_n is also written as a function of its place in sequence, i.e., t_n may also be written as $t(n)$. We solve this problem, here, by **forward chaining method.** The same problem will be solved by **characteristic equation method** in the next section of this chapter.

$$\begin{aligned} t(2) &= t(2^{1/2}) + c \log_2 2 \\ &= t(2^{1/4}) + c \log_2 2^{1/2} + c \log_2 2 \\ &= t(2^{1/8}) + c \log_2 2^{1/4} + c \log_2 2^{1/2} + c \log_2 2 \\ &= t(2^{1/2n}) + c \log_2 2^{1/n} + \ldots + c \log_2 2^{1/4} + c \log_2 2^{1/2} + c \\ &\quad \log_2 2 \end{aligned}$$

Since t is defined for n = 1 only, we have to let $t(2^{1/2n}) \rightarrow t(1)$
This is possible when $n \rightarrow \infty$

i.e., when $n \to \infty \Rightarrow \dfrac{1}{2n} \to 0 \Rightarrow 2^{\frac{1}{2n}} \to 1$

Therefore, $t(2) = t(1) + [c \log_2 2 + c \log_2 2^{1/2} + c \log_2 2^{1/4} + ...]$

$$= t(1) + c \log_2 2 \left[1 + \frac{1}{2} + \frac{1}{4} + \frac{1}{8} + ...\right] = t(1) + c$$

$$\log_2 2 \left[\frac{1}{1 - \frac{1}{2}}\right]$$

$= t(1) + 2c \log_2 2$

i.e., $t(2) = t(1) + 2c \log_2 2$

|| *ry* $t(3) = t(1) + 2c \log_2 3$

$t(4) = t(1) + 2c \log_2 4$

$\vdots$

$t(n) = t(1) + 2c \log_2 n$

Therefore $t(n) = t(1) + 2c \log_2 n.$

3.3 Characteristic Equation Method

In principle, this method can be used to solve **any** constant order linear recurrence equation with constant coefficient. This recurrence relation may be homogeneous or non-homogeneous. Before attempting to solve any such problem, let us first, understand what is characteristic equation for a given recurrence equation and how to find it.

A given recurrence equation of the mentioned type can be arranged in standard form as:

$$A_n + c_1 A_{n-1} + c_2 A_{n-2} + c_3 A_{n-3} = \textbf{RHS} \text{ ———————— } (1)$$

Where c_1, c_2, c_3 are constant coefficients and **RHS** has one of the following forms:

Table 3.2

Form	Examples
Homogeneous	0
A constant to the nth power	2^n, π^{-s}, 2^{-n}, $\sqrt{2^n}$
A polynomial in *n*	3, n^2, $n^2 - n$, $n^3 + 2n - 1$
A product of a constant to the *n*th power and a polynomial in *n*	$2^n (n^2 + 2n - 1)$, $(n - 1)n^6$, $n6^n$
A linear combination of any of the above	$(2^n + 3^{n/2})(n^2 + 2n - 1) + 5$

In the recurrence equation (1), assigning RHS = 0, we get

$$A_n + c_1 A_{n-1} + c_2 A_{n-2} + c_3 A_{n-3} = 0 \qquad (2)$$

This equation (2) gives the homogeneous part of the given recurrence equation. **Every recurrence equation has a homogeneous part.** If the recurrence relation is homogeneous then it has only homogeneous part and solving such equation is **one step process** (to be discussed in this section). On the other hand, if the given recurrence equation is non-homogeneous then its homogeneous part is obtained by assigning **RHS equal to zero. A characteristic equation corresponds to homogeneous part of the given recurrence relation.** The characteristic equation of (2) is given as:

$$x^3 + c_1 x^2 + c_2 x + c_3 = 0 \qquad (3)$$

This has been obtained by the following procedure:

- Find the order of the recurrence equation. Here it is 3.
- Take any variable (say x) and substitute A_n, A_{n-1}, A_{n-2}, by x^3, x^2, x respectively in the homogeneous part of the recurrence equation.

Equation so obtained is called **characteristic equation** of the given recurrence equation.

Example 1: The characteristic equation of the recurrence equation

$c_n = 3\, c_{n-1} - 2\, c_{n-2}$; is given by

$x^2 - 3x + 2 = 0$

Example 2: The characteristic equation of the recurrence equation

$f_n = f_{n-1} + f_{n-2}$ is given by

$x^2 - x - 1 = 0$

Example 3: The characteristic equation of the recurrence equation

$A_n - 5\,A_{n-1} + 6\,A_{n-2} = 2^n + n$; is given by

$x^2 - 5x + 6 = 0$

Now, we shall discuss two theorems which are the basis for finding the solution for the homogeneous part of any linear recurrence equation.

Theorem 1: If the characteristic equation $x^2 - r_1x - r_2 = 0$ of the recurrence equation $a_n = r_1a_{n-1} + r_2a_{n-2}$ has two distinct roots s_1 and s_2 then

$$a_n = us_1^{\,n} + vs_2^{\,n}$$

is the closed form formula for the sequence where us and v depend on the initial condition.

Proof: Since s_1 and s_2 are roots of

$$x^2 - r_1x - r_2 = 0 \qquad (1)$$

We have,

$$s_1^{\,2} - r_1s_1 - r_2 = 0 \qquad (2)$$

$$s_2^{\,2} - r_1s_2 - r_2 = 0 \qquad (3)$$

Since u and v are dependent on the initial conditions, we have

$$a_1 = us_1 + vs_2 \;\&\; a_2 = us_1^{\,2} + vs_2^{\,2}$$

Now,

$$\begin{aligned} a_n &= us_1^{\,n} + vs_2^{\,n} \\ &= us_1^{\,n-2}s_1^{\,2} + vs_2^{\,n-2}s_2^{\,2} \\ &= us_1^{\,n-2}[r_1s_1 + r_2] + vs_2^{\,n-2}[r_1s_1 + r_2] \quad \text{[from equation (2) \& (3)} \\ &= r_1us_1^{\,n-1} + r_2us_1^{\,n-2} + r_1vs_2^{\,n-1} + r_2vs_2^{\,n-2} \end{aligned}$$

$$= r_1[us_1^{n-1} + vs_2^{n-1}] + r_2[us_1^{n-2} + vs_2^{n-2}]$$

$$= r_1a_{n-1} + r_2a_{n-2}$$

i.e. $a_n = us_1^n + vs_2^n$ *is an explicit formula for the given recurrence equation.*

Proved.

Theorem 2: If the characteristic equation $x^2 - r_1x - r_2 = 0$ of the recurrence equation $a_n = r_1a_{n-1} + r_2a_{n-2}$ has *a* single root *s* then

$$a_n = us^n + vns^n$$

is the closed form formula for the sequence where *u* and *v* depend on the initial condition.

Proof: Since *s* is roots of

$$x^2 - r_1 x - r_2 = 0 \qquad (1)$$

We have,

$$s^2 - r_1s - r_2 = 0 \qquad (2)$$

Now,

$$a_n = us^n + vns^n = usn + v(n-1)s^n + vs^n$$

$$= us^{n-2}s^2 + v(n-1)s^{n-2}s^2 + vs^n$$

$$= [us^{n-2} + v(n-1)s^{n-2}]\, s^2 + vs^n$$

$$= [us^{n-2} + v(n-1)s^{n-2}]\, (r_1s + r_2) + vs^n$$

$$= r_1[us^{n-1} + v(n-1)s^{n-1}] + r_2[us^{n-2} + v(n-1)s^{n-2}] + vs^n$$

$$= r_1[us^{n-1} + v(n-1)s^{n-1}] + r_2[us^{n-2} + v(n-2)s^{n-2}] + r_2vs^{n-2} + vs^n$$

$$= r_1a_{n-1} + r_2a_{n-2} + vs^{n-2}[r_2 + s^2] \qquad (3)$$

From equation (2) we have

$$s = \frac{r_1 \pm \sqrt{r_1^2 + 4r_2}}{2} \qquad \left[\because x = \frac{-b \pm \sqrt{b^2 - 4ac}}{2a}\right]$$

$\because$ *s is an equal root we have,*

$$\sqrt{r_1^2 + 4r_2} = 0 \text{ or, } s = \frac{r_1}{2} \text{ or, } r_1 = 2s$$

Now substituting $r_1 = 2s$ *in equation* (2), *we get*

$$s^2 - 2s^2 - r_2 = 0$$

$$\text{or,} \quad - s^2 - r = 0$$

$$\text{or,} \quad s^2 + r_2 = 0 \qquad \text{———} (4)$$

Using the result from equation (4) in equation (3), we have

$$a_n = r_1 a_{n-1} + r_2 a_{n-2}$$

Therefore the explicit formula for the recurrence equation $a_n = r_1 a_{n-1} + r_2 a_{n-2}$ is

$$a_n = us^n + vns^n$$

With initial conditions

$$a_1 = us + vs \text{ and } a_2 = us^2 + 2vs^2$$

Proved.

Note:

1. **In general,** if $s_1, s_2, s_3, \ldots s_r$ are r distinct roots of the characteristic equation of a rth order recurrence equation then its explicit formula is given by

$$a_n = u_1 s_1{}^n + u_2 s_2{}^n + u_3 s_3{}^n + \ldots + u_1 s_1{}^n;$$

   Where $u_1, u_2, u_3, \ldots u_r$ depend on initial conditions.
2. If s is r times equal root of the characteristic equation of a rth order recurrence equation then its explicit formula is given by

$$a_n = u_1 s^n + u_2 ns_2{}^n + u_3 n^2 s^n + \ldots + u_r n^{r-1} s^n;$$

   where $u_1, u_2, u_3 \ldots u_r$ depend on initial conditions.
3. A combination of 1 and 2 is also possible, i.e., some roots of *a* characteristic equation are distinct and some are equal. In that case, for distinct roots we use method outlined in 1 and for repeated roots we use method mentioned in 2.

Having discussed the theorem, we shall now learn the way to solve a given recurrence equation using the characteristic equation method. This is based on the method of characteristic equation with

undetermined constants used to solve the related continuous function in differential calculus. It relies on the **principle of uniqueness, i.e., any answer which works is the correct answer since there can be only one correct answer.**

There are two parts to the total solution. The **homogeneous** part of the solution depends only on what is on the left of the recurrence equation when expressed in the standard form. The **particular part** of the total solution depends on what is on the RHS and has the same form as the RHS. We will calculate the two parts separately and add them to form the total solution. There are four steps in the process as listed below.

Step 1: Find the homogeneous solution to the homogeneous equation. This results when you set the RHS zero. **If it is already zero**, skip the next two steps and go directly to the step 4. Your answer will contain one or more "undetermined coefficients" whose values cannot be determined until **step 4.**

Step 2: Find the particular solution by guessing a **form similar to the RHS.** This step does not produce any additional undetermined coefficient, nor does it eliminate those from the **step 1.**

Step 3: Combine the homogeneous and particular solutions.

Step 4: Use boundary or initial conditions to eliminate the undetermined constants from the step 1.

This method is straightforward and very effective. Its major *limitation* in that it only works with a specific subset of recurrence equations. However, by using the change of variable method, a wider range of recurrence equations can be solved than is immediately apparent. In fact, the method presented here is useful for solving many recurrence equations that occur in the analysis of algorithm. Let us now demonstrate the procedure by solving some problems.

Example 4: Solve the recurrence equation

$a_r - 7a_{r-1} + 10a_{r-2} = 2^r$ with initial condition $a_0 = 0$ and $a_1 = 6$.

Solution: The general solution, also called homogeneous solution, to the problem is given by homogeneous part of the given recurrence equation. The homogeneous part of the equation

$$a_r - 7a_{r-1} + 10a_{r-2} = 2^r \quad \text{------} \quad (1)$$

is

$$a_r - 7a_{r-1} + 10a_{r-2} = 0 \text{ ——— (2)}$$

The characteristic equation (2) is given as

$$x^2 - 7x + 10 = 0$$

or, $(x-2)(x-5) = 0$

or, $x = 2$ *and* $x = 5$.

Since the two roots 2 & 5 of characteristic equation are distinct, the homogeneous solution is given by

$$a_r = A2^r + B5^r \text{ ——— (3)}$$

Now, particular solution is given by

$$a_r = rC2^r$$

The reader must be wondering how this guess has been made. Please wait till the final solution of this problem. Thereafter we shall discuss the way to deal with particular solution before proceeding to the next example.

Substituting the value of a_r in equation (1), we get

$$C[r2^r - 7(r-1)2^{r-1} + 10(r-2)2^{r-2}] = 2^r$$

or, $C[4r - 7(r-1)2 + 10(r-2)]2^{r-2} = 2^r$

or, $C[4r - 14(r-1) + 10r - 20] = 4$

or, $C[4r - 14r + 14 + 10r - 20] = 4$

or, $-6C = 4$ or, $C = -\frac{2}{3}$

$\therefore$ *Particular solution is* $a_r = -\frac{2}{3}r2^r$

The **complete**, also called **total,** solution is obtained by combining the homogeneous and particular solutions. This is given as

$$a_r = A2^r + B5^r - \frac{2}{3}r2^r \text{ ——— (4)}$$

The equation (4) contains two undetermined coefficients. A and B, which are to be determined. To find this, we use the given initial conditions for $r = 0$ and $r = 1$. Since values of a_0 and a_1 are given, putting $r = 0$ and $r = 1$ in equation (4) we get the following equations (5) and (6), respectively

$$a_0 = A2^0 + B5^0 - \frac{2}{3} * 0 * 2^0$$

or, $0 = A + B$ ——————————(5) [$\because a_0 = 0$ *is given*]

And,

$$a_1 = A2^1 + B5^1 - \frac{2}{3} * 1 * 2^1$$

or, $6 = 2A + 5B - \frac{4}{3}$ [$\because a_1 = 6$ *is given*]

or, $2A + 5B = \frac{2}{3}$ ——————————(6)

Solving equations (5) and (6), we get

$$A = -\frac{22}{9}, \text{ and } B = \frac{22}{9}$$

Replacing A and B in equation (4) by its respective values, we get the closed form formula for the given recurrence equation. Thus,

$$a_r = -\frac{22}{9}2^r + \frac{22}{9}5^r - \frac{2}{3}r2^r$$

or, $a_r = \frac{22}{9}\left[5^r - 2^r\right] - \frac{2}{3}r2^r$

Ans.

As assured before, we shall now discuss the way to find a particular solution before proceeding further. **The form of a particular solution has nothing to do with the order of the recurrence relation.** It only **depends** on the form of the RHS of recurrence equation expressed in standard form. Guess a solution of the same form but with undetermined coefficients, which have to be calculated. We find their values by substituting the **guessed** particular solution into the recurrence equation. You may find the following table useful for guessing a particular solution.

Table 3.3

RHS	***Guessed Particular solution***
17(constant)	C(constant)
π^n (Constant to the nth Power)	$C\pi^n$ (Constant to the nth Power)
$2^n + 5^n + 3$ (Linear combination)	$D2^n + B5^n + F$ (Same linear combination)
$5n^3$ (Polynomial in n)	$An^3 + Bn^2 + Cn + D$ (Decreasing Polynomial)
$5n^3 - 1$ (Polynomial in n)	$An^3 + Bn^2 + Cn + D$ (Decreasing Polynomial)
$3n^2 5^n$ (Linear combination)	$5^n (Bn^2 + Cn + D)$ (Linear combination)

Notice how a polynomial in n produces a decreasing polynomial with all orders; down to, and including the constant term. Without those extra terms, usually the coefficients cannot be determined successfully.

When the RHS is of the form of homogeneous solution there is slight complication. You must have noticed that in example 4, the RHS was of the form 2^r so we should have guessed $C2^r$, instead we guessed

$$a_r = rC2^r$$

We cannot try a particular solution of the form:

$$a_r = C2^r$$

because that is already a homogeneous solution. So the left side will sum to zero. This is demonstrated by substituting

$$a_r = C2^r$$

in equation (1) of example 4. Thus, we have

$$C2^r - 7C2^{r-1} + 10C2^{r-2} = 2^r$$

or, $[4C - 14C + 10C]2^{r-2} = 2^r$

or, $0 = 4$ *which is impossible.*

This implies that coefficient C could not have been determined if we would have guessed the above way.

Example 5: Solve the recurrence equation

$A_n - A_{n-1} - A_{n-2} = 2_n$ with $A_n = 0$ and $A_{n-1} = 1$.

Solution: The homogeneous part of the equation

$$A_n - A_{n-1} - A_{n-2} = 2_n \qquad (1)$$

is

$$A_n - A_{n-1} - A_{n-2} = 0 \qquad (2)$$

The characteristic equation of (2) is given as

$$x^2 - x - 1 = 0$$

$$\therefore x = \frac{1 \pm \sqrt{1+4}}{2} \quad or, \quad x = \frac{1+\sqrt{5}}{2}, \; and \; \frac{1-\sqrt{5}}{2}$$

Thus, two roots of the characteristic equation are distinct. So, the homogeneous solution is given by

$$A_n = A\left[\frac{1+\sqrt{5}}{2}\right]^n + B\left[\frac{1-\sqrt{5}}{2}\right]^n \qquad (3)$$

The particular solution is given by the guess

$$A_n = Cn + D$$

Substituting this value in equation (1), we get

$$Cn + D - C(n-1) - D - C(n-2) - D = 2n$$

or, $(Cn - Cn - Cn) + D + C - D + 2C - D = 2n$

or, $-Cn + 3C - D = 2n$

or, $-C = 2, \quad 3C - D = 0$

or, $C = -2, \quad D = -6$

$\therefore A_n = -2n - 6$ *is the particular solution.*

Combining homogeneous and particular solutions, we get total solution as

$$A_n = A\left[\frac{1+\sqrt{5}}{2}\right]^n + B\left[\frac{1-\sqrt{5}}{2}\right]^n - 2n - 6 \quad \text{———} (4)$$

The equation (4) contains two undetermined coefficients A & B, which are to be determined. To find this we use the given initial conditions for $n = 0$ and $n = 1$, since values of A_0 and A_1 are given Putting $n = 0$ and $n = 1$ in equation (4), we get the following equations (5) and (6), respectively.

$$A_0 = A\left[\frac{1+\sqrt{5}}{2}\right]^0 + B\left[\frac{1-\sqrt{5}}{2}\right]^0 - 2*0 - 6$$

or, $0 = A + B - 6$ ——— (5) [$\because A_0 = 0$ *is given*]

and

$$A_1 = A\left[\frac{1+\sqrt{5}}{2}\right]^1 + B\left[\frac{1-\sqrt{5}}{2}\right]^1 - 2*1 - 6$$

or, $1 = A\left[\frac{1+\sqrt{5}}{2}\right] + B\left[\frac{1-\sqrt{5}}{2}\right] - 8$ –(6) [$\because A_1 = 1$ *is given*]

Solving equation (5) and (6) we get,

$$B - \frac{1-\sqrt{5}}{1+\sqrt{5}}B = 6 - \frac{18}{1+\sqrt{5}}, \text{ or}$$

$$B\frac{1+\sqrt{5}-1+\sqrt{5}}{1+\sqrt{5}} = \frac{6+6\sqrt{5}-18}{1+\sqrt{5}}$$

or, $B = \frac{3(\sqrt{5}-2)}{\sqrt{5}}$, *and* $A = \frac{3(\sqrt{5}+2)}{\sqrt{5}}$

Replacing A and B in equation (4) by its respective values, we get the closed form formula for the given recurrence equation. Therefore, the final solution is:

$$A_n = \frac{3(\sqrt{5}+2)}{\sqrt{5}}\left[\frac{1+\sqrt{5}}{2}\right]^n + \frac{3(\sqrt{5}-2)}{\sqrt{5}}\left[\frac{1-\sqrt{5}}{2}\right]^n - 2n - 6$$

Ans.

Example 6: Solve the recurrence equation

$$a_n - 8a_{n-1} + 16a_{n-2} = 0 \quad with \quad a_2 = 16 \quad and \quad a_3 = 80.$$

Solution: The given equation is homogeneous, so only the homogeneous part of the solution is required. In this case, we do not have to try for the particular solution and hence no combining of the particular part with the homogeneous part is needed. Here, we will go to the step 4 from the step 1, skipping the steps 2 & 3.

The characteristic equation of the given recurrence equation is

$$x^2 - 8x + 16 = 0$$

$$or, \quad (x-4)^2 = 0 \Rightarrow x = 4, 4$$

Since the two roots of characteristic equation are equal, the homogeneous solution is given by .

$$a_n = (u + vn)4^n \qquad (1)$$

The equation (1) contains two undetermined coefficients, u & v, which are to be determined. To find this, we use the given initial conditions for $n = 2$ and $n = 3$, since values of a_2 are given. Putting $n = 2$ and $n = 3$ in equation (1), we get the following equations (2) and (3), respectively.

When n = 2,

$$a_2 = (u + 2v)4^2, \text{ or, } 16 = 16(u + 2v)$$

$$or, \; u + 2v = 1 \qquad (2)$$

And,

When n = 3,

$$a_3 = (u + 3v)4^3, \text{ or, } 80 = 64(u + 3v)$$

$$or,\ u + 3v = \frac{5}{4} \text{ ———————————— } (3)$$

Solving equation (2) and (3) we have

$$u = \frac{1}{2}, \quad and \quad v = \frac{1}{4}$$

Replacing u and v by theirs respective values in (1), we have the solution in the form of closed formula. Therefore,

$$a_n = \left[\frac{1}{2} + \frac{n}{4}\right]4^n$$

Ans.

Example 7: Solve the recurrence equation

$a_n = a_{n-3} + 2(n - 1)$ with $a_0 = 1$.

Solution: The given equation is non-homogeneous of order **one.** This can also be solved **either** by **backtracking** or, by **summation method.** Here, we shall solve it by characteristic method. The characteristic equation of the given recurrence equation is

$x - 1 = 0$, *or*, $x = 1$

The characteristic equation has only one root, so the homogeneous solution is given by

$$a_n = A * 1^n \text{ ———————————— } (1)$$

Now to find particular solution, we guess

$a_n = B_n + D$

Now substituting this guess for a_0 and a_{0-1} in the given recurrence equation, we get

$Bn + D - B(n - 1) - D = 2n - 2$

or, $B = 2n - 2 \Rightarrow 0 = 2$, *This is impossible.*

Notice coefficient of n in LHS is zero and in RHS it is 2

This has happened because of our assumption (guess). Now we will modify our guess and take a **polynomial in *n* of a degree one higher than what we took earlier** as a guess to the particular solution. Let

$a_n = Bn^2 + Cn + D$

Now substituting this guess for a_n and a_{n-1} in the given recurrence equation, we get

$$Bn^2 + Cn + D - B(n-1)^2 - C(n-1) - D = 2n - 2$$

or, $2Bn + (C - B) = 2n - 2, \Rightarrow 2B = 2$, and $C - B = -2$

$\Rightarrow B = 1$ and $C = -1$ and $D = 0$

$\therefore a_n = n^2 - n$, *is the particular solution.*

Here, we have assigned $D = 0$. Readers may try assigning any finite value to D and may test the end result. The shall remain the same.

The complete solution is then given by combining homogeneous and particular solution as

$$a_n = A + n^2 - n \quad \text{———} (2)$$

To determine the value of A, we use the initial condition $a_0 = 1$. We have from equation (2), $a_0 = A$ or $A = 1$ when $n = 0$. therefore,

$$a_n = n^2 - n + 1$$

Ans.

In example 7, initially, we guessed a particular solution that failed and then we guessed another that succeeded. Reader may ask, then **where to begin and where to stop**. Some people may find it easier to solve polynomial particular solutions by beginning at the top (at the order of RHS, or above if the RHS is of the form of homogeneous solution) and then by eliminating term until a solution is found. On the other hand, some may proceed in reverse direction and begin at the bottom (try a constant) and then work on first power of n, second power of n and on until a solution is found. Pick whichever may work best for you. In any case, if you are able to find a set of coefficients which work then that is the solution. If you cannot find a consistent set of coefficients, your guess was poor. There will be no undetermined coefficients in the particular solution. If you have any, then what you have found is not a particular solution.

Example 8: Solve the recurrence equation

$a_n - sa_{n-1} + 6a_{n-2} = 2^n + n$ *with initial condition* $a_1 = 0$ and $a_2 = 10$.

Solution: The general to the problem is given by the homogeneous part of the given recurrence equation. The homogeneous part of the equation

$$a_n - 5a_{n-1} + 6a_{n-2} = 2^n + n \text{ ——— (1)}$$

is

$$a_n - 5a_{n-1} + 6a_{n-2} = 0 \text{ ——— (2)}$$

The characteristic equation of (2) is given as

$$x^2 - 5x + 6 = 0$$

$$or, \quad (x-2)(x-3) = 0$$

$$or, \quad x = 2 \text{ and } x = 3$$

The two roots, 2 & 3, or characteristic equation are distinct. So the homogeneous solution is given by

$$a_n = A2^n + B3^n \text{ ——— (3)}$$

Let particular solution be

$$a_n = Cn2^n + Dn + E$$

Substituting this particular solution in equation (1) for a_n a_{n-1} and a_{n-2}, we get

$$Cn2^n + Dn + E - 5[C(n-1)2^{n-1} + D(n-1) + E]$$
$$+ 6[C(n-2)2^{n-2} + D(n-2) + E] = 2^n + n$$

$$or, \quad Cn2^n - 5(n-1)2^{n-1} + 6(n-2)2^{n-2}] + D[n - 5(n-1)$$
$$+ 6(n-2] + E[1 - 5 + 6] = 2^n + n$$

$$or, \quad C[4n - 10(n-1) + 6(n-2)]2^{n-2} + D[n - 5 + 5 + 6n - 12)]$$
$$+ 2E = 2^n + n$$

$$or, \quad C[4n - 10n + 10 + 6n - 12)]2^{n-2} + D[2n - 7)] + 2E = 2^n + n$$

$$or, \quad C[(-2)]2^{n-2} + D[2n - 7)] + 2E = 2^n + n$$

$$or, \quad -2C = 4,\ 2D = 1 \text{ and } -7D + 2E = 0$$

$$\text{or,} \quad C = -2, \quad D = \frac{1}{2} \quad \text{and} \quad E = \frac{7}{4}$$

Therefore, particular solution is

$$a_n = -2n2^n + \frac{n}{2} + \frac{7}{4}$$

Total solution is given by

$$a_n = A2^n + B3^n - 2n2^n + \frac{n}{2} + \frac{7}{4} \quad \text{———————} (4)$$

Using initial conditions for $n = 1$ and $n = 2$, we get the following two equations (5) and (6), respectively.

$$a_1 = A2^1 + B3^1 - 2 * 1 * 2 + \frac{1}{2} + \frac{7}{4}$$

$$or, \quad 0 = 2A + 3B - 4 + \frac{1}{2} + \frac{7}{4} \qquad [\because a_1 = 0]$$

$$or, \quad 2A + 3B = \frac{7}{4} \quad \text{———————} (5)$$

And,

$$a_2 = A2^2 + B3^2 - 2 * 2 * 2^2 + \frac{2}{2} + \frac{7}{4}$$

$$or, \quad 10 = 4A + 9B - 16 + 1 + \frac{7}{4} \quad [\because a_2 = 10]$$

$$or, \quad 4A + 9B = \frac{93}{4} \quad \text{———————} (6)$$

Solving equations (5) and (6), we get

$$A = -9, \quad and \quad B = \frac{79}{12}$$

Replacing the values of A and B in equation (4), we get the closed form formula for the given recurrence equation.

$$a_n = -9 * 2^n + \frac{79}{12}3^n - 2n2^n + \frac{n}{2} + \frac{7}{4}$$

Ans.

Example 9: Solve the recurrence equation

$$c_n = 3c_{n-1} - 2c_{n-2} \quad for\ n \geq 3 \quad with \quad c_1 = 5 \quad and\ c_2 = 3$$

Solution: The given equation is homogeneous so only homogeneous part of the solution is required.

The characteristic equation of the given recurrence equation is

$$x^2 - 3x + 2 \quad = 0$$

$$or, \quad (x-1)(x-2) = 0 \Rightarrow x = 1,\ and\ x = 2$$

The two roots of the characteristic equation are distinct. So the homogeneous solution is given by

$$c_n = u * 1^n + v * 2^n \text{ ——————— } (1)$$

The equation (1) contains two undetermined coefficients u & v which are to be determined. To find this, we use the given initial conditions for $n = 1$ and $n = 2$, since the values of c_1 and c_2 are given. Putting $n = 1$ and $n = 2$ in equation (1), we get the following equations (2) and (3), respectively.

When $n = 1$

$$c_1 = u * 1 + v * 2 \ or,\ 5 = u + 2v$$

$$or \quad u + 2v = 5 \text{ ——————— } (2)$$

And,

When $n = 2$

$$c_2 = u * 1^2 + v * 2^2 \ or,\ 3 = u + 4v$$

$$or, \quad u + 4v = 3 \text{ ——————— } (3)$$

By solving equations (2) and (3), we have

$$u = 7, \quad \text{and} \quad v = -1$$

Replacing u and v by theirs respective values in (1), we have the solution in the form of closed form formula. Therefore.

$$c_n = 7 - 2^n$$

Ans.

Example 10: Solve the recurrence equation

$$T_n = T_{\sqrt{n}} + C \log_2 n \text{ with } T_1 = 1$$

by characteristic equation method.

Solution: The given equation can be written is standard form as

$$T_n - T_{\sqrt{n}} = C \log_2 n \qquad \text{———— (1)}$$

The homogeneous solution of (1) is given by the equation

$$T_n - T_{\sqrt{n}} = 0 \qquad \text{———— (2)}$$

The characteristic equation of (2) is

$$x - \sqrt{x} = 0 \quad or, \quad x^2 - x = 0 \ \ or, \quad x\,(x-1) = 0$$

$$\therefore x = 0, \text{ and } x = 1$$

The homogeneous solution is then given by

$$T_n = A * 0^n + B * 1^n \qquad \text{———— (3)}$$

Guess the particular solution as

$$T_n = ZC \log_2 n$$

Using this particular solution in equation (1), we get

$$ZC \log_2 n - ZC \log_2 n^{\frac{1}{2}} = C \log_2 n$$

$$or, \quad \left[Z - \frac{Z}{2}\right] C \ \log_2 n = C \ \log_2 n$$

$$Z - \frac{Z}{2} = 1 \qquad or, \quad Z = 2$$

Therefore, particular solution is $T_n = 2C \log_2 n$.

Combining homogeneous and particular solution, we get the complete solution, which is:

$$T_n = A * 0^n + b * 1^n + 2C \log_2 n$$

$$or, \quad T_n = B + 2C \log_2 n$$

For $n = 1$ the initial condition is given. Putting $n = 1$ in the above equation, we can find the value of undetermined coefficient B as below.

When $n = 1,\ T_1 = B + 2C \log_2 1,$

or, $1 = B + 2C * 0$ $[\because T_1 = 1 \text{ and } \log_2 1 = 0]$

or, $B = 1$

Replacing B by 1 in the complete solution, we have

$T_n = 1 + 2C \log_2 n$

Ans.

Before winding up this section, I would like to introduce—**the technique of change of variables**. This is very useful and effective for solving a range of recurrence equations, which are otherwise perceived to be unsolvable.

Example 11: Solve the recurrence equation

$$a_r = \sqrt{a_{r-1} + \sqrt{a_{r-2} + \sqrt{a_{r-3} + \sqrt{\cdots}}}} \quad with \quad a_0 = 4.$$

Solution: **Wow!** Is it solvable by any of the above methods? At first, it seems no. But squaring the given equation we have

$$a_r^{\ 2} = a_{r-1} + \sqrt{a_{r-2} + \sqrt{a_{r-3} + \sqrt{a_{r-4} + \sqrt{\cdots}}}}$$

or,

$$a_r^{\ 2} = a_{r-1} + a_{r-1} \qquad \left[\because a_{r-1} = \sqrt{a_{r-1} + \sqrt{a_{r-2} + \sqrt{a_{r-4} + \sqrt{\cdots}}}}\right]$$

or, $a_r^2 = 2a_{r-1} \Rightarrow a_r^2 - 2a_{r-1} = 0$ ——————— (1)

Now, what about equation (1)? Is it solvable? Again it seems no. Let us replace a_r with a new sequence term. Let

$b_r = \log_2 a_r \Rightarrow a_r = 2^{b_r}$

Substituting the value of a_r in terms of the new sequence variable b_r in equation (1), we get

$$\left(2^{b_r}\right)^2 - 2 * 2^{b_{r-1}} = 0$$

or, $$2^{2b_r} - 2^{b_{r-1}+1} = 0$$

This is possible only if

$$2b_r - b_{r-1} + 1 = 0 \text{ ——————————— } (2)$$

with initial condition $b_0 = \log_2 a_0 = \log 4 \Rightarrow b_0 = 2$

Now, our job has been reduced to solve the equation (2) which is of **first order, linear and non-homogeneous.** The characteristic equation of (2) is

$$2x - 1 = 0 \quad \Rightarrow \quad x = \frac{1}{2}$$

Therefore, homogeneous solution is

$$b_r = A\left(\frac{1}{2}\right)^r \text{ ——————————— } (3)$$

For particular solution, we guess $b_r = B$. Substituting this value for sequence term b_r and b_{r-1} in equation (2), we have

$$2B - B = 1 \Rightarrow B = 1$$

$\therefore$ *particular solution is* $b_r = 1$

Then, the total solution is

$$b_r = A\left(\frac{1}{2}\right)^r + 1$$

Using the initial condition $b_0 = 2$ we get $A = 1$. Therefore the solution to the changed equation is

$$b_r = \left(\frac{1}{2}\right)^r + 1,$$

Since, $b_r = \log_2 a_r$, $a_r = 2^{b_r}$

Therefore, final solution is given by

$$a_r = 2^{\left(\frac{1}{2}\right)^r + 1}$$

Ans.

Skill for solving problem by substitution method can be improved by rigorous practice only. There is no short cut to this. Readers are advised to work out the problems given in the exercises.

Example 12: Solve the recurrence equation

$a_n + 9a_{n-1} = 0$ with initial condition $a_0 = 0$ and $a_1 = 1$.

Solution: The characteristic equation for the given recurrence equation is given by

$x^2 + 9 = 0 \Rightarrow x^2 = -9 \Rightarrow x = \pm 3i$

Therefore the homogeneous solution is

$a_n = A(-3i)^n + B(3i)^n$

Don't worry if the roots are imaginary or complex. When the values of the coefficients are determined at the very end, the resulting sequence values a_n will be real. Using the initial condition for $n = 0$ and $n = 1$, we get the following two equations.

$$0 = A(-3i)^0 + B(3i)^0$$

or, $A + B = 0$ ——— (1)

and,

$$1 = A(-3i)^1 + B(3i)^1$$

or, $-3iA + 3iB = 1$ ——— (2)

Now solving equation (1) and (2) we get values of A and B which are

$$A = \frac{-1}{6i} = \frac{i}{6} \quad \textit{and} \quad B = \frac{-i}{6}$$

Substituting the value of A and B in homogeneous solution we get

$$a_n = \frac{i}{6}(-3i)^n + \frac{-i}{6}(3i)^n$$

You can verify that **all sequence terms will be real.**

3.4 Generating Function Method

One of the uses of generating function (**GF**) method is to find the closed form formula for a recurrence equation. Before using this method, **ensure that the given recurrence equation is in linear form.** A non-linear recurrence equation cannot be solved by the **GF** method. Use **substitution of variable** technique to convert a non-linear recurrence equation into linear. Solving a recurrence equation using **GF** method involves two steps process:

Step 1: Find generating function for the sequence for which the general term is given by recurrence equation.

Step 2: Find coefficient of x^n or $x^n/n!$ depending upon whether the GF is binomial or exponential to get a_n, the general term of the sequence.

The value so obtained will be an algebraic formula for a_n, expressed in terms of n which is the position of a_n in sequence. Our goal is to find the same while solving a recurrence equation. The process of doing this is illustrated with the following examples.

Example 1: Solve the following recurrence equation using generating function.

$$a_n = a_{n-1} + 2(n-1);\ a_0 = 1$$

Solution: We will solve this using binomial generating function. Let $f(x)$ be the binomial generating function for the sequence a of which general term is given by the given recurrence. Therefore. We have

$$f(x) = \sum_{n=0}^{\infty} a_n x^n = a_0 + \sum_{n=1}^{\infty} a_n x^n$$

$$or,\ f(x) = a_0 + \sum_{n=1}^{\infty} [a_{n-1} + 2(n-1)]x^n$$

$$= a_0 + \sum_{n=1}^{\infty} a_{n-1}x^n + 2\sum_{n=1}^{\infty}(n-1)x^n$$

$$= a_0 + x\sum_{n=1}^{\infty} a_{n-1}x^{n-1} + 2x\sum_{n=1}^{\infty}(n-1)x^{n-1}$$

$$= a_0 + x\sum_{m=0}^{\infty} a_m x^m + 2x\left[0 * x^0 + 1 * x^1 + 2 * x^2 + \ldots\right] \quad [n-1=m]$$

$$= a_0 + xf(x) + 2x^2\left[1 + 2x + 3x^2 + \ldots\right] \quad \left[\because f(x) = \sum_{m=0}^{\infty} a_m x^m\right]$$

or, $(1-x)f(x) = a_0 + \dfrac{2x^2}{(1-x)^2} = 1 + \dfrac{2x^2}{(1-x)^2} \quad [\because a_0 = 1]$

or, $f(x) = \dfrac{1-2x+3x^2}{(1-x)^3}$

or, $f(x) = \dfrac{1}{(1-x)^3} - \dfrac{2x}{(1-x)^3} + \dfrac{3x^2}{(1-x)^3}$ *is the GF of the equation.*

The value of a_n is given by the coefficient of x^n in $f(x)$. Therefore,

Coeff. of x^n = *Coeff. of* x^n *in* $\dfrac{1}{(1-x)^3} - 2x$ * *Coeff. of* x^{n-1} *in* $\dfrac{1}{(1-x)^3}$

$+ 3x^2$ * *Coeff of* x^{n-2} *in* $\dfrac{1}{(1-x)^3}$

$$= {}^{3+n-1}C_n - 2 * {}^{3+(n-1)-1}C_{n-1} + 3 * {}^{3+(n-2)-1}C_{n-2}$$

or, $a_n = {}^{n+2}C_n - 2 * {}^{n+1}C_{n-1} + 3 * {}^{n}C_{n-2}$

$$a_n = \frac{n^2+3n+2}{2} - (n^2+n) + \frac{3(n^2-n)}{2} =$$

or, $$\frac{n^2+3n+2-2n^2-2n+3n^2-3n}{2}$$

or, $$a_n = \frac{2n^2-2n+2}{2} = n^2-n+1 \qquad \text{i.e.} \qquad a_n = n^2-n+1$$

Ans.

Example 2: Solve the following recurrence equation using generating function

$$f_n = f_{n-1} + f_{n-2} \; for\; n \geq 2 \quad and \; f_0 = f_1 = 1$$

Solution: We will solve this using binomial generating function. Let $f(x)$ be the binomial generating function for the sequence f of which general term is given by the above recurrence equation. Then, we have

$$f(x) = \sum_{n=0}^{\infty} f_n x^n = f_0 + f_1 x + \sum_{n=2}^{\infty} a_n x^n$$

or, $$f(x) = f_0 + f_1 x + \sum_{n=2}^{\infty}[f_{n-1} + f_{n-2}]x^n$$

$$= f_0 + f_1 x + \sum_{n=2}^{\infty} f_{n-1}x^n + \sum_{n=2}^{\infty} f_{n-2}x^n$$

$$= f_0 + f_0 x + \sum_{n=2}^{\infty} f_{n-1}x^n + \sum_{n=2}^{\infty} f_{n-2}x^n \quad [\because f_0 = f_1]$$

$$= f_0 + x\sum_{n=1}^{\infty} f_{n-1}x^{n-1} + x^2\sum_{n=2}^{\infty} f_{n-2}x^{n-2}$$

$$= f_0 + x\sum_{n-1=0}^{\infty} f_{n-1}x^{n-1} + x^2\sum_{n-2=0}^{\infty} f_{n-2}x^{n-2}$$

$$or,\ f(x) = f_0 + xf(x) + x^2 f(x) = 1 + xf(x) + x^2 f(x)$$

$$\text{or, } f(x) = \frac{1}{(1 - x - x^2)} \text{ is the GF of the equation.}$$

Using partial fraction, we can write $f(x)$ as

$$f(x) = \frac{1}{\sqrt{5}}\left[\frac{1}{1 - x_1 x} - \frac{1}{x - x_2 x}\right]$$

$$where,\ x_1 = \frac{1+\sqrt{5}}{2},\ and\ \ x_2 = \frac{1-\sqrt{5}}{2}$$

$$Therefore,\ coefficient\ of\ \ x^n = \frac{1}{\sqrt{5}}\left[\left(\frac{1+\sqrt{5}}{2}\right)^n - \left(\frac{1-\sqrt{5}}{2}\right)^n\right]$$

$$i.e.,\ f_n = \frac{1}{\sqrt{5}}\left[\left(\frac{1+\sqrt{5}}{2}\right)^n - \left(\frac{1-\sqrt{5}}{2}\right)^n\right]$$

Ans.

By this time, you must have noticed that the hidden rule is to arrange the Σ so that it conforms to the definition of the general term i.e., if a_n is defined for say, $n \geq 4$, then before replacing a_n by its definition we must arrange Σ so that its starting point begins from $n = 4$. Any early replacement will result in wrong result.

Example 3: Solve the following recurrence equation using generating function.

$$a_n - 8a_{n-1} + 16a_{n-2} = 0 \quad for\ n \geq 4 \quad and \quad a_2 = 16, a_3 = 80$$

Solution: We will use binomial generating function to solve this problem. Let $f(x)$ be the binomial generating function for the sequence ***a*** of which general term is given by the above recurrence equation. One thing to be noticed here is that the sequence term start from $n = 2$, so summation will be used accordingly. $f(x)$ is given as:

$$f(x) = \sum_{n=2}^{\infty} a_n x^n = a_2x^2 + a_3x^3 + \sum_{n=4}^{\infty} a_n x^n$$

$$\text{or, } f(x) = a_2x^2 + a_3x^3 + \sum_{n=4}^{\infty} [8a_{n-1} - 16a_{n-2}]x^n$$

$$= a_2x^2 + a_3x^3 + 8\sum_{n=4}^{\infty} a_{n-1}x^n - 16\sum_{n=4}^{\infty} a_{n-2}x^n$$

$$= a_2\,x^2 + a_3\,x^3 + 8[a_3\,x^4 + a_4\,x^5 + \ldots] - 16[a_2\,x^4 + a_3\,x^5 + \ldots]$$

$$= a_2\,x^2 + a_3\,x^3 + 8[a_2\,x^2 + a_3\,x^3 + \ldots]x - 8a_2\,x^3 - 16[a_2\,x^2 + a_3x^3 + \ldots]\,x^2$$

$$= a_2\,x^2 + a_3\,x^3 + 8xf(x) - 8a_2\,x^3 - 16x^2 f(x)$$

or, $(1 - 8x + 16x^2) f(x) = a_2\,x^2 + a_3\,x^3 - 8a_2\,x^3$

Now, substituting the value of a_2 and a_3 in the above equation, we get

$$(1 - 8x + 16x^2) f(x) = 16x^2 + 80x^3 - 8 * 16x^3$$

$$\text{or, } f(x) = \frac{16x^2 - 48x^3}{1 - 8x + 16x^2} = 16x^2 \frac{1 - 3x}{(1 - 4x)^2}$$

$$\text{Therefore, } f(x) = 16x^2 \frac{1 - 3x}{(1 - 4x)^2}$$

To find the general term a_n, we will have to find the coefficient of x^n from $f(x)$. Using Partial fraction method, we can write

$$\frac{1 - 3x}{(1 - 4x)^2} = \frac{A}{(1 - 4x)} + \frac{Bx}{(1 - 4x)^2}$$

Solving the above equation for A and B, we get

$A = 1$, $B = 1$ *therefore, the GF can be written as*

$$f(x) = 16x^2 \left[\frac{1}{1 - 4x} + \frac{x}{(1 - 4x)^2}\right]$$

Hence, coefficient of x^n in $f(x)$ is given as 16 * (coefficient of x^{n-2} in first term + coefficient of x_{n-3} in second term). This is equal to

$$16 * (4^{n-2} + (n-2)4^{n-3}$$

$$\text{or,} \quad a_n = 16 * 4^{n-2}\left[1 + \frac{n-2}{4}\right]$$

$$\text{or,} \quad a_n = 4^n\left[\frac{1}{2} + \frac{n}{4}\right]$$

Ans.

Example 4: Solve the following recurrence equation using generating function.

$$c_n = 3c_{n-1} - 2c_{n-2} = 0 \text{ for } n \geq 3 \text{ and } c_1 = 5, c_2 = 3.$$

Solution 4: We will again use binomial generating function to solve this problem. Let $f(x)$ be the binomial generating function for the sequence ***c*** of which general term is given by the above recurrence equation. Since the sequence term starts from $n = 1$, so the summation will be used accordingly. $f(x)$ is given by;

$$f(x) = \sum_{n=1}^{\infty} c_n x^n = c_1 x + c_2 x^2 + \sum_{n=3}^{\infty} c_n x^n$$

$$\text{or,} \quad f(x) = c_1 x + c_2 x^2 + \sum_{n=3}^{\infty}(3c_{n-1} - 2c_{n-2})x^n \quad [\text{by def n of } c_n]$$

$$= c_1 x + c_2 x^2 + 3\sum_{n=3}^{\infty} c_{n-1} x^n - 2\sum_{n=3}^{\infty} c_{n-2} x^n$$

$$= c_1 x + c_2 x^2 + 3[c_2 x^3 + c_3 x^4 + c_4 x^5 + \ldots] - 2[c_1 x^3 + c_2 x^4 + c_3 x^5 + \ldots]$$

$$\text{or,} \quad f(x) = c_1 x + c_2 x^2 + 3[c_1 x + c_2 x^2 + c_3 x^3 + \ldots]x - 3c_1 x^2 - 2[c_1 x + c_2 x^2 + c_3 x^3 + \ldots]x^2$$

$$\text{or,} \quad f(x) = c_1 x + c_2 x^2 + 3xf(x) - 3c_1 x^2 - 2x^2 f(x)$$

$$\text{or,} \quad (1 - 3x + 2x^2) f(x) = c_1 x + c_2 x^2 - 3c_1 x^2 = 5x + 3x^2 - 3 * 5x^2$$

$$Therefore, f(x) = \frac{5x - 12x^2}{(1-x)(1-2x)} = x\frac{5-12}{(1-x)(1-2x)}$$

To find the general term a_n, we will have to find the coefficient of x^n from $f(x)$. Using Partial fraction method, we can write

$$\frac{5-12x}{(1-x)(1-2x)} = \frac{A}{(1+x)} + \frac{B}{(1-2x)}$$

Solving the above equation for A and B we get

$A = 7, B = -2$ *Therefore, the GF can be written as*

$$f(x) = x\left[\frac{7}{1-x} - \frac{2}{(1-2x)}\right]$$

Hence, coefficient of x^n in $f(x)$ is given as (coefficient of x^{n-1} in first term + coefficient of x^{n-2} in second term). This is equal to

$$7 - 2 * 2^{n-1}$$

or, $a_n = 7 - 2^n$

Ans.

Now let us solve another problem by generating function method in which RHS is an exponential function. This problem has been solved using the binomial generating function.

Example 5: Solve the following recurrence equation using generating function.

$$a_n - 7a_{n-1} + 10a_{n-2} = 2^n \textit{ for } n \geq 2 \textit{ and } a_0 = 0, a_1 = 6.$$

Solution: Let $f(x)$ be the binomial generating function for the sequence a of which general term is given by the above recurrence equation. Then, $f(x)$ is given by

$$f(x) = \sum_{n=0}^{\infty} a_n x^n = a_0 x + a_1 x + \sum_{n=2}^{\infty} a_n x^n$$

$$or,\ f(x) = a_0 x + a_1 x + \sum_{n=2}^{\infty}(7a_{n-1} - 10a_{n-2} + 2^n)x^n \text{ [by def n of } a_n]$$

$$= a_0x + a_1x + 7\sum_{n=2}^{\infty} a_{n-1}x^n - 10\sum_{n=2}^{\infty} a_{n-2}x^n + \sum_{n=2}^{\infty} 2^n x^n$$

$$= a_0x + a_1x + 7x\sum_{n=2}^{\infty} a_{n-1}x^{n-1} - 10x^2\sum_{n=2}^{\infty} a_{n-2}x^{n-2} + \sum_{n=2}^{\infty} 2^n x^n$$

or, $f(x) =$

$$a_0x + a_1x + 7xf(x) - 10x^2 f(x) + \frac{4x^2}{1-2x}\left[\because \sum_{n=2}^{\infty} 2^n x^n = \frac{4x^2}{1-2x}\right]$$

or, $(1-7x+10x^2)f(x) = a_0x + a_1x + \frac{4x^2}{1-2x} = 6x + \frac{4x^2}{1-2x}$

Therefore, $f(x) = \dfrac{6x - 8x^2}{(1-2x)^2(1-5x)} = x\dfrac{6-8x}{(1-2x)^2(1-5x)}$

As in the previous example, we will have to find the coefficient of x^n form $f(x)$. Using Partial fraction method, we can write

$$\frac{6x-8x^2}{(1-2x)^2(1-5x)} = \frac{A}{(1-5x)} + \frac{B}{(1-2x)} + \frac{Cx}{(1-2x)^2}$$

Solving the above equation for A, B and C, we get

$A = \frac{22}{9}, B = -\frac{22}{9},$ *and* $C = -\frac{4}{3}$ *Therefore, the GF can be written as:*

$$f(x) = \frac{22}{9}\left[\frac{1}{1-5x}\right] - \frac{22}{9}\left[\frac{1}{1-2x}\right] - \frac{4}{3}\left[\frac{x}{(1-2x)^2}\right]$$

Hence, coefficient of x^n in $f(x)$ is given as (coefficient of x^n in first term + coefficient of x^n in second term + coefficient of x^n in third term). This is equal to

$$\frac{22}{9}5^n - \frac{22}{9}2^n - \frac{4}{3}n2^{n-1}$$

$$\text{or,} \quad a_n = \frac{22}{9}\left[5^n - 2^n\right] - \frac{2}{3}n2^n$$

Ans.

A given recurrence equation can also be solved by using exponential generating function. If the problem is related to combination, we always use binomial generating function, whereas, if a problem is related to permutation, we use exponential generating function. The following problem is not solvable by using binomial generating function as it is related to permutation of numbers. Try to solve the following using binomial GF.

Example 6: Solve the following recurrence equation using exponential generating function.

$d_n = (n-1)(d_{n-1} + d_{n-2})$ *for* $n \geq 3$ *and* $d_1 = 0$, $d_2 = 1$.

Solution: Let $f(x)$ be the exponential generating function of the above recurrence relation, then $f(x)$ is given as

$$f(x) = \sum_{n=0}^{\infty} d_n \frac{x^n}{n!} \quad \text{———} (1)$$

Differentiating (1) with respect to x, we get,

$$f(x) = \sum_{n=0}^{\infty} d_n \frac{x^{n-1}}{(n-1)!} \quad \text{———} (2)$$

Replacing d_n by its definition, in the equation (2), we get

$$f(x) = \sum_{n=0}^{\infty} (n-1)(d_{n-1} + d_{n-2}) \frac{x^{n-1}}{(n-1)!}$$

$$\text{or,} \quad = \sum_{n=0}^{\infty} (n-1)d_{n-1} \frac{x^{n-1}}{(n-1)!} + \sum_{n=0}^{\infty} (n-1)d_{n-2} \frac{x^{n-1}}{(n-1)!}$$

$$or, \quad f(x) = \sum_{n=0}^{\infty} d_{n-1} \frac{x^{n-1}}{(n-2)!} + \sum_{n=0}^{\infty} d_{n-2} \frac{x^{n-1}}{(n-2)!}$$

$$= x\sum_{n=0}^{\infty} d_{n-1} \frac{x^{n-2}}{(n-2)!} + x\sum_{n=0}^{\infty} d_{n-2} \frac{x^{n-2}}{(n-2)!}$$

$$= x\left[d_0 . \frac{x^{-1}}{(-1)!} + d_1 . \frac{x^0}{(0)!} + d_2 . \frac{x^1}{1!} + d_3 . \frac{x^2}{2!} + \ldots\right]$$

$$+x\left[d_0 + d_1 . x + d_2 . \frac{x^2}{2!} + d_3 . \frac{x^3}{3!} + \ldots\right]$$

or, $f(x) = xf(x) + xf(x)$

or, $(1-x)f(x) = xf(x) \Rightarrow \quad \frac{f'(x)}{f(x)} = \frac{x}{1-x} \Rightarrow \frac{f'(x)}{f(x)} = \frac{1}{1-x} - 1$

or, $\frac{f'(x)}{f(x)} = \frac{1}{1-x} - 1$ ——————————(3)

Integrating both sides of equation (3) we get,

$$\log f(x) = -\log(1-x) - x$$

or, $\log f(x), (1-x) = -x$

or, $f(x).(1-x) = e^{-x}$

or $f(x) = \frac{1}{1-x}$

given recurrence equation.

To find the general term from this generating function we will have to find the coefficient of $x^n/n!$. Which is given as

$$d_n = n!\left(1 - \frac{1}{1!} + \frac{1}{2!} - \ldots\ldots..(-1)^n \frac{1}{n!}\right)$$

Ans.

Readerss are suggested to give emphasis on understanding the technique to solve a recurrence equation rather than simply following the examples given in the text. If there is only one method to solve a particular problem, one has no option but to follow it. This has the advantage of skipping the selection of method process. However, when there are multiple methods to solve the same problem, selection of choice always gives an edge to those who selects better method. A better method for one problem may be equally bad for the other. The selection of choice is perfected by experience.

Exercise 3

1. Solve the following recurrence equations.
   (a) $a_r - 7a_{r-1} + 10a_{r-2} = 0$ with $a_0 = 0$ *and* $a_1 = 3$.
   (b) $a_r - 4a_{r-1} + 4a_{r-2} = 0$ *with* $a_0 = 1$ *and* $a_1 = 6$.
   (c) $a_r - 7a_{r-1} + 10a_{r-2} = 3^r$ *with* $a_0 = 0$ *and* $a_1 = 1$.
   (d) $a_r + 6a_{r-2} + 9a_{r-2} = 3$ *with* $a_0 = 0$ *and* $a_1 = 1$.
   (e) $a_r + a_{r-1} + a_{r-2} = 0$ *with* $a_0 = 0$ *and* $a_1 = 2$.
   (f) $a_r - a_{r-1} - a_{r-2} = 0$ *with* $a_0 = 0$ *and* $a_1 = 1$.
   (g) $a_r - 2a_{r-1} + 2a_{r-2} - a_{r-3} = 0$ *with* $a_0 = 2$, $a_0 = 1$ *and* $a_2 = 1$.
2. Given that $a_0 = 0$, $a_1 = 1$, $a_2 = 4$, and $a_3 = 12$, satisfy the recurrence equation

   $a_r + C_1 a_{r-1} + C_2 a_{r-2} = 0$ *and*

   determine a_r.

   In Exercises 3 through 8, identify whether the given recurrence equation is linear homogeneous or not. If the equation is linear homogeneous equation, give its degrees.

3. $a_n = 2.5 * a_{n-1}$
4. $b_n = -3b_{n-1} - 2b_{n-2}$
5. $c_n = 2^n * c_{n-1}$
6. $d_n = nd_{n-3}$
7. $e_n = 5e_{n-1} + 3$
8. $g_n = \sqrt{g_{n-1} + g_{n-1}}$

   In Exercises 9 through 14, use the backtracking method to find an explicit formula for the sequence defined by recurrence equation and initial condition(s).

9. $a_n = 2.5 * a_{n-1}$ *with* $a_1 = 4$
10. $b_n = b_{n-1} - 2$ *with* $b_1 = 0$
11. $c_n = c_{n-1} + n$ *with* $c_1 = 4$
12. $d_n = -1.1 * d_{n-1}$ *with* $d_1 = 5$
13. $e_n = 5e_{n-1} + 3$ *with* $e_1 = 2$
14. $g_n = ng_{n-1}$ *with* $g_1 = 6$

In Exercises 15 through 20, solve each of the recurrence equations.

15. $a_n = 4a_{n-1} + 5a_{n-2}$ *with* $a_1 = 2$ *and* $a_2 = 6$.
16. $b_n = -3b_{n-1} - 2b_{n-2}$ *with* $b_1 = -2$ *and* $b_2 = 4$
17. $c_n = -6c_{n-1} - 9c_{n-2}$ *with* $c_1 = 2.5$ *and* $c_2 = 4.7$
18. $d_n = 4d_{n-1} - 4d_{n-2}$ *with* $d_1 = 1$ *and* $d_2 = 7$
19. $e_n = 2e_{n-2}$ *with* $e_1 = \sqrt{2}$ *and* $e_2 = 6$
20. $g_n = 2g_{n-1} - 2g_{n-2}$ *with* $g_1 = 1$ *and* $g_2 = 4$
21. Develop a general explicit formula for a non-homogeneous recurrence equation of the form
$$a_n = ra_{n-1} + s$$
where, r and s are constants.
22. Find the closed formula for the Fibonacci sequence
$$f_n = f_{n-1} + f_{n-2}$$
23. The solution of the recurrence relation $C_0 a_r + C_2 a_{r-1} + C_2 a_{r-2} = f(r)$ is
$$3^r + 4^r + 2$$
If $f(r) = 6$ for all r, determine C_0, C_1, and C_2
24. Solve the difference equation
$$a_r - ra_{r-1} = r! \text{ for } r \geq 1 \text{ and given that } a_0 = 2$$
[Hint: Use change of variable method. Let $b_r = a_r/r!$]
25. Solve the difference equation
$$ar^2 - 2a_{r-1}{}^2 = 1 \text{ for } r \geq 1 \text{ and given that } a_0 = 2$$
[Hint: Use change of variable method. Let $b_r = a_r^2$
26. Solve the difference equation
$$ra_r + ra_{r-1} - a_{r-1} = 2_r \text{ for } r \geq 1 \text{ and given that } a_0 = 273$$
[Hint: Use change of variable method. Let $b_r = ra_r$]

27. Let d_r denote the number of ways of permuting r integers $\{1, 2, 3, 4, \ldots r\}$ so that the integer will not be in the ith position for $1 \leq i \leq r$.

(a) Use combinatorial a_r gument show that

$$d = (r-1)(a_{r-1} - b_{r-1})$$

(b) Show that

$$d_r = r!\left(1 - \frac{1}{1!} + \frac{1}{2!} - \frac{1}{3!} \ldots\ldots\ldots (-1)^r \frac{1}{r!}\right)$$

satisfies the recurrence equation (a).

28. The solution of the recurrence equation

$$a_r = Aa_{r-1} + B3^r \; for \; r \geq 1$$

is

$$a_r = C2^r + D3^{r+1} \; for \; r \geq 0$$

Given that $a_0 = 19$ and $a_1 = 50$, determine the constants A, B, C, and D.

29. Consider the multiplication of bacteria in a controlled environment. Let a_r denote the number of bacteria there are on rth day. If $a_r - 2a_{r-1}$ be the rate of growth of rth day and that the rate of growth doubles every day, determine a_r given that $a_0 = 1$.

30. How many r bits long binary sequences are having no adjacent 0's?

31. Let a_r denote that total assets of a bank at the end of the rth month. This is equal to the sum of the total deposit in the rth month and 1.1 times the total assets at the end of the previous month. Given that the total deposit is a constant 100 (in thousand Rs.) determine a_r if $a_0 = 0$.

32. n circular discs are slipped onto a peg with the largest disc at the bottom. These rings are to be transferred one at a time onto another pet, and there is a third peg available on which discs can be left temporarily. During the course of transfer, no disc can be placed on top of the smaller one. Find how many moves are required to transfer all n discs from one peg to other.

33. A particle is moving in the horizontal direction. The distance it travels in each second is equal to twice of the distance in previous second. Let a_n denote the position of the particle in nth second, determine a_n given that $a_0 = 3$ and $a_3 = 10$.

34. Consider a sorting algorithm for an array containing $n \geq 2$ numbers.

    (i) Use $2n - 1$ comparisons to determine the largest and the second largest of the n numbers.

    (ii) Recursively, sort the remaining $r - 2$ numbers.

    Let a_n denotes the number of comparisons used for sorting r numbers. Determine a_n.

35. Let a_n denotes the numbers of partitions of a set of n elements. Show that

$$a_{r+1} = \sum_{i=0}^{n} {}^{n}C_i a_i$$

where $a_0 = 1$.

36. Let a_n denotes that total assets of a company in Rupees in the nth year. Clearly, $a_n - a_{n-1}$ is the increase in assets during nth year. If the increase in assets during each year is always five times the increase during the previous year, what are the total assets in the nth year? Given that $a_0 = 3$ and $a_1 = 7$.

37. Let a_n the number of subsets of the set $\{1, 2, 3, \ldots . n\}$ that do not contain two consecutive numbers. Determine a_n.

38. Consider the operation of a factory whose average profit in every two successive month is equal to the average new order in that period. Let a_n denotes the new order received and b_n denotes the monthly profit in the month. Determine b_n when $a_n = 2n$ for all $n \geq 0 =$ and $b_0 = 0$.

Determine b_n when $b_0 = 0$ and

$$a_n = \begin{cases} 2^n & 0 \leq n \leq 9 \\ 2^{10} & n \geq 10 \end{cases}$$

39. Determine the particular solution for the difference equations

(a) $a_n - 3a_{n-1} + 2a_{n-2} = 2^n$ *and* (b) $a_n - 4a_{n-1} + 4a_{n-2} = 2^n$

40. Determine the particular solution for the difference equations

(a) $a_n - 2a_{n-1} = 7n^2$ *and* (b) $a_n - 2a_{n-2} = 7n$

Solve the following recurrence equations 41 through 50 using generating function method.

41. $A_n = A_{n-1} + A_{n-2} + 2n$ given that $A_0 = 0$ and $A_1 = 1$
42. $a_n + a_{n-1} = 3n2^n$ *given that* $a_0 = 0$
43. $a_n = 5a_{n-1} - 3$ *given that* $a_0 = 1$
44. $h_n = 4h_{n-2}$ *given that* $h_0 = 0$ *and* $h_1 = 1$
45. $a_n = \begin{cases} 0 & \text{for } n = 0 \\ 2a_{n-1} + 3 & \text{for } n > 3 \end{cases}$
46. $a_n = \begin{cases} 16 & \text{for } n - 2 \\ 60 & \text{for } n - 3 \\ 8a_{n-1} - 16a_{n-2} & \text{for } n > 3 \end{cases}$
47. $a_n - 2a_{n-1} - 4a_{n-2} = 4_n$ given that $a_0 = 1$ and $a_1 = 7$
48. $a_n - 5a_{n-1} + 6a_{n-2} = 2^n + n$ *given that* $a_1 = 0$ and $a_2 = 10$
49. $3a_{n+2} - 8a_{n+1} - 3a_n = 3^n - 2n + 1$
50. $a_{n+2} - 6a_{n+1} + 8a_n = n^2 + 2 - 5 * 3^n$
51. Solve the following recurrence equations.

(i) $a_n = 2a_{\frac{n}{2}}$ *with* $a_1 = 1$

(ii) $a_n = 2a_{\frac{n}{2}} + n$ *with* $a_1 = 1$

(iii) $a_n = 2a_{\frac{n}{2}} + 3$ *with* $a_0 = 0$.

4

Combinatorics

Proofs are the heart of mathematics. If you are studying mathematics either as a major or as a subject to be used extensively in other major fields of study, then you must come to the terms with proofs. You must be able to read, understand and write them (proofs). **What is the secret?** What magic do you need to know? The straight answer is **there is no secret, no mystery and no magic.** All that is needed is some common sense and a basic understanding of a few trusted and easy to understand techniques.

The **permutation** and **combination,** as a method, has been used to prove many counting laws and to solve many counting problems. We shall discuss the basic concepts of permutation and combination and use of **generating functions** to solve problems related to it in this chapter. Whenever and wherever we discuss the topics of **combinatorics** the list of solution techniques seems to be incomplete without discussing the **pigeonhole principle.** We will learn about this principle and its applications in this chapter.

4.1 Methods of Proofs

The **basic structure** of a proof is easy. It is just a series of statements. Each statement being either

- An assumption or
- A conclusion, clearly following from an assumption or previously proved result.

And that is all. Sometimes there will be some clarifying remarks, but that is for the reader and has no logical bearing on the structure of the proof. A well-written proof will flow. That is, the reader should feel as though they are being taken on a ride that takes them directly and inevitably to the desired conclusion without any distraction about irrelevant details. Each step should be clear or at least clearly justified. A good proof is easy to understand. When you are finished with the proof, apply the simple test listed below to every step (sentence): is it

- Clearly an assumption, or
- A justified conclusion?

If the sentence fails the test, may be it is not needed in the proof. In this section, we will discuss the following techniques one by one.

- Direct proofs
- Proof by contradiction
- Proof by contra-positive
- If, and Only If
- Mathematical Induction
- Counter Examples
- Proof by Exhaustion
- Constructive Verses Existential Proofs

Direct Proofs

If a and b are two natural numbers, we say that ***a* divides *b*** if ∃ a natural number $\boldsymbol{k}$ such that $\boldsymbol{b = a*k}$. For example, 3 divides 21 because ∃ a natural number k (=7) such that 21 = 3 k. Now let us proof the following theorem.

Theorem 1: If a divides b and b divides c then prove that a divides c.

Proof: By our assumption, and the definition of divisibility, ∃ natural numbers k_1 and k_2 such that $\boldsymbol{b = a\, k_1}$ and $\boldsymbol{c = b\, k_2}$. Consequently, $\boldsymbol{c = b\, k_2 = a\, k_1\, k_2}$. Let $\boldsymbol{k = k_1\, k_2}$. Since $\boldsymbol{k}$ is a natural number such that $\boldsymbol{c = a\, k}$, so by the definition of divisibility a divides c.

Proved.

Most theorems that we want to prove are either explicitly or implicitly in the form "**If *P*, Then *Q***". In the theorem proved above, ***P*** is "If a divides b and b divides c" and ***Q*** is "a divides c". **This is the standard form of a theorem.** A **direct proof** should be thought of as a flow of implications beginning with "***P***" and ending with "***Q***" i.e., $\boldsymbol{P \rightarrow \ldots \rightarrow Q}$.

Most of the proofs are direct proofs. Always try direct proofs first, while proving any theorems or statements, unless you have a good reason not to do so. If you find a simple proof, and you are convinced of its correctness, then do not feel shy about. Many times proofs are simple and short. See the theorem below.

Theorem 2: Every odd integer is the difference of two perfect squares.

Proof: Let $2n + 1$ is an odd integer for any $n \in I$. Then

$$2n + 1 = (n + 1)^2 - n^2$$

Proved.

Wow! What is this? This is the proof and that too of one line.

Example 1: Prove that the number 1000…001 (with $3n - 1$ zeroes for $n > 0$) is a composite number.

Solution: We can rewrite the given number as

$$\begin{aligned} 1000\ldots011 &= 10^{3n} + 1 \textit{ where } n > 0 \textit{ and } n \in N \\ &= (10^n)^3 + 1 \\ &= (10^n + 1)(10^{2n} - 10^n + 1) \end{aligned}$$

In the above expression, both the factors are integers and are greater than 1. Thus the given number is a composite number.

Example 2: If r_1 and r_2 are two distinct roots of the polynomial $p(x) = x^2 + bx + c$, then show that $r_1 r_2 = -b$ and $r_1 + r_2 = c$.

Solution: If r_1 and r_2 are two distinct roots of the polynomial $p(x)$, then we can write $p(x) = (x - r_1)(x - r_2)$. Upon expansion of the right hand side, we get

$$p(x) = x^2 - (r_1 + r_2)x + r_1 r_2 \text{ ——————————— } (1)$$

We know that two polynomial are said to be equal if the corresponding coefficient are equal. Comparing the coefficients of the given polynomial with that of equation (1), we get

$$r_1 + r_2 = -b \text{ and } r_1 r_2 = c$$

Proved.

Proof by contradiction

In a proof by contradiction we assume, along with the hypothesis, the **logical negation** of the result to be proved. Then we reach some kind of contradiction. That is, if we want to prove "If ***P*** and ***Q***", we assume ***P*** and Not ***Q***. The contradiction we arrive at could be some conclusion contradicting one of our assumptions, or something obviously untrue like 6 = 2. Proof by contradiction is often used to prove the impossibility of something. You assume it is possible and arrive at some contradiction. Read the proof of the irrationality of the square root of 2 in the following theorem.

Theorem 3: Prove that the square root of 2 is an irrational number.

Proof:

Let $\sqrt{2} = s$

So $s^2 = 2$ ——————————————————(1)

If s were a rational number, we can write

$$s = \frac{p}{q} \text{ ——————————————————(2)}$$

Where p is an integer and q is any positive integer. We can assure that the **gcd** $(p, q) = 1$ by driving out common factor between p and q, if any. Now using equation (2), in equation (1), we get

$p^2 = 2q^2$ ——————————————————(3)

Conclusion: From equation (3), **2** must be a prime factor of $\boldsymbol{p^2}$. Since **2** itself is a prime, 2 must be a factor of $\boldsymbol{p}$ itself. Therefore, 2*2 is a factor of $\boldsymbol{p^2}$. This implies that 2*2 is a factor of $2*\boldsymbol{q^2}$.

$\Rightarrow$ 2 is a factor of $\boldsymbol{q^2}$

$\Rightarrow$ 2 is a factor of $\boldsymbol{q}$

Since, 2 is factor of both $\boldsymbol{p}$ and $\boldsymbol{q}$ **gcd** $(p, q) = 2$. This is contrary to our assumption that **gcd** $(p, q) = 1$. Therefore, $\sqrt{2}$ is an irrational number.

Proved.

Many such theorems are proved using the method of contradictions. One of the first proofs by contradiction is the following theorem attributed to Euclid.

Theorem 4: Prove that there are infinite many prime numbers.

Proof: Assume to the contrary that there are only finite numbers, of prime numbers and all of them can be listed as

$$p_1, p_2, p_3, \cdots, p_n$$

Now, consider the number

$$q = p_1 p_2 p_3 \cdots p_n + 1$$

This number is not divisible by any of the listed primes since if we decided q by p_r for each $r = 1, 2, 3, \ldots, n$, we get remainder as 1. Well then, we must conclude that q is a prime number different from the primes listed above. This is contrary to the assumption that all primes are in the list

$$p_1, p_2, p_3, \cdots, p_n$$

Hence, there are infinite many prime numbers.

Proved.

Example 3: Prove that there are no rational number solution to the equation

$$x^3 + x + 1 = 0$$

Solution: There is formula for solving the general cubic equation

$$ax^3 + bx^2 + cx + d = 0$$

But, in this example, we wish to prove there is no rational solution (roots) to the given cubic equation without using the general cubic formula. Let us assume that there is a rational number $\boldsymbol{p/q}$, in reduced form, with $\boldsymbol{p} \neq 0$ that satisfies the equation. Then, we have

$$\frac{p^3}{q^3} + \frac{p}{q} + 1 = 0 \quad or, \; p^3 + pq^2 + q^3 = 0$$

There are, now, three cases to consider. *Case 1:* If both p and q are odd, then the LHS of the above equation is odd. *Case 2:* If p is odd and q is even, then also the LHS is odd. *Case 3:* If p is even and

q is odd, then again LHS is odd. The fourth case, p and q both being even does not arise because of our assumption that p/q is in reduced form. Since 0 is not an odd number, we arrive at a contradiction that **0** is an odd number. This contradiction is because of our assumption that the given equation has a rational root. Hence, there are no rational solution to given cubic equation.

Proved.

Proof by Contra-Positive

Proof by contra-positive takes advantage of the logical equivalence between "***P* implies *Q***" and "**Not *Q* implies Not *P***". For example, the assertion "If it is my car, then it is red" is equivalent to "If that car is not red, then it is not mine". So, to prove "If *P*, then *Q*" by the method of contra-positive means to prove "If Not *Q*, then Not *P*". At first it seems that both contra-positive method and contradiction method are similar, but there is subtle difference between them.

- Contradiction: Assume *P* and Not *Q* and prove some sort of contradiction.
- Contra-positive: Assume not *Q* and prove Not *P*.

Read the proof of the following theorem to understand the contra-positive method.

Theorem 5: If n is a positive integer such that $n \equiv_4 2$ or $n \equiv_4 3$, then n is not a perfect square. Prove it.

Proof: We will prove the contra-positive version of the given statement. That is if n is perfect square, then n mod (4) is neither 2 nor 3, i.e., if n is a perfect square then n mod (4) is either 0 or 1. If n is a perfect square, then $\exists$ an integer k such that $n \equiv k^2$. There are four possible cases to consider, corresponding to four possible remainders 0, 1, 2 and 3 when, k is divided by 4.

Case 1: If k mod (4) = 0, then $k = 4q$, for some integer q. Then, $n = k^2 = 16q^2 = 4(4q^2) \Rightarrow n$ mod (4) = 0

Case 2: If k mod (4) = 1, then $k = 4q + 1$, for some integer q. Then, $n = k^2 = 16q^2 + 8q + 1 = 4(4q^2 + 2q) + 1 \Rightarrow n$ mod (4) = 1

Case 3: If k mod (4) = 2, then $k = 4q + 2$, for some integer q. Then, $n = k^2 = 16q^2 + 16q + 4 = 4(4q^2 + 4q + 1) \Rightarrow n$ mod (4) = 0

Case 4: If k mod (4) = 3, then $k = 4q + 3$, for some integer q. Then, $n = k^2 = 16q^2 + 24q + 9 = 4(4q^2 + 6q + 2) + 1 \Rightarrow n$ mod (4) = 1

Hence, it is proved that if n is perfect square, then n mod (4) cannot be equal to 2 or 3. That is if n mod (4) is 2 or 3, then n is not a perfect square.

Proved.

If, and Only If

Many theorems are stated in the form "P, if, and only if Q". Another way to say the same thing is: "Q is necessary, and sufficient for P". This means two things: one "If P, Then Q" and "If Q, Then P". So to prove an "If, and Only If" theorem, you must prove two implications. See the following example.

Example 4: Prove that if $\boldsymbol{a}$ is an integer, then $\boldsymbol{a}$ is not evenly divisible by 3 if, and only if, $\boldsymbol{a^2 - 1}$ is evenly divisible by 3.

Solution: Since this an "if, and only if" problem, we must prove two implications. (**"If"**) We must prove that "a is not evenly divisible by 3 if $a^2 - 1$ is evenly divisible by 3". So we assume that 3 evenly divides

$$a^2 - 1 = (a - 1)(a + 1)$$

Since 3 is a prime number therefore, 3 must evenly divide either $\boldsymbol{a - 1}$ or $\boldsymbol{a + 1}$. In either case, it should be apparent that 3 cannot evenly divide $\boldsymbol{a}$.

(**"Only If"**) We must prove "$\boldsymbol{a}$ is not evenly divisible by 3 only if $a^2 - 1$ is evenly divisible by 3". This means "If $\boldsymbol{a}$ is not evenly divisible by 3, then $a^2 - 1$ is evenly divisible by 3". We can write $\boldsymbol{a = 3q + r}$, where $r = 0$, 1 or 2. Our assumption that $\boldsymbol{a}$ is not divisible by 3 implies that r cannot be 0. If $r = 1$, then $a - 1 = 3q$ and so 3 evenly divides $a^2 - 1 = (a - 1)(a + 1)$. Similarly, if $r = 2$, then $a + 1 = 3q + 3$ therefore 3 evenly divides $a^2 - 1$.

Proved.

Example 5: A positive integer n is evenly divisible by 3 if, and only if, the sum of the digits of n is divisible by 3.

Solution: Let n is a positive integer whose digit representation is $a_0, a_1, \ldots, a_k$. That means

$n = a_0 + 10a_1 + 10a_2 + \ldots\ldots + 10^k a_k$ and digit sum is

$$s = a_0 + a_1 + a_2 + \ldots\ldots + a_k$$

Now, we have

$$n - s = a_0 + 10a_1 + 10^2 a_2 + \ldots\ldots + 10^k a_k - (a_0 + a_1 + a_2 + \ldots\ldots + a_k)$$

$$= 9a_1 + 99a_2 + \ldots\ldots + (999\ldots9)a_k$$

Where the last term has k 9's. So, clearly $n - s$ is divisible by 3. It follows that n is divisible by 3 if, and only if, s is divisible by 3.

Proved.

Mathematical Induction

Suppose we have to prove a theorem or statement of the form "For all integers n greater than equal to $\boldsymbol{n}$, $\boldsymbol{P(n)}$ is true". $\boldsymbol{P(n)}$ must be an assertion that we wish to be true for all $\boldsymbol{n = k, k+1, k+2, \ldots}$; like a mathematical formula. We first verify the **initial step.** That is we must verify that $\boldsymbol{P(k)}$ is true. Next comes the inductive step. Here we must prove "if there is a $\boldsymbol{m}$, greater than or equal to $\boldsymbol{k}$, for which $\boldsymbol{P(k)}$ is true, then for this same $\boldsymbol{k}$, $\boldsymbol{P(k+1)}$ is true".

Since, we have verified $\boldsymbol{P(k)}$, it follows from inductive step that $\boldsymbol{P(k+1)}$ is true, and hence $\boldsymbol{P(k+2)}$ is true, and hence $\boldsymbol{P(k+3)}$ is true, and so on. In this way the theorem is proved using mathematical induction. See the following examples to understand the working procedure of mathematical induction in proving any formula (theorem or statement).

Example 6: For any positive integer n,

$$1 + 2 + 3 + \ldots\ldots\ldots + n = \frac{n(n+1)}{2}$$

Prove this using mathematical induction method.

Solution: Let the given formula be $P(n)$. Then we have to prove that $P(n)$ is true for all positive integer n i.e., for $n = 1, 2, 3, \ldots m, \ldots$; the prove consists of two steps: the **initial step** and **inductive step.**

Initial Step: For $n = 1$, we have $P(1) = 1(1+1)/2$, which is clearly true. Thus the $P(1)$ is verified.

Inductive Step: Here we must prove the assertion "if there is a m such that $P(m)$ is true, then for this same m, $P(m+1)$ is true". Thus, we assume that there is a m such that

$$1 + 2 + 3 + \ldots\ldots\ldots + m = \frac{m(m+1)}{2}$$

We must prove, for this same m, the formula

$$1 + 2 + 3 + \ldots\ldots\ldots + m + (m+1) = \frac{(m+1)(m+2)}{2}$$

We have

$$1 + 2 + 3 + \ldots\ldots\ldots + m = \frac{m(m+1)}{2} \quad [\textit{assumption that } P(m) \textit{ is true}]$$

Adding $(m + 1)$ *to the both sides of the above equation, we get*

$$1 + 2 + 3 + \ldots\ldots\ldots + m + (m+1) = \frac{m(m+1)}{2} + (m+1)$$

$$= \frac{m^2 + m + 2m + 2}{2}$$

$$= \frac{(m+1)(m+2)}{2}$$

$$\textit{i.e., } P(m+1) = \frac{(m+1)((m+1)+1}{2} \quad \textit{is true.}$$

Therefore, by induction $P(n)$ is true for all positive integer

Proved.

Example 7: Prove by mathematical induction that for all positive $n \in N$

$$3^{2^n} - 1 \textit{ is not divisible by } 2^{n+3}$$

Solution: Let the given statement be ***P*(*n*)**.

Initial Step: Let $n = 1$, then, we have

$3^{2^n} - 1 = 3^2 - 1 = 8$ *and*

$2^{n+3} = 2^{1+3} = 16$

Since 8 is not divisible by 16, the statement is true for $n = 1$ i.e., ***P*(1)** is true.

Inductive Step: Let the given statement is true for $n = k$ i.e., ***P*(*k*)** is true. Then we have to prove that for the same ***k*, *P*(*k*+1)** is also true. Now,

$$3^{2^{k+1}} - 1 = 3^{2^k \cdot 2} - 1 = (3^{2^k})^2 - 1$$

$$= \left(3^{2^k} - 1\right)\left(3^{2^k} + 1\right)$$

In the above expression, the RHS has two factors. The first factor is ***P*(*k*)**, which is assumed to be true. Further, the two factors are two consecutive even numbers. Therefore, when divided by **2**, one of the factors becomes an odd number, and thereafter this factor cannot be further divided by any power of **2**. Since ***P*(*k*)** is true, we have

$3^{2^k} - 1$ *is not divisible* 2^{k+3}

and also any odd multiple of $3^{2^k} - 1$ *is not divisible by* 2^{k+3}.

Now, if we prove that the second factor is twice of an odd number then the task will be over.

$$3^{2^k} - 1 = (3^2)^{2^k - 1} - 1$$

$$= (3^2 - 1)\left[(3^2)^{2^k - 2} + (3^2)^{2^k - 3} + \ldots\ldots\ldots + 1\right]$$

$$= 8\left[(3^2)^{2^k - 2} + (3^2)^{2^k - 3} + \ldots\ldots\ldots + 1\right]$$

Since the first factor has **8** as a factor, so it is not twice of an odd number. Hence, the second factor is twice of an odd number. Therefore,

$3^{2^{k+1}} - 1$ *is not divisible by* $2^{(k+1)+3}$

This implies that ***P*(*k*+1)** is true. Therefore, by induction ***P*(*n*)** is true for all positive $n \in$ N.

Proved.

Example 8: Prove by mathematical induction that any integer composed of **3**^{n} identical digits is divisible by **3**^{n} for $n \geq 1$.

Solution: Let ***P*(*n*)** be the statement that that any integer composed of **3**^{n} identical digits is divisible by **3**^{n}.

Initial Step: Let $n = 1$. Then, the number consists of three identical digits. The sum of these digits divisible by 3. Hence, the number is divisible by 3. This implies that ***P*(1)** is true.

Inductive Step: Let the given statement is true for $n = k$ i.e., ***P*(*k*)** is true. Then we have to prove that for the same ***k*, *P*(*k*+1)** is also true. Before proceeding for proof, let us give a look on the following facts.

$$\begin{bmatrix} & 222 = 2(10^{2.3^0} + 10^{1.3^0} + 1) = 2.(111) & \textit{for} \ \ n=1 \\ \| ry, & 222,222,222 = 222(10^{2.3^1} + 10^{1.3^1} + 1) = 222(1001001) & \textit{for} \ \ n=2 \\ \| ry, & 222...27 \text{ times} = 222,222,222\,(10^{2.3^2} + 10^{1.3^2} + 1) & \\ & = 222(1000000001000000001) & \textit{for} \ \ n=3 \end{bmatrix}$$

Let x is any integer consisting of $3n + 1$ identical digits. Then, x can be written as

$$x = y \times z$$

Where y is an integer consisting of same digit 3^n times and z is of the form

$$z = \left(10^{23^k} + 10^{18^k} + 1\right)$$

We have assumed that y is divisible by 3^n. Since, z contains exactly 3 **ones** and remaining **zeroes**, z is divisible by 3. Therefore, the product $y \times z$ is divisible by 3^{n+1}. This implies that ***P*(*k*+1)** is also true. Therefore, by induction ***P*(*n*)** is true for all positive $n \geq 1$.

Proved.

Example 9: Define *a* sequence $\boldsymbol{a} = a_0, a_1, a_2, \ldots a_n, \ldots$ by the recurrence equation

$$a_{n+1} = 2a_n - a_n^2$$

Then a closed form formula for a_n is given by

$$a_n = 1 - (1 - a_0)^{2^n} \textit{ for } n = 0, 1, 2, 3, \ldots$$

Prove this by mathematical induction.

Solution: Let the given statement be ***P*(*n*)**.

Initial Step: Let $n = 0$, then, we have from recurrence equation

$$a_1 = 2a_0 - a_0^2 = 1 - (1 - a_0)^{2^1}$$

Therefore, ***P*(0)** is true.

Inductive Step: Let the given statement is true for $n = k$ i.e., ***P*(*k*)** is true i.e., closed form formula for a_k is given as below.

$$a_k = 1 - (1 - a_0)2^k$$

Then we have to prove that for the same ***k*, *P*(*κ*+1)** is also true. Now from recurrence equation, we have

$$a_{k+1} = 2a_k - a_k^2 \textit{ substituting the value of } a_k\textit{, we get}$$

$$a_{k+1} = 2\left[1 - (1 - a_0)^{2^k}\right] - \left[1 - (1 - a_0)^{2^k}\right]^2$$

$$or,\ a_{k+1} = \left[1 - (1 - a_0)^{2^k}\right]\left[2 - \left[1 - (1 - a_0)^{2^k}\right]\right]$$

$$or,\ a_{k+1} = \left[1 - (1 - a_0)^{2^k}\right]\left[1 - (1 - a_0)^{2^k}\right] = \left[1 - \left\{(1 - a_0)^{2^k}\right\}\right]$$

$$or,\ a_{k+1} = \left[1 - (1 - a_0)^{2^{k+1}}\right]$$

This implies that ***P*(*k*+1)** is also true. Therefore, by induction ***P*(*n*)** is true for all positive $n \geq 0$.

Proved.

Counter Examples

Counter Examples play an important role in mathematics. A complicated proof may be the only way to demonstrate the validity of a particular theorem, whereas a single counter example is enough to refute the validity of a proposed theorem. For example, number of the form $2^{2^n} + 1$, where n is positive integer, were once thought to be prime. Those numbers are prime for $n = 1, 2, 3$ and 4. When $n = 5$, we get $2^{2^5} + 1 = 4294967297 = (641)(6700417)$, a composite number, thus, we conclude that when faced with a number in the form $2^{2^n} + 1$, we are not to assume it either prime or composite, unless we know for sure for some other reason.

A natural place for counter examples to occur is when the converse of a known theorem comes into question. The **converse** of an assertion in the form "**If *P*, Then *Q***" is "**If *Q*, then *P***".

Example 10: State the converse of "if a and b are even integers then $a + b$ is an even integer". Show that converse is not true by giving a counter example.

Solution: The converse of the given statement is "if $a + b$ is an integer then a & b are even". The counter example to disprove this hypothesis can be given by $a = 3$ and $b = 5$. Here, $a + b = 8$ is even but a and b are odd.

Proved.

Example 11: In calculus we have learnt that if a function is differentiable at a point, then it is continuous at that point. What would the converse assert? It would state that if a function is continuous at a point, then it is differentiable at that point. But we know that this is not true. The counter example is $f(x) = |x|$. This function is continuous at $x = 0$, but it is not differentiable at $x = 0$. This counter example is all we need to disprove the converse of a theorem.

Proof by Exhaustion

Sometimes the most straightforward, if not the most elegant, way to construct a proof is by checking all possible cases and verifying the result in all such cases. This method of proof is called **proof by exhaustion** or **case by case proof.** Let us see the following example.

Example 12: Show that if n is a positive integer then $n^7 - n$ is divisible by 7.

Solution: The given expression can be written as

$$n^7 - n = n(n^6 - 1) = n(n^3 - 1)(n^3 + 1) = n(n-1)(n^2 + n + 1)(n+1)(n^2 - n + 1)$$

Now, when n is divided by 7, there are seven possible remainders depending on $n = 7q + r$ where $r = 0, 1, 2, 3, 4, 5$ or 6. We will examine in all these cases to ensure that whichever positive integer n, we take, $n^7 - n$ is divisible by 7.

Case 1: When $r = 0$, i.e., $n = 7q$. Then $n^7 - n$ is divisible by 7 since n is a factor.

Case 2: When $r = 1$, i.e., $n = 7q + 1$. Then $n - 1$ is divisible by 7. Since $(n - 1)$ is a factor of $n^7 - n$, $n^7 - n$ is divisible by 7.

Case 3: When $r = 2$, i.e., $n = 7q + 2$, then the factor $(n^2 + n + 1) = (7q + 2)^2 + (7q + 2) + 1 = 49q^2 + 35q + 7$ is clearly divisible by 7 and hence $n^7 - n$ is divisible by 7.

Case 4: When $r = 3$, i.e., $n = 7q + 3$, then the factor $(n^2 - n + 1) = (7q + 3)^2 - (7q + 3) + 1 = 49q^2 + 35q + 7$ is clearly divisible by 7 and hence $n^7 - n$ is divisible by 7.

Case 5: When $r = 4$, i.e., $n = 7q + 4$, then the factor $(n^2 + n + 1) = (7q + 4) + (7q + 4) + 1 = 49q^2 + 63q + 21$ is clearly divisible by 7 and hence $n^2 - n$ is divisible by 7.

Case 6: When $r = 5$, i.e., $n = 7q + 5$, then the factor $(n^2 - n + 1) = (7q + 5)^2 - (7q + 5) + 1 = 49q^2 + 63q + 21$ is clearly divisible by 7 and hence $n^7 - n$ is divisible by 7.

Case 7: When $r = 6$, i.e., $n + 1 = 7q + 7$ is clearly divisible by 7 and hence $n^7 - n$ is divisible by 7.

Therefore, we have proved that in all cases of n, $n^7 - n$ is divisible by 7.

Proved.

Constructive Verses Existential Proofs

How would you prove that $2^{99} + 1$ is a composite number? Certainly, we will try to exhibit a factorization: $2^{99} + 1 = (2^{33})^3 + 1 = (2^{33} + 1)(2^{66} + 2^{33} + 1)$. That is to say that we have constructed a factorization to show that $2^{99} + 1$ is a composite number. We call such a method of proof as **proof by construction.**

Example 13: **A Pythagorean triple** is a triplet of positive integers (a, b, c) that satisfies the equation $a^2 + b^2 = c^2$. For example, (3, 4, 5) is a Pythagorean triple because $3^2 + 4^2 = 5^2$. Are there any more? Yes, there are infinitely many. The triplet $(3m, 4m, 5m)$ for all positive integer m are Pythagorean triples. We call $(3m, 4m, 5m)$ **one parameter** family of solutions. Here m is a parameter. *Similarly prove that there is a two parameters family of Pythagorean triples.*

Solution: Let us construct the solution. Let $a = m^2 - n^2$ and $b = 2\,m\,n$ where m and n are positive integers with $m > n$. Then $a^2 + b^2 = (m^2 - n^2)^2 + (2\,m\,n)^2 = (m^2 + n^2)^2$. Thus, $(m^2 - n^2, 2\,m\,n, m^2 + n^2)$, for $m > n$, is two parameters family of Pythagorean triples.

Proved.

Example 14: There is rational number that lies strictly between the square root of 10^{100} and the square root of $10^{100} + 1$. Prove this using constructive proof method.

Solution: The square root of 10^{100} is 10^{50}. Let us assume that $x = 10^{50} + 10^{-51}$. Clearly, x is a rational number greater than 10^{50}. Now we have to prove that x is less than the square root of $10^{100} + 1$. Let us compute square of x, which is $(10^{50} + 10^{-51})^2 = 10^{100} + (2).10^{-1} + 10^{-102}$, which is clearly less than $10^{100} + 1$.

Proved.

Sometimes it is possible to prove the existence of something mathematical without actually constructing it. The reason could be that you cannot think of a constructive proof or that a constructive proof is very long and tedious. In any case, existential proofs are another important method for providing a theorem. Let us see the following example.

Example 15: Prove that the polynomial $p(x) = x^3 + x - 1$ has exactly one real root.

Solution: We have to prove this in two parts: First there exists a real root and second there is not more than one real root to the given polynomial.

(**First Part:** Direct Existential proof.) We know from Intermediate value theorem that if $p(x)$ changes sign in an interval $[a, b]$ then there is a real number c in $[a, b]$ such that $p(c) = 0$ i.e., $p(x)$ has a real root in the interval $[a, b]$. Let $a = 0$ and $b = 1$, then $p(a) = -1 < 0$ and $p(b) = 1 > 0$, thus there exists a real number c in $[0, 1]$ such that $p(c) = 0$. Hence, we have proved that there is a real root for the given polynomial $p(x)$.

(**Second Part:** Only one real root) This we shall prove by contradiction method. Let us assume that there are more than one real root to the polynomial $p(x)$. Let c_1 and c_2 be two real roots to the polynomial $p(x)$, then $p(c_1) = p(c_2) = 0$. From Mean Value Theorem, there must exist a number c between c_1 and c_2 such that

$$p'(c) = \frac{p(c_2) - p(c_1)}{c_2 - c_1} = 0$$

But a direct calculation shows that $p(x) = 3x + 1$, which can never be zero as $x^2 \geq 0$. Therefore, we have proved that there exists not more than one real roots.

Proved.

Theorem 6: If p is a prime number, then show that

$(a + b)^p = a^p + b^p$ *+multiple of p*

Proof: Using binomial expansion, we can write

$$(a + b)^p = a^p + {}^pC_1 a^{p-1}b + {}^pC_2 a^{p-2}b^2 + {}^pC_3 a^{p-3}b^3 + \ldots + {}^pC_p b^1$$

or,

$$(a + b)^p =$$

$$a^p + {}^pC_1 a^{p-1}b + {}^pC_2 a^{p-2}b^2 + {}^pC_3 a^{p-3}b^3 + \ldots + {}^pC_{p-1} ab^{p-1} + b^p$$

or,

$$(a + b)^p =$$

$$a^p + b^p + \left\{ {}^pC_1 a^{p-1}b + {}^pC_2 a^{p-2}b^2 + {}^pC_3 a^{p-3}b^3 + \ldots + {}^pC_{p-1} ab^{p-1} \right\}$$

Here, every term inside {} has coefficient of the form ${}^pC_r$. It is also given that p is a prime number and $1 \leq r \leq p - 1$. Now the general coefficient ${}^pC_r$ can be expanded as

$${}^pC_r = \frac{p(p-1)(p-2)\cdots(p-r+1)}{r!}$$

Since p is a prime number it is not divisible by any of the number from 2 to r. Further r is a number less than p, so p cannot divide r. This implies that p is factor of every term inside {} in the above expansion. Thus we can write.

$$(a + b)^p = a^p + b^p + \textit{multiple of } p$$

Proved.

In general, if p is a prime number then, for $a_1, a_2, a_3, \ldots, a_n$, we can write

$$(a_1 + a_2 + \ldots + a_n)^p = a_1^p + a_2^p + \ldots + a_n^p + \textit{multiple of } p$$

In particular, if $a_1 = a_2 = a_3 = \ldots = a_n = 1$ then, we have from above generalization $n^p = n +$ multiple of p i.e., $n^p - p$ is divisible by p.

Example 16: Show that $23^{47} - 23$ is divisible by 47.

Solution: Readers may apply the above theorem to prove this. This is left as an exercise to the reader.

4.2 Permutation

1 and 2 are two digits. The numbers 12 and 21 are formed with these two digits. Further, no other numbers can be formed with only these two digits. The only difference in 12 and 21 is the order of arrangement of digit 1 and 2. Next, let us suppose that we have three letters *a*, *b*, and *c*. Then, all possible arrangements of any two letters out of these three letters can be enumerated as: *ab*, *ac*, *bc*, ba, *ca* and *cb*. If we make an arrangement of all three letters out of these three, then we have *abc*, *acb*, *bac*, *bca*, *cab* and *cba* as possible arrangements. Each of the distinct order of arrangements of a given set of distinct objects, taking some or all of them at a time (with or without repetition), is called a **permutation** of the objects.

It is obvious from the example that the total number of permutations of n distinct objects, taken r at a time ($r \leq n$), is equal to the total number of ways of placing n objects in r boxes. This is denoted as $\boldsymbol{P(n, \mathrm{r})}$ or $^n\boldsymbol{P}_r$. This number is equal to $n(n-1)(n-2)\ldots(n-r+1)$. Since the first box can be filled by any of the n objects, so we have n choices. Once an object is selected, we are left with $n - 1$ objects and $r - 1$ empty boxes to fill. The second time we can select any of the $n - 1$ objects, so we have $n - 1$ choices and so on. This is the direct result of the **multiplication principle of counting**.

Theorem 1: **(multiplication principle of counting)** Suppose that two tasks T_1 and T_2 are to be performed in sequence. If T_1 can be performed in n_1 ways, and for each of these n_1 ways. T_2 can be performed in n_2 ways, the sequence T_1 T_2 can be performed in n_1 n_2 ways.

Proof: Each choice of a method of performing T_1 will result in a different way of performing the task sequence. There are n_1 ways to perform T_1, and for each of these wé may choose n_2 ways of performing T_2. Therefore, there will be, in total, $n_1 n_2$ ways to complete the tasks sequence T_1 T_2.

Proved.

This principle can be extended to the finite sequence of $m > 2$ tasks. Suppose that tasks T_1, T_2, T_3, ..., T_m are to be performed in sequence. If T_1 can be performed in n_1 ways and for each of n_1 ways. T_2 can be performed in n_2 ways and for each of n_1 n_2 ways of performing task sequence T_1, T_2, T_3 can be performed in n_3 ways and so on, then the tasks sequence T_1 T_2 T_3 ...T_m can be performed in n_1 n_2 n_3...n_m ways. Readers may recall **the principle of inclusion and exclusion,** discussed in the chapter 1. This principle is also known as **addition principle of counting.** It is, sometimes, used in permutation of objects also. Let there are three flight services and four train services between two places ***A*** and ***B***. A person can complete his (or her) journey between these two places in seven possible ways. Notice the use of **addition principle of counting**.

Example 1: A gentleman has 6 friends to invite. In how many ways can he send invitation cards to them, if he has three servants to carry the cards?

Solution: A card can be send to any one friend by any one of the three servants. Let us take the tasks of sending cards to six friends as T_1, T_2, T_3, T_4, T_5 and T_6. Each of the tasks can be completed in three distinct ways according to the number of servant to carry the cards. Thus, by the multiplication principle of counting the tasks T_1 T_2 T_3 T_4 T_5 T_6 can be performed in $3\times3\times3\times3\times3\times3 = 729$ ways. **Ans.**

Example 2: A telegraph has 5 arms and each arm is capable of 4 distinct positions, including the position of rest. What is the total number of signals that can be made?

Solution: There are five arms say T_1, T_2, T_3, T_4, and T_5. Each arm can be in any one of the four positions. For each of the position of arm T_1, there are four possible position for the second arm T_2, for each of the possible positions for $T_1 T_2$, there are four possible positions for the third arm T_3 and so on. Thus, by the multiplication principle of counting the total possible for $T_1 T_2 T_3 T_4 T_5$ is 4×4×4×4×4= 1024. Since each distinct position is a distinct signal, so total number of possible signals is 1024 including the signal (meaningless) corresponding to the situation when all arms are in rest position. Therefore, the total number of signals that can be generated is 1024 – 1 = 1023.

Ans.

Example 3: How many numbers of three digits can be formed with the digits 1, 2, 3, 4 and 5 if the digits in the same numbers are not repeated? How many such numbers are possible between 100 and 10000?

Solution: Here we have to find the number of permutation of 5 distinct objects (digits) taken 3 at a time. This given by $^5P_3$ = 5×4×3 = 60.

Ans.

Second part: The numbers between 100 and 10000 are numbers of three digits and of four digits. The total number of three digits numbers, formed with the given digits, is calculated above and it is equal to 60. Similarly, the total number of four digits numbers, formed with 1, 2, 3, 4 and 5, is given by $^5P_4$ = 5×4×3×2 = 120. Thus, the required number is 60 + 120 = 180.

Ans.

Notation ^nP_r

We know that ^nP_r is the number of permutation of a distinct objects taken r at a time, and this is equal to $n(n-1)(n-2)\ldots(n-r+1)$. Therefore,

$$^nP_r = \frac{n(n-1)(n-2)\cdots(n-r+1) * (n-r)!}{(n-r)!} = \frac{n!}{(n-r)!}$$

Therefore, $${}^nP_r = \frac{n!}{(n-r)!}$$

Meaning of 0! and 1/(–r)!

According to the definition of factorial, 0! is meaningless. But, permutation of 1 object taken 1 at a time is obviously equal to 1, i.e., ${}^1P_1 = 1$ Using the above formula, we have

$${}^1P_1 = \frac{1!}{(1-1)!} \quad or, \quad 1 = \frac{1}{0!} \quad or, \quad 0! = 1$$

Thus, for consistency, we take 0! = 1. Similarly, in the definition ${}^n\boldsymbol{P}_r$, if we put $n = 0$, we have the following result.

$$n(n-1)(n-2)\cdots(n-r+1) = \frac{n!}{(n-r)!}$$

$$\Rightarrow 0(0-1)(0-2)\cdots(0-r+1) = \frac{0!}{(0-r)!} \Rightarrow 0 = \frac{0!}{(-r)!}$$

In fact, factorial of a negative integer does not exist. However, we write 0 (zero) in place of $1/(-r)!$, wherever $1/(-r)!$ occurs in a product.

Permutation of objects when all are not distinct

Suppose there are $\boldsymbol{n}$ objects in which $\boldsymbol{p}$ objects are of one type, $\boldsymbol{q}$ objects are of second type, $\boldsymbol{r}$ objects are of third type and rest all are distinct. Let us further assume that the required number of permutations in this case is $\boldsymbol{x}$. If we replace $\boldsymbol{p}$ objects of the same type with $\boldsymbol{p}$ distinct objects, then we shall have total number of permutations equal to $\boldsymbol{x*p!}$. Similarly, if we replace all equal $\boldsymbol{q}$ objects and all equal $\boldsymbol{r}$ objects with the $\boldsymbol{q}$ numbers of distinct objects and $\boldsymbol{r}$ numbers of distinct objects respectively, then we shall have total number of permutations equal to $\boldsymbol{x*p!q!r!}$. Further, all these replacement will make all the $\boldsymbol{n}$ objects distinct. We know that the number of permutation of $\boldsymbol{n}$ distinct objects, taken $\boldsymbol{n}$ at a time, is given by ${}^n\boldsymbol{P}_n = n!$. Therefore, we have

$$x * p!q!r! = n! \Rightarrow x = \frac{n!}{p!q!r!}$$

Example 4: How many distinguishable permutations of the letters in the word BANANA are there?

Solution: The word 'BANANA' has 6 letters. All the letters are not distinct. Let us use subscript to distinguish them temporarily. Let the letters be B. A_1, N_1, A_2, N_2, A_3. Thus, the number of permutations are 6! = 720. Some of the permutations are identical like $A_1 A_2 A_3 BN_1 N_2$ and $A_1 A_2 A_3 BN_2 N_1$ except the order in which the N's appear. This means that if we drop the subscripts, the total number of permutations will be 6!/2! = 360. Similarly, if we drop subscript with A's then total number of distinguishable permutations are 360/3! = 60.

Ans.

In general, the above situation can be described as: "***The number of distinguishable permutations that can be formed from a collection of n objects, taken all n at a time, which the first object appears k_1 times, the second object k_2 times, and so on, is given by***

$$\frac{n!}{k_1!k_2!k_3!\cdots k_r!}$$

Where $k_1 + k_2 + ... k_r = n$'

Example 5: Find the number of positive integers greater than a million that can be formed with the digits 2, 3, 0, 3, 4, 2 and 3.

Solution: The numbers greater than a million must be of 7 digits. In the given set of digits, 2 appear twice, 3 appear thrice and all others are distinct. Thus, the total number of seven digits numbers that can be formed with given digits is

$$\frac{7!}{2!3!} = 420$$

The set of these 420 positive integers include some numbers begin with 0. Clearly, these numbers are less than a million and they must not be counted in our answer. The number of such numbers is given by the permutations of 6 non-zero digits and is equal to

$$\frac{6!}{2!3!} = 60$$

Therefore, the number of positive integers greater than a million that can be formed with given digits is equal to 420 – 60 = 360. **Ans.**

Permutation when objects can be repeated

Suppose we are given five digits: 1, 2, 3, 4 and 5. A 3-digits number, that can be formed using these digits, may be 123 or 111 or 122 or 533 or 551 etc. Repetitions of digits are allowed here. To form a 3-digits number, we have to fill three places with digits, we have five digits. The first place can be filled in 5 ways. For each of these ways, the second place can be filled in 5 ways so we have 5×5 ways to fill first two places. Similarly, for each of these 25 ways, the third place can be filled in 5 ways. Thus we have 5×5×5 = 125 ways to fill three places. This implies that we can form 125 only 3-digits numbers in this case. In general this result can be summarized as "*The number of permutations of n distinct objects, taken r at a time, when repetitions are allowed, is given by* n^r."

Example 6: There are 10 stalls for animals in an exhibition. Three animals: lion, pussycat and horse are to be exhibited. Animals of each kind are not less than 10 in number. What is the possible number of ways of arranging the exhibition.

Solution: There are three types of animals and 10 stalls. One stall can be filled by any of the three animals. Once the first stall is filled, the second stall can be filled again in three ways by placing any of the three animals in it. We have to fill 10 such stalls and number of each animals is greater than equal to 10, so we have 3^{10} = 59049 ways to fill the stalls. Thus, we can arrange the exhibition in 59049 ways.

Ans.

Circular Permutation

A permutation is called a **linear permutation** if the objects are arranged in a row, i.e., there is a first object and a last object in the arrangement. For example, $G_1G_2G_3G_4$, $G_2G_3G_4G_1$, $G_3G_4G_1G_2$ and $G_4G_1G_2G_3$ are all different arrangements, if we take them as linear permutations of four objects. However, if the same four arrangements are viewed as a circular arrangement as shown in the following diagram, all the four are same. The reason is there is nothing like the first object or the last object in a **circular permutation**. Any object could be the first and then the last is determined. Thus *in a circular permutation, we consider one object as fixed and the remaining objects are arranged as in any linear permutation.* The number of ways of allowing four people to sit in a row is given by 4!, whereas the number of ways of allowing the same four people to sit in a circle is (4 – 1) ! = 3!.

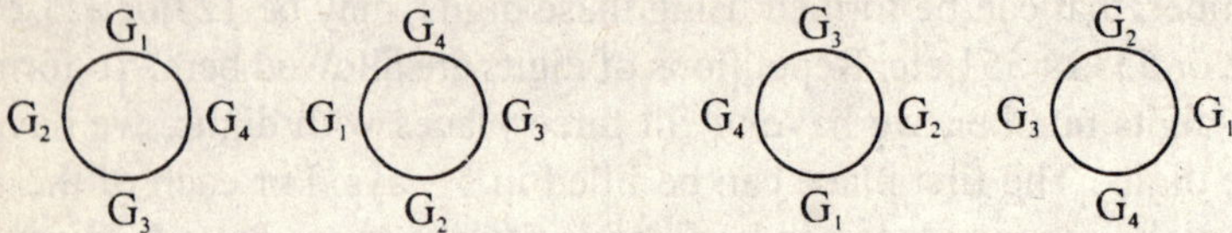

In circular permutation, **clockwise** and **anti-clockwise** arrangements of objects are possible and both are distinguishable. In the diagram shown below, the first is the permutation $G_1G_2G_3G_4$ -anti-clockwise, and $G_1G_2G_3G_4$ -clockwise. Therefore, when distinction is made between anti-clockwise and clockwise arrangements then the number of circular permutations of n objects taken all n at a time is $(n - 1)!$. However, when there is no distinction between the directions of traversal along the circle, the number of circular permutations of n objects taken all n at a time is **half of the $(n - 1)!$.**

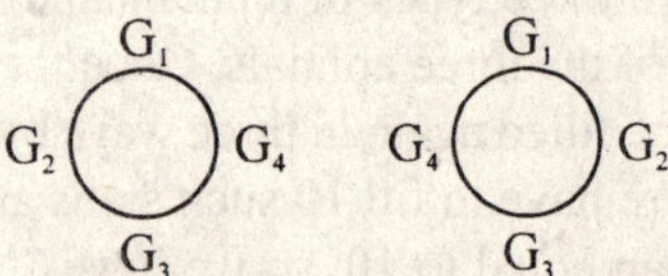

Example 7: In how many different ways can 5 men and 5 women sit around a table, if

(i) there is no restriction;

(ii) no two women sit together?

Solution: The problem is related to circular permutation of 10 objects (5 men and 5 women). If there is no restriction then the number of permutations is (10 – 1)! = 9! = 362880. Notice here the difference in arrangement between clockwise and anti-clockwise.

In the second case, there is a restriction that no two women are allowed to sit side by side. To meet this restriction each women should occupy a sit between two men. The number of ways five men can sit around a table is 4! = 24. Once these five men have sat on alternate chairs, the five women can occupy the 5 empty chairs in 5! Ways. Thus, total number of ways, in this case, will be 24*5! = 24*120 = 2880.

Ans.

Restricted Permutation

We have learned to find the number of permutations of given number of objects when choice of selection is arbitrary. In real life application, we encounter many situations in which counting is required under some constraints (restriction). For example, to find the number of positive even integers of 3 digits (distinct) that can be formed with digits 1, 2, 3, 4 and 5. Here, to make a 3-digit number, we have to fill 3 places by the given digits. The unit place can be filled either by 2 or by 4 to make the number even. The remaining two places can be filled by any of the remaining four digits. Thus, count of 3-digits positive even integers, having all distinct digits, = $2 * {}^4P_2$ = 2*12 = 24. Let us see some examples to understand the concept of restricted permutation.

Example 8: In how many ways can 10 different examination papers be scheduled so that

(i) the best and the worst always come together;
(ii) the best and the worst never come together?

Solution: (i) Let us combine the best and the worst paper together and consider them as one object. We have, now, 9 objects (papers). These 9 objects can be arranged in 9! Ways. And in each of these 9! arrangements, the best and the worst paper can be arranged in 2! ways. Therefore, the number of ways in which the 10 papers can be scheduled in this situation = 2!*9! = 725760.

(ii) Without any restriction, the 10 papers can be scheduled in 10! ways. We have just calculated in part (i) that total number of ways in which the 10 papers can be scheduled so that the best and the worst always come together = 725760. Therefore, the number of ways of scheduling 10 papers so that the best and the worst never come together = 10! – 725760 = 3628800 – 725760 = 2903040.

Ans.

In the example 8, we have used the notation that "*the number of ways of arranging object under some restriction = the number of arrangements of the same number of object without restriction–the number of arrangements of the same number of arrangements with the opposite restriction*". Let us see a few more examples on restricted permutation.

Example 9: Find the sum of all the four-digit numbers that can be formed with the digits 3, 2, 3 and 4.

Solution: One thing is worth noticing here that a four-digit number so formed does not contain a repeated digit except the digit 3. This is implied from the question, because if it were not so, 3 should not have been repeated in the list of the digits.

To find the sum of the four-digit numbers formed with 3, 2, 3 and 4, we have to calculate the sum of digits at unit place in all such numbers. The sum of the digits at ten, hundred and thousand place will be the same, only theirs place value will change.

The number of four-digit numbers in which 2 appears at unit place is determined by the number of permutations of digits 3, 3 and 4 to fill ten, hundred and thousand place. And, this is = 3!/2! = 3.

Similarly, the number of four-digit numbers in which 3 appears at unit place is = 3! = 6.

The number of four-digit numbers in which 4 appears at unit place is 3!/2! = 3. Therefore, sum of the digits in the unit place of all the numbers

$= 3\times2 + 6\times3 + 3\times4 = 36.$

As stated above, the sum of the digits in all numbers at ten, hundred and thousands places is 36 each. Thus the sum of all such numbers

$= 36\times1000 + 36\times100 + 36\times10 + 36 = 39996$

Ans.

Example 10: How many binary sequences of r-bits long have even number of 1's?

Solution: There will be 2^r possible binary sequences of r-bits long. This can be verified by the permutation of objects when repetitions are allowed. There are r places and two objects (0 and 1). The first place can be filled in 2 ways, for each of these, the second place can be filled in 2 ways, so we have 2×2 ways to fill first two places. Extending the sequence up to the rth steps, we have 2^r possible arrangements and 2^r possible binary sequences.

We can now make pairs of binary sequences in such a way that two sequences differ only r^{th} place. There will 2^{r-1} such pairs, and in each pair one sequence will have even number of 1's. Thus, number of binary sequences of r bits long having even number of 1's = 2^{r-1}.

Ans.

Example 11: We are asked to make slips for all five-digit numbers. Since the digits 0, 1, 6, 8 and 9 can be read as 0, 1, 9, 8 and 6 when they are read upside down, there are pairs of numbers that can share same slip if the slips are read upside down or right side up (e.g., 89166 and 99168). Find the number of slips required for all five-digit numbers.

Solution: We have 10 digits. We have to make all five-digits numbers. The total such numbers is equal to 10^5. Here we have to make slips for these many numbers. The number made of digits 0, 1, 6, 8 and 9 can be read upside down or right side up. And, there are 5^5 many such five-digit numbers (all those five-digit numbers made of digits 0, 1, 6, 8 and 9). Out of these 5^5 many numbers, however, there are some numbers that read the same either upside down or right side up. For example, 91816, and there are 3×5^2 such numbers (center place filled with 0, 1 or 8). Thus, there are $5^5 - 3 \times 5^2$ numbers that can be read upside down or right side up. And, for these numbers we need only

$$\frac{1}{2}\left(5^5 - 3 * 5^2\right)$$

number of slips. Therefore number of slips required to be made, is

$$10^{10} - \frac{1}{2}\left(5^5 - 3 * 5^2\right)$$

Ans.

Generating function for Permutation

Recall the exponential function that we have learnt in the chapter 2. The coefficient of $x^r/r!$ in a polynomial $p(x)$ is ${}^nP_r$. And ${}^nP_r$ is the number of permutations of n objects taken r at a time. Generating function is a very hardy technique for solving many counting pròblems. The general procedure for finding permutations is based on the exponential generating function.

Example 12: Find the generating function, also called enumerator, for permutations of n objects with the following specified conditions:

(a) Each object occurs at the most twice.
(b) Each object occurs at least twice.
(c) Each object occurs at least once and at the most k times.

Solution: (a) Each object occurs at the most twice implies that an object may occur 0, 1 or 2 times. The exponential generating function for an object under this condition is given as

$$1 + x + \frac{x^2}{2!}$$

There are n objects, so the generating fimction for the problem is written as

$$\left(1 + x + \frac{x^2}{2!}\right)^n$$

(b) In this case, an object occurs at least twice. This implies that an object may appear 2 or more times. The exponential generating function for an object under this condition is

$$\frac{x^2}{2!} + \frac{x^3}{3!} + \frac{x^4}{4!} + \ldots\ldots\ldots$$

Therefore, for the problem dealing with n objects, the generating function can be written as

$$\left(\frac{x^2}{2!} + \frac{x^3}{3!} + \frac{x^4}{4!} + \ldots\ldots\ldots\right)^n$$

(c) Here, each object occurs at least once and at the most k times. That is to say that an object may occur 1 or 2 or 3 or ...or k times. The exponential generating function for an object under this condition is

$$x + \frac{x^2}{2!} + \frac{x^3}{3!} + \ldots\ldots + \frac{x^k}{k!}$$

Since the problem is for n objects, the generating function for the problem is written as

$$\left(x + \frac{x^2}{2!} + \frac{x^3}{3!} + \ldots\ldots + \frac{x^k}{k!}\right)^n$$

Ans.

Example 13: A fair six-sided die is tossed four times and the numbers shown are recorded in a sequence. How many different sequences are there?

Solution: Let us assume, here, that each toss is an object and the number appearing on the face of a die is the number of times it occurs. Thus each object occurs at least once and at the most 6 times. Also the order of appearing of different 1's (conceptually) to make it 2 or 3 or 4 or 5 or 6 is fixed and is one and only one way. The generating function for first toss is given as

$$x + x^2 + x^3 + \ldots + x^6$$

The sum of all the coefficient, here, is 6 and number of possible sequences of numbers, on face of a die in one toss, is 6 only. Die is tossed four times, therefore generating function for the problem is

$$\{x + x^2 + x^3 + \ldots + x^6\}^4$$

The number of sequence is the sum of coefficient of all the terms in this generating function. This is equal to 6^4. Reader may verify this.

Ans.

Example 14: How many different words can be made from letters of the word '**committee**'.

Solution: The word committee contains letters *c*, *o* & *i* once and *m*, *t* & *c* twice, when a word is formed from these letters, a letter may appear at the most the number of times it appear in the word **committee** or not at all. So, generating function for *c*, *o* and *i* is given by $(1 + x)$ each, whereas, for *c*, *m* and *c* is given by $(1 + x + x^2/2!)$ each. Thus generating function for the problem is then given by

$$(1+x)^3\left(1+x+\frac{x^2}{2!}\right)^3 = (1+3x+3x^2+x^3)\left(1+x+\frac{x^2}{2!}\right)^3$$

If words are to be formed taking all the letters at once, then the numbers of such words is given by the coefficient of $x^9/9!$, and this is equal to $9!/2!2!2!$.

Ans.

Obviously, it is cumbersome to use generating functions to solve counting problems by hand calculation, because it's no fun multiplying polynomials. But it is very easy for computers to do the calculations, so generating functions provide a powerful tool for solving counting problems by machine. In addition, generating functions provide an excellent tool for studying counting problems in the abstract.

4.3 Combination

You may recall the difference in arrangement of terms in a sequence and that in a set. In the former, order of terms is important, whereas in the latter, order is **not important**. Generally, when order matters, we count the number of sequences or permutations and when order is not important, we count number of subsets or combinations.

Let A be a set such that $|A| = n$. Then count of all r-element $(r \leq n)$ subsets of A is the combination of n elements taken r at a time. This is denoted as ${}^nC_r$. We know that ${}^nP_r$ is the number of permutations of n elements taken r at a time. This includes the different orders in which r elements can be arranged. It is also known that the r elements can be arranged in $r!$ different ways. Thus, if we divide the ${}^nP_r$ by $r!$, then the order consideration is done away with. The result so obtained is the number of arrangements of n elements taken r at a time and that too without the consideration of order. This is, precisely, the definition of ${}^nC_r$. Therefore,

$$^nC_r = \frac{1}{r!}{}^nP_r = \frac{n!}{r!(n-r)!}$$

Example 1: A person has 8 children of which he takes 3 at a time to a circus. He does not take the same three children twice to the circus. How many times he will have to go to circus to ensure that every three children have seen the circus together? In this case find the number of times a particular child has visited the circus.

Solution: Here we have to find the number of combinations of 8 children taken 3 at a time. Note that order of selection of child is not important in this case. This selection can be made in ${}^8C_3$ ways. We can make ${}^8C_3 = 56$ distinct groups of three children, and for each such group, the person will have to go to circus once. Therefore, the person will have to visit circus 56 times.

Second Part: A particular child goes to circus with every possible pair of two children out of remaining 7 children. Number of such possible pair is ${}^7C_2 = 21$. Therefore, a particular child goes to circus 21 times.

Ans.

Example 2: From 8 men and 4 women a team of 5 is to be formed. In how many ways can this be done so as to include at least one woman?

Solution: This is a case of restricted combination. Total number of persons = 8 men + 4 women = 12 persons. A team of 5 has to be made and this can be made in ${}^{12}C_5$ ways. This count includes the case of teams containing all five men (i.e. no women in the team) which is equal to ${}^8C_5$. Thus, number of ways in which the specified team can be selected $= {}^{12}C_5 - {}^8C_5 =$

$$\frac{12!}{5!7!} - \frac{8!}{5!3!} = \frac{12 \times 11 \times 10 \times 9 \times 8}{5 \times 4 \times 3 \times 2} - \frac{8 \times 7 \times 6}{3 \times 2} = 792 - 56 = 736$$

Ans.

Example 3: In a party of 30 people, each shakes hand with others. Hoe many handshakes took place in the party?

Solution: In a normal case, a handshake involves two persons. This case is of counting 2-elements subsets of a set containing 30 elements. And this count is ${}^{10}C_2 = 30\times29/2 = 435$.

Ans.

Example 4: What is the number of diagonals that can be drawn in a polygon n sides?

Solution: A polygon having n sides has n vertices. A diagonal is a line between two points, which are not adjacent to each other. The total number of lines that can be drawn in a polygon of n vertices = ${}^{n}C_2$. There are already n sides (lines) which are not diagonal. The remaining lines will be diagonals. Therefore, the number of diagonals that can be drawn in a polygon of side n is equal to ${}^{n}C_2 - n =$

$$\frac{n!}{2!(n-2)!} - n = \frac{n(n-1)}{2} - n = \frac{n^2 - n - 2n}{2} = \frac{n(n-3)}{2}$$

Ans.

Example 5: There are 10 points in a 2-D plane. Four of these are co-linear. Find the number of different straight lines that can be drawn by joining these points.

Solution: Any two points are always co-linear. So, a line can be drawn between any two points. If there are three **non-co-linear points** (a single line cannot be drawn joining all these three points), we can draw $3 = {}^{3}C_2$ distinct lines. A triangle is an example of this. Thus, we can draw ${}^{10}C_2$ distinct lines joining 10 points. Out of these 10 points, 4 are co-linear. So ${}^{4}C_2$ lines will be same and we consider them as one line. Therefore, actual number of lines that can be drawn = ${}^{1}C_2 - {}^{4}C_2 + 1 = 45 - 6 + 1 = 40$.

Ans.

Combination when replacement is allowed

Consider the situation of placing 10 balls of different colors in 3 pots, say A, B and C, with the flexibility that whenever required any number of balls of any color are available. A person places a red ball in pot A and is willing to place a red ball again in pot B and in pot C. Other balls may be placed according to the different possible combinations. After placing a red ball in pot A, only 9 balls are left and none is of red color. To place a red ball in pot B, the person needs a red ball, that is provided making the total number of balls 10 again. After placing this new ball in pot B, again a new red ball is required to place it in pot C. Once provided the number of balls is again 10. Once this new ball is placed in pot C, the person is left with 9 balls and 3 balls are in 3 pots. This implies that 3 balls have been selected from 9 + 3 balls. The number of combinations in this case is ${}^{9+3}C_3 = {}^{10+3-1}C_3$. This situation can be generalized in the following words.

"*Suppose that k selections are to be made from n objects without regard to order. Also assume that repeats are allowed and at least k copies of each of the n objects are available. Then, the number of ways selections of k objects can be made from n object is* ${}^{n+k-1}C_k$."

Example 6: A bookstore allows the recipient of a gift coupon to choose 6 books from the combined list of 10 best-selling fiction books and 10 best-selling non-fiction books. In how many different ways can the selection of 6 books be made?

Solution: The number of different types of books is 10 + 10 = 20. A gift coupon recipient may select any 6 books, possibly 6 copies of a single book. This is a case of selection of 6 objects from 20 objects with repetitions allowed. The number of ways the selection can be made is ${}^{20+6-1}C_6 = {}^{24}C_6 = 177100$.

Ans.

Example 7: If three dice are rolled, and we make a set of numbers shown on the three dice. How many different sets are possible?

Solution: Rolling three dice is equivalent to selecting three numbers from the list of six numbers 1, 2, 3, 4, 5 and 6 with repetition allowed. Because sequence 111, 121 etc. are possible. Thus the different possible combinations is ${}^{6+3-1}C_3 = {}^{8}C_3 = 56$.

Ans.

Open Selection

Suppose a coin is tossed n times. Obviously there will be 2^n sequences of H (Heads) and T (Tails) each of the form $X_1X_2X_3X_4 \ldots X_n$. Each X_1

is either H or T. If we are asked to find number of sequences having at least one head or sequences having one or more or all heads, then the count is $2^n - 1$. This is the result of the following sum

$${}^nC_1 + {}^nC_2 + {}^nC_3 + \ldots + {}^nC_n$$

Similarly, if we are asked to find the number of sequences having at least 5 heads and at the most 10 heads ($n \geq 10$), then the number is given by the sum ${}^nC_5 + {}^nC_6 + {}^nC_7 + \ldots + {}^nC_{10}$. In general, combinations of n items taking r at a time where r lies between a and b where $1 \leq a$ and $b \leq n$, is given by the sum

$${}^nC_a + {}^nC_{a+1} + {}^nC_{a+2} + \ldots + {}^nC_b$$

Example 8: In an election, there are four candidates contesting for three vacancies an elector can vote for any number of candidates not exceeding the number of vacancies. In how many ways can one cast his votes?

Solution: An elector may vote for any one or, any two or, any three candidates out of total 4. Therefore, an elector may vote in ${}^4C_1 + {}^4C_2 + {}^4C_3 = 4 + 6 + 4 = 14$ different possible ways.

Ans.

Example 9: In an election the number of candidates is one more than the number of vacancies. If a voter can vote in 30 different ways, find the number of candidates.

Solution: Let the number of candidates be x. An elector may vote for any one or any two or, any three up to maximum of any $x - 1$ candidates from total of 4, because number of vacancies is $x - 1$ only. Therefore, number of ways in which an elector can cast his vote is ${}^xC_1 + {}^xC_2 + {}^xC_3 + \ldots + {}^xC_{x-1}$ and this value is given to be 30. Thus,

$${}^xC_1 + {}^xC_2 + {}^xC_3 + \ldots + {}^xC_x - 1 = 30$$

or, $${}^xC_1 + {}^xC_2 + {}^xC_3 + \ldots + {}^xC_{x-1} + {}^xC_x - {}^xC_0 - {}^xC_x = 30$$

or, $$2^x - 2 = 30 \Rightarrow x = 5$$

Therefore, the number of candidates is 5.

Ans.

Example 10: In an examination a candidate has to pass in each of the 5 papers. How many different combinations of papers are there so that a student may fail?

Solution: For a student to pass the examination, he/she will have to pass in each of the five papers. To fail, student may fail in any one or, in any two or so on including in all the five papers. Thus, a student may fail in as many as

$$^5C_1 + {^5C_2} + {^5C_3} + {^5C_4} + {^5C_5} = 2^5 - 1 = 31 \text{ ways.}$$

Ans.

Example 11: Find the total number of selections of at least one red ball from 4 red balls and 3 green balls, if

(a) The balls of the same colour are different,

(b) The balls of the same colour are identical.

Solution: (a) From 4 different red balls and 3 different green balls, we have to find number of selections taking at least one red ball and any number of (including 0) 3 green balls. The total number of ways of selecting at least one red ball from 4 different red balls = $^4C_1 + {^4C_2} + {^4C_3} + {^4C_4} = 15$. Corresponding to each of these selections, the number of ways of selecting green balls = $^3C_0 + {^3C_1} + {^3C_2} + {^3C_3} = 8$. Therefore, total number of different ways of selection = $15 \times 8 = 120$.

Ans.

Distributing different objects into groups

Suppose 12 different books are to be equally distributed between two persons A and B. In how many different ways this task can be accomplished. When half of the books have been given to A, the rest 6 books have already been selected for B. Since the order of selection is not important, the task of selecting 6 books from 12 books can be done in $^{12}C_6$ different ways. If the task is to make two sets of 6 books each, then it can be performed in $^{12}C_6 \div 2!$ different ways. In the last case, interchange of two sets does not give any new combination, whereas, in the previous case it is important.

Now consider the equal distribution of the same 12 books among 4 persons. We can select 3 books from 12 books in $^{12}C_3$ ways for first person. For each of these selections, 3 books can be selected from the remaining 9 in ${}^{9}C_3$ ways for second person. Similarly, we can select 3 books in ${}^{6}C_3$ ways for third person and in ${}^{3}C_3$ ways for last person. Therefore, total number of ways in which 12 books can be distributed equally among 4 persons is ${}^{12}C_3\,{}^{9}C_3\,{}^{6}C_3\,{}^{3}C_3 =$

$$\frac{12!}{3!9!} \times \frac{9!}{3!6!} \times \frac{6!}{3!3!} \times \frac{3!}{3!0!} = \frac{12!}{(3!)^4}$$

The above result is to be divided by 4!, if we have to find the number of ways to make 4 sets of 3 books each. Instead of divided equally, if A has to get 7 books and B has to get 5 books, then the number of ways = ${}^{12}C_7$ = 12!/(7!5!). This discussion, in general, can be summarized as follows.

- *The number of ways of distributing p + q different objects between two distinguishable groups in such a ways that one group gets p objects and other gets q objects is given by*

$$\frac{(p+q)!}{p!q!}$$

- *If the two groups are indistinguishable in the above case then, the number of ways of distribution is given by*

$$\frac{(p+q)!}{2!p!q!}$$

- *In general, if n different objects are to be distributed among m distinguishable groups containing* $p_1, p_2, p_3, \ldots, p_m$ objects, where $p_1 + p_2 + p_3 + \ldots + p_m = n$. The number of ways in which this tasks can be completed is given by

$$\frac{n!}{p_1!\,p_2!\,p_3!\ldots p_m!}$$

- *In the previous result if the m groups are indistinguishable, the number of ways of distribution is given by*

$$\frac{n!}{m!\,p_1!\,p_2!\,p_3!\ldots p_m!}$$

Example 12: In how many ways can a pack of 52 cards be equally divided into four groups? If the cards are to be distributed equally among four players, then find the number of ways of this distribution.

Solution: First Part: When 52 cards are distributed equally among four groups, each group contains 13 cards. Since groups are indistinguishable, the number of ways of distribution is given by

$$\frac{52!}{4!13!13!13!13!}$$

Second Part: Here four groups (players) are distinguishable, thus number of ways of distribution is given by

$$\frac{52!}{13!13!13!13!}$$

Ans.

Generating Function for Combinations

Recall the notion of a binomial generating function that we have studied in the chapter 2. The coefficient of x^r in $(1 + x)^n$ is ${}^nC_r$ is the number of combinations of n objects taken r at a time. This implies that if we can find a binomial generating function for a given combination problem, the task of finding the number of combinations is reduced to getting the relevant coefficient. We shall use binomial generating function to solve a few problems here.

Example 13: A library has 5 indistinguishable black books, 4 indistinguishable red books and 3 indistinguishable yellow books. In how many distinguishable ways can a student take home (a) 6 books? (b) 6 books taking at least 1 of each colour? (c) 6 books taking 2 of each colour?

Solution: Here all books of a particular colour are indistinguishable. A black (or red or yellow) book can be selected or, not selected. If selected then maximum number of black that can be selected is 5, i.e., possible ways of selections for black books are 0, 1, 2, 3, 4 or 5. Similarly, the possible ways of selections red books are 0, 1, 2, 3 or 4 and for yellow books are 0, 1, 2 or 3. If we take three variables x, y and z for the numbers of black, red and yellow books selected by a student, respectively then the number of solutions to the equation

$x + y + z - 6$ where $0 \le x \le 5$, $0 \le y \le 4$ and $0 \le z \le 3$

is the required number of ways in which a student can take home 6 books. The generating function for x is $(1 + x + x^2 + x^3 + x^4 + x^5)$, for y is $(1 + y + y^2 + y^3 + y^4)$ and for z is $(1 + z + z^2 + z^3)$. If we replace y and z by x, we get the generating function for the above problems as

$$f(x) = (1 + x + x^2 + x^3 + x^4 + x^5)\,(1 + x + x^2 + x^3 + x^4)\,(1 + x + x^2 + x^3)$$

The coefficient of x^6 in $f(x)$ is the required number, and this number is **18.**

In the second part, at least one book of each colour has to be selected, so generating function $f(x)$ is given as

$$f(x) = (x + x^2 + x^3 + x^4 + x^5)\,(x + x^2 + x^3 + x^4)\,(x + x^2 + x^3)$$

The coefficient of x^6 in $f(x)$ is the required number, and this number is **9.**

In the third part, two books of each colour are to be selected. So generating function $f(x)$ is given as $f(x) = x^2 * x^2 * x^2$

The coefficient of x^6 in $f(x)$ is the required number, and this number is 1.

Ans.

Example 14: A library has 5 black books, 4 red and 3 yellow books, all with different titles. How many distinguishable ways can a student take home 6 books, 2 of each colour?

Solution: Notice that this problem is different from that in the example 6. In this case, the books of a particular colour are distinguishable by their title. Each book can be either selected or not, giving the possible number of selection for each book as 0 or 1. This can be written, in polynomial form, as $1 + x$. Thus, the generating function for 5 black is $(1 + x)^5$. Similarly, the generating function for 4 red books is $(1 + y)^4$ and for 3 yellow books is $(1 + z)^3$. The count of number of ways 6 books, 2 of each colour, can be selected, we take coefficient of $x^2 y^2 z^2$ in the generating function

$$f(x) = (1 + x)^5\,(1 + y)^4\,(1 + z)^3$$

The coefficient of $x^2 y^2 z^2$ is ${}^5C_2\,{}^4C_2\,{}^3C_2 = 10 \times 6 \times 3 = 180$.

Ans.

Example 15: In how many ways can one choose n pieces of fruit, assuming there are an infinitely large number of apples, bananas, oranges and pears, and he (she) wants an even number of apples, an odd number of bananas, not more than 4 oranges and at least two pears?

Solution: The generating functions for the selection of apple can be written as $(1 + x^2 + x^4 + x^6 + \ldots)$, for the selection of bananas can be written as $(x + x^3 + x^5 + x^7 + \ldots)$, for the selection of oranges can be written as $(1 + x + x^2 + x^3 + x^4)$ and for pears it can be written as $(x^2 + x^3 + x^4 + \ldots)$. Therefore, the generating function for this problem is

$$f(x) = (1 + x^2 + x^4 + x^6 + \ldots)\ (x + x^3 + x^5 + x^7 + \ldots)$$
$$(1 + x + x^2 + x^3 + x^4)\ (x^2 + x^3 + x^4 + \ldots)$$

The answer is simply the nth coefficient of this infinite series. *Here, although we cannot see the numerical answers, this function contains all the data: a computer could readily provide us with a table of the answers for value of n as large as we please, of course, within the limits of computer memory.*

Ans.

Example 16: In how many ways $2n + 1$ items can be distributed among 3 persons so that the sum of the number of items received by any two persons should exceed the number of items received by the other?

Solution: Let us take three variables x, y and z for the number of items received by the three persons. Then number of ways $2n + 1$ items can be distributed among three persons is equal to the number of positive integer solutions to the equation

$$x + y + z = 2n + 1 \text{ where } 1 \leq x, y, z \leq n$$

To ensure that the sum of the variables must exceed the third one, it is important to define the range of values for each variable as above. Therefore, the generating function for the problem is written as

$$f(x) =$$
$$(x + x^2 + x^3 + \ldots + x^n)(x + x^2 + x^3 + \ldots + x^n)(x + x^2 + x^3 + \ldots + x^n)$$

$$f(x) = (x + x^2 + x^3 + ... + x^n)^3$$

$$= \left\{\frac{x(1-x^n)}{1-x}\right\}^3$$

$$= x^3(1 + 3x^n + 3x^{2n} + x^{3n})(1 + x)^3$$

In this function $f(x)$, the coefficient of x^{2n+1} is the count we are looking for. And this coefficient is equal to the coefficient of x^{2n-2} in $(1 - 3x^n + 3x^{2n} - x^{3n})(1 - x)^{-3}$. This value is

$$^{3+2n-2-1}C_{2n-2} - 3\times{}^{3+n-2-1}C_{n-2}$$

$$or, \quad ^{2n}C_{2n-2} - 3\times{}^{n}C_{n-2} = \frac{2n(2n-1)}{2!} - 3\frac{n(n-1)}{2!}$$

$$= \frac{n}{2}(4n - 2 - 3n + 3) = \frac{n(n+1)}{2}$$

Ans.

Mixed Problem

Some combinatorial problems require that the counting of permutations and combinations be combined or supplemented by the direct use of the addition or the multiplication principles of counting. See the following example.

Example 17: A valid password consists of seven symbols. Symbols are chosen from digits and Roman capital alphabets. The first symbol of the password must be a Roman capital alphabet. How many different passwords are possible?

Solution: Number of symbols = 26 + 10 = 36. The first place of seven characters password can be chosen in 26 ways. The remaining 6 places can be filled in 36 ways each i.e., the second symbol for a password can be chosen in 36 ways, and for each of this, the third place can be filled in 36 ways and so on. Thus, the number of different possible password = $26\times(36)^6$ = 56596340736.

Ans.

4.4 Pigeonhole Principle

"If **five** men (pigeonholes) are married to **six** women (pigeons) then at least one men has more than one wife."

Or

"If **five** men (pigeons) are married to **four** women (pigeonholes) then at least one women has more than one husband."

Or

"If we have **six** pigeons in **five** pigeonholes then at least one pigeonholes would have more than one pigeon."

Arguments of this kind are used in solution to many mathematical problems and quite a few beautiful theorems have been proven with their help. All these arguments share the name "**Pigeonhole Principle**". In this text we shall use this theorem to prove problems related to numerical, counting, integer division and geometry.

Theorem 1: If $\boldsymbol{n}$ objects (pigeons) are placed in $\boldsymbol{m}$ places (pigeonholes) for $m < n$, then one of the places (pigeonholes) must contains at least

$$\left\lfloor \frac{n-1}{m} \right\rfloor + 1$$

objects (pigeons).

Proof: The proof is very simple. Let us assume that none of the places (pigeonholes) contains more than **floor((*n* – 1)/*m*)** objects (pigeons). Then there are at the most m * **floor((*n* – 1)/*m*)** $\boldsymbol{\leq m * (n-1)/m = n-1}$ objects. This is contrary to the given fact that there are $\boldsymbol{n}$ object. This contradiction is because of our assumption that "none of the places contains more than **floor((*n* – 1)/*m*)** objects". Thus our assumption is wrong. Therefore, one of the pigeonholes must contains at least

$$\left\lfloor \frac{n-1}{m} \right\rfloor + 1$$

objects.

Proved.

This so called **'pigeonhole principle'** is very simple. Its application however, is very **intelligence intensive**. It needs skill, which can be developed by attempting on different types of examples. The core is to identify the **pigeons, pigeonholes** and theirs respective numbers. Once you have these figures, you have almost found the solution to the problem. Let us see some examples to understand the application of this principle.

Example 1: 15 students wrote a dictation. Tarun made 13 errors. Each of the other students made less than that number of errors. Prove that at least two students made equal number of errors.

Solution: Since maximum numbers of errors committed by any student is 13, there are students who might have made no errors, some may have made one error and so on. But, none of the students has made more than 13 errors. Let there are 14 chairs (pigeonholes) marked 0, 1, 2, ... 13. We let the students be seated in the chair having mark equal to the number of errors committed by the student. After 14 students have been seated, there is one student left who will be sharing chair with any of the 14 students. This means that there are at least two students who have made same number of errors. Here, n = number of pigeons = number of students = 15. m = number of pigeonholes = 14 (count of number from 0 to 13, possible number of errors by different student). Thus using the pigeonhole principle we have at least one pigeonhole containing at least

$$\left\lceil \frac{n-1}{m} \right\rceil + 1 = \left\lceil \frac{15-1}{14} \right\rceil + 1 = 1 + 1 = 2$$

students.

Proved.

Example 2: There are 500 boxes with apples. Each contains no more than x apples. Find the maximum possible value of x, such that one can for sure find 3 boxes containing equal number of apples.

Solution: Let us place these 500 boxes (pigeons) in pigeonholes marked with number according to the **number of apples** in a box. Let there are pigeonholes marked with 0 for empty boxes, 1 for boxes containing one apple, 2 for boxes containing two apples... and so on up to x. Thus we have $x + 1$ pigeonholes. Applying the pigeonhole principle, we get

$$\left[\frac{500-1}{x+1}\right]+1=3 \quad Or, \quad \left[\frac{500-1}{x+1}\right]=2 \quad Or, \quad 499=2x+2$$

$$Or, \quad x=\left[\frac{499-2}{2}\right]=248.$$

Therefore, the maximum possible value for x is 248.

Ans.

Example 3: There are 33 students in the class, and the sum of their ages is 430 years. Is it true that for some 20 students the sum of their ages is more than 260 years?

Solution: Consider the average age of students which is 430/33 = 13.03 years. Now some students may be of age less than this average, some may be of average age and some other may be of age greater than average. Let there be three pigeonholes ***L***, ***A*** and ***G*** for less than, average and greater than ages respectively. If we put the 33 pigeons (students) in these three pigeonholes, then there are three possibilities.

- All three pigeonholes get equal number of pigeons, or
- The distribution is skewed toward ***L***, or
- The distribution is skewed toward ***G***.

In the first case, we can always pick up 20 students (all from ***G*** and 9 from ***A***) so that their ages will total more than 260 years. In the third case too we can pick up 20 students (all from ***G***, remaining from ***A*** and, if any more required then from ***L***) so that the ages will total more than 260 years. In second case, we use the fact that **if any number is less than the average then there exists a number which will be greater than the average.** Thus if there are students in ***L***, then there must be some students in **G** whose age(s) will maintain the average. By picking up 20 students as in the cases of first and third, theirs total age will exceed 260 years. Thus, we can always find 20 students such that the sum of their ages is more than 269 years.

Ans.

Example 4: There are pencils in the box: 10 red ones, 8 blue, 8 green, 4 yellow. Let us take, with eyed closed, some number of pencils from the box. What is the least number of pencils we have to take in order to ensure that we get at least 4 pencils of the same colour?

Solution: Let us consider 4 pigeonholes corresponding to four colours. Then, our problem is to find the number of pigeons (pencils) so that when placed in holes, one of the holes must contain at least four pigeons (pencils). In the theorem above, $m = 4$, then we have to find n such that

$$\left\lceil \frac{n-1}{4} \right\rceil + 1 = 4 \quad or, \quad n - 1 = 12 \quad or, \quad n = 13$$

Therefore, at least 13 pencils are to be picked up to ensure that we get at least 4 pencils of same colour.

Ans.

The examples discussed above are related to counting problems. This principle can be used to solve many such counting problems. Let us now consider a few examples related to some other area of applications.

Example 5: There are 50 people in a room. Since of them are acquainted with each other, some not. Prove that there are two persons in the room who have equal numbers of acquaintances.

Solution: If there is a person in the room who has no acquaintances at all then each of the other persons in the room may have either 1, or 2, or 3, ..., or 48 acquaintances, or do not have acquaintances at all. Therefore we have 49 pigeonholes numbered 0, 1, 2, 3, ..., 48. If we put 50 people one by one in the pigeonholes according to the number of acquaintances the person has, we have one pigeonhole containing at least

$$\left\lceil \frac{50-1}{49} \right\rceil + 1 = 1 + 1 = 2$$

pigeons. Thus, there are at least two persons in the room who have equal number of acquaintances.

Proved.

Example 6: Prove that, given any 12 natural numbers, we can chose two of them such that their difference is divisible by 11.

Solution: "If two numbers a and b have the same remainder upon division by a number c then the difference a – b is divisible by c." We shall use this property of integers to solve this type of problems. There are 11 possible remainder upon division by 11. These are 0, 1, 2, ..., 10. Let these be the pigeonholes. We have 12 pigeons (natural numbers). If we put these pigeons in the 11 pigeonholes, then we have one pigeonhole containing at least

$$\left[\frac{12-1}{11}\right]+1=2$$

integers yielding same remainder when divided by 11. Using the property of integers, as stated above, we conclude that we can choose two of the 12 natural numbers such that their difference is divisible by 11.

Proved.

Example 7: We are given five arbitrary natural numbers $\boldsymbol{a}$, $\boldsymbol{b}$, $\boldsymbol{c}$, $\boldsymbol{d}$, $\boldsymbol{e}$. Prove that either one of them is divisible by 5, or the sum of several numbers in a row is divisible by 5.

Solution: Let us consider five sums $\boldsymbol{a}$, $\boldsymbol{a+b}$, $\boldsymbol{a+b+c}$, $\boldsymbol{a+b+c+d}$, $\boldsymbol{a+b+c+d+e}$. When a number is divided by 5, the possible remainders are 0, 1, 2, 3, 4. Let the number of pigeonholes is 5. If we place the five sums one by one in these holes then either each hole will get one pigeon **or** some of holes will get more than one and some none. In the first case, the sum corresponding to holes **0** is divisible by 5. And, in the second case, the difference of the two sums falling in the same hole is divisible by 5. Also note that the difference of the sums is again one of the natural number or sum of one or more given natural numbers in row. Thus, we have proved that earlier one of the numbers or the sum of several numbers in a row is divisible by 5.

Proved.

Example 8: Prove that there exists a multiple of 1997 whose decimal expansion contains only digits 1 and 0.

Solution: Let us write 1998 numbers like **1, 11, 111, 1111, ..., 111...111**. Each of these numbers yields 0 or 1 or 2 or ...1996 as remainder when divided by 1997. Let these 1997 possible remainders are the pigeonholes and the above 1998 written numbers are pigeons. If we place the pigeons one by one in the holes then we have at least one hole containing at least

$$\left[\frac{1998-1}{1997}\right]+1=2$$

numbers from the 1998 numbers, listed as above. The difference of these two numbers is a number containing only digits 1 and 0.

Proved.

Example 9: 51 points were placed in an arbitrary way into the square of side 1. Prove that some 3 of these points can be covered by a circle of radius 1/7.

Solution: Let us divide the square into 25 squares of side 1/5 each. Then at least one of these small squares will contain at least

$$\left[\frac{51-1}{25}\right]+1=3$$

points (pigeons). Now the circle circumvented around the square with three points also contains these three points and has radius.

$$r=\sqrt{\left(\frac{1}{10}\right)^2+\left(\frac{1}{10}\right)^2}=\sqrt{\frac{2}{100}}=\sqrt{\frac{1}{50}}<\sqrt{\frac{1}{49}}=\frac{1}{7}$$

Hence, there are at least three points inside a circle of radius 1/7.

Proved.

Now, we shall use this theorem to solve a problem that is based on constructive prove method. See the example below and before looking at the solution just try to solve on your own.

Example 10: A chess player wants to prepare for a championship match by playing some practice games in 77 days. She wants to play at least one game a day but no more than 132 games altogether. Show that there is a period of consecutive days within which she plays exactly 21 games, irrespective of the scheduling of her practice games.

Solution: Let a_i be the total number of games she plays up through the ith days. Obviously, the sequence a_1, a_2, a_3, ..., a_{77} is a monotonically increasing sequence, where $a_1 \geq 1$ and $a_{77} \leq 132$. Now let us take another sequence b_1, b_2, b_3, ...b_{77} such that $b_i = a_i +$ 21 and $b_{77} \leq 153$ (Notice, here, the construction of sequence ***b***). This sequence ***b*** is also monotonically increasing. We have, now, 154 numbers (77 from sequence ***a*** and 77 from sequence ***b***) and the values of these numbers range from 1 to 153. Therefore from pigeonhole principle, we have at least two numbers having same value. Since both ***a*** and ***b*** are monotonically increasing, two numbers of ***a*** and two numbers of ***b*** cannot be equal. So, there exists a number in the sequence ***a***, which is equal to a number in the sequence ***b***. Let a_i is equal to $= b_j$ i.e., $a_i = a_j + 21$. This proves that there are consecutive days ***i*** through ***j*** during which she played exactly 21 games.

Proved.

Example 11: Prove that some integral power of 2 has the decimal expansion which starts with the digits 1999 i.e., there exists integer n such that

$2^n = 1999...$

Solution: Let there is a number the decimal expansion of which starts with the digits 1999.... If we prove that there exists a number starting with the digits 1999..., such that $\log_2$ 1999... is an integer, then the problem is solved. Readers may try to solve this. You may use the concept that $\log_2 10$ is an irrational number.

Exercise

1. Population of New Delhi is above 600, 000.00. Each has no more than 100, 000 hairs on his or her head. Prove that some 599 residents of New Delhi have equal number of hairs.
2. Assume that there are $n > 1$ people in the room. Prove that at least two of them have equal number of acquaintances.

3. Ten football teams play for a trophy. Every two of them have to meet in a game. Prove that at any given time there are two teams having played equal number of games.
4. Prove that, given three natural numbers, we can select two of them with even sum.
5. Prove that of any 100 natural numbers, there is one number, or the sum of several numbers, which is divisible by 100.
6. Prove that there is a number of the form 19971997... 199700...00 divisible by 1998.
7. Prove that for any natural number n there is a number written only by 5's and 0's and divisible by n.
8. Prove that of any 52 natural numbers one can find two numbers m and n such $m + n$ or $m - n$ is divisible by 100. Is the same statement true for 51 arbitrary natural numbers?
9. Given n integers, prove that there are several of them (may be one) with the sum divisible by n.
10. Prove that there is a natural number of the form 19971997... 1997 divisible by 1999.
11. We are given 7 straight lines on the plane. No two of them are parallel. Prove that there are two lines with the angle between them less than 26^0. Can we claim the same for 25^0?
12. Five points are positioned inside of the equilateral triangle of side 1 unit length. Prove that there are two of them at the distance less than 0.5 from each other.
13. A city has 10, 000 different telephone lines numbered by 4-digit numbers. More than half of these are in suburban. Show that there are two telephone numbers in the sub-urban whose sum is again the number of a sub-urban telephone line.
14. Ten people volunteer for a three-person committee. Every possible committee of three that can be formed from these 10 names is written on a slip of paper, one slip for each possible committee, and the slips are put in 10 hats. Show that at least one hat contains 12 or more slips of paper.
15. A store has introductory sales on 12 types of candy bars. A customer may choose one bar of any five different types and will be charged no more than Rs. 175. Show that although different choices may cost different amounts, there must be at

least two different ways to choose so that the cost will be the same for both choices.

16. Show that there must be at least 90 ways to choose six numbers from 1 to 1 so that all the choices have the same sum.
17. How many friends must you have to guarantee that at least five of them will have birthdays in choices same month?
18. Prove that if any 14 numbers from 1 to 25 are chosen, then one of them is a multiple of another.
19. If a is an integer divisible by 4, then a is the difference of two perfect squares.
20. Prove that the sum of two rational numbers is a rational number.
21. If r_1, r_2 and r_3 are three distinct roots of polynomial $p(x) = x^3 + bx^2 + cx + d$, then show that $r_1 r_2 + r_1 r_3 + r_2 r_3 = c$.
22. Prove that the cube root of 2 is irrational.
23. Prove tnat there are no positive integer solutions to the diophantine equation $x^2 - y^2 = 10$.
24. Show that there is no rational number solution to the equation $x^5 + x^4 + x^3 + x^2 + 1 = 0$.
25. Show that the sum of a rational number ***a*** and an irrational number ***b*** is an irrational number.
26. Show that if product of two integers is even then at least one of them must be even. Show that if product of two integers is odd then both of then must be odd.
27. Show that if product of two real numbers is an irrational number then one of them must be an irrational number.
28. Prove that an integer ***a*** is not evenly divisible by 5 if, and only if, $n^4 - 1$ is evenly divisible by 5.
29. A **Diophantine equation** is an equation for which you seek integer solutions. For example, the triples (x, y, z) are positive integer solutions to the equation $x^2 + y^2 = z^2$. Shows that there are no positive integer solution to the diophantine equation $x^2 - y^2 = 1$.
30. Two integers are said to have same **parity** if they both are odd or both are even. If x and y are two integers for which $x + y$ is even, then show that x and y have the same parity.
31. Prove that a positive integer n is evenly divisible by 9 if and only if, the sum of the digits of n is divisible by 9.

32. Show that a positive integer n is evenly divisible by 11 if, and only if, the difference of the sum of the digits in the even and odd position in n is divisible by 11.
33. How many distinguishable permutations of the letters in the word (a) BANANA (b) ASSOCIATIVE (c) REQUIREMENTS are there?
34. In how many ways can seven people be seated in a circle?
35. In a psychological experiment, a person must arrange a square, a cube, a circle, a triangle and a pentagon in a row. How many different arrangements are possible?
36. A bookshelf is to be used to display six new books. Suppose that there are eight computer science books and five French books from which to choose. If we decide to show four books of computer science and two French books and we are required to keep the books in each subject together, how many different displays are possible?
37. Most of the programming languages allow variables names to be of eight letters or digits with restriction that first character must be a letter. How many eight character variable names are possible?
38. At a certain college, the housing office has decided to appoint, for each floor, one male and one female residential advisor. How many different pairs of advisors can be selected for a seven-story building from 12 male candidates and 15 female candidates?
39. In how many ways can a committee of 6 people be selected from a group of 10 people if one person is to be designated as chairperson of the committee?
40. How many ways can you choose three of seven fiction books and two of six nonfiction books to take with you on your vacation?
41. In an experiment, n fair six sided dice are tossed and numbers showing on top are recorded. (a) How many sequences are possible? (b) How many sequences contains exactly one six? (c) How many sequences contain exactly four twos, assuming $n \geq 4$?
42. There are 10 points in a plane of which only 4 are collinear. How many different triangles can be formed with these points as vertices?

43. Find the number of ways of distributing 16 pens among 4 persons in such a way that every one must get at least three pens.
44. How many integers between land 100000 have the sum of their digits equal to 16?
45. In how many ways can 52 cards be distributed amongst four persons sitting in a circle to play a game of cards?
46. A library has 5 indistinguishable black books, 4 indistinguishable red books and indistinguishable yellow books. How many distinguishable ways can a student take home 6 books? Solve it using generating function.
47. In the exercise 46 above, how many ways can a student take home 6 books, 2 of each colour?
48. In how many ways can two adjacent squares be selected from 8×8 chessboard?
49. In how many ways can two integers be selected from integers 1 to 100 so that their difference is (a) exactly seven? (b) less than equal to 7?
50. There are 10 pairs of shoes in a shoe rack. If eight shoes are chosen at random, what is the probability that no complete pair of shoes is chosen? That exactly one complete pair of shoes is chosen?
51. Two numbers are chosen one after the other from numbers between 1 and 100. What is the probability that the sum of the two numbers is divisible by 3?
52. Ten passengers get into an elevator on the ground floor of a 20-story building. What is the probability that they will all get off at different floors?
53. How many permutations of 10 digits 0, 1, 2, ..., 9 are there in which the first digit is greater than 1 and the last digit is less than 8?
54. How many permutations of 26 letters $a, b, c, \ldots, z$ are there in which the final letter is not a, b, or c and the last letter is not w, x, x, y, or z?
55. In how many ways can two integers be selected from 1, 2..., $n - 1$ so that their sum is larger than n?

56. Find the number of ways to place $2n + 1$ balls in three different boxes so that any two boxes together should contain more balls than the other one. Use generating function method.

Prove the following using mathematical induction

57. $1 + 2 + 3 + \cdots + n = \dfrac{n(n+1)}{2}$

58. $\overline{\bigcap_{i=1}^{n} A_i} = \bigcup_{i=1}^{n} \overline{A_i}$

59. $1^2 + 2^2 + 3^2 + \ldots + n^2 = \dfrac{n(n+1)(2n+1)}{6}$

60. $1^3 + 2^3 + 3^3 + \ldots + n^3 = \left[\dfrac{n(n+1)}{2}\right]^2$

61. $a + ar + ar^2 + \ldots + ar^{n-1} = \dfrac{a(1 - r^n)}{1 - r}$ *for* $r \neq 1$

62. $1 + 2^n < 3^n$ for $n \geq 2$

63. $n < 2^n$ for $n \geq 0$

64. $\left(\bigcap_{i=1}^{n} A_i\right) \cup B = \bigcap_{i=1}^{n} (A_i \cap B)$

65. $\dfrac{1}{1*2} + \dfrac{1}{2*3} + \dfrac{1}{3*4} + \ldots\ldots + \dfrac{1}{n*(n+1)} - \dfrac{n}{n+1}$

66. Use induction method to show that if p is a prime and $p \mid a^n$ for $n \geq 1$ then $p \mid a$.

67. Find smallest positive integer n such that $2^n > n^2$ and show that $2^m > m^2$ for $m > n$.

68. Using induction method, prove that $10^{2n-1} + 1$ is divisible by 11 for all $n \in N$.

69. $2^{5n+5} - 31n - 32$ is divisible by 961 for all $n \in N$. prove this by mathematical induction.

70. Prove that $7^{2n} + 2^{3n-3} * 3^{n-1}$ is divisible by 25 for all $n \in N$. Use mathematical induction method to prove this.

71. Using mathematical induction method, prove that $11^{n+2} + 12^{2n+1}$ is divisible by 133 for any $n \in N$.

72. Using induction or otherwise, prove that for any non-negative integer m, n, r and k.

$$\sum_{m=0}^{k}(n-m)\frac{(r+m)!}{m!} = \frac{(r+k+1)!}{k!}\left[\frac{n}{r+1} + \frac{k}{r+2}\right]$$

Prove the following by induction method.

73. $\frac{1}{2}, \frac{3}{4}, \frac{5}{6} \cdots \frac{2n-1}{2n} < \frac{1}{\sqrt{2n+1}}$ *for all* $n \in N$

74. $10^{n-2} > 81n$ for $n \in N$ and $n \geq 5$

75. $\frac{n^5}{5} + \frac{n^3}{3} + \frac{7n}{15}$ *is a natural number for all* $n \in N$.

76. $p^{n+1} + (p+1)^{2n-1}$ is divisble by $p^2 + p + 1, p \in N$.

77. Let $s_1 = 2, s_2 = 0, s_3 = -14$, and $s_{n+1} = 9s_n - 23s_n - 1 + 15s_{n-2}$, *for* $n \geq 3$, then show that

$$s_n = 3^{n-1} - 5^{n-1} + 2 \textit{ for } n \geq 1.$$

78. *Let* $u_1 = u_2 = 5$, *and* $u_{n+1} = u_n + 6u_{n-1}$, *for* $n \geq 2$, *then show that*

$$u_n = 3^n - (-2)^n \textit{ for } n \geq 1.$$

5

Group

Let G be a non-empty set. If $f = G \times G \rightarrow G$ be a function such that for any two elements a and b, the ordered pair (a, b) is in G, then we say that G is closed with respect to this binary function f and f is called a binary operation on G. If f is any binary operations like +, *, / etc, then G is said to be closed with respect to a binary operation * if and only if for any two elements a and b of G, $a*b$ is in G. For example, addition is a binary operation on the set of natural number N, because, the sum of any two natural numbers is a natural number. However, subtraction is not a binary operation on N as the $5 - 9 = -4 \notin N$. A binary operations on a set G is also called binary composition in a set G. In group theory, we are concerned with binary operations only. Therefore the phrase 'binary operation' and the word 'operation' are interchangeably used.

A non-empty set G is said to be equipped with a binary operation * if * is an operation on G, and this is denoted as $(G, *)$. A non-empty set G equipped with one or more binary operations is called an algebraic structure. Thus $(G, *)$ is a representation of an algebraic structure with one operation. For example, $(I, +)$ $(N, +)$, $(Q, -)$ etc, are algebraic structure with one operations, whereas, $(R, +, *)$ is an algebraic structure with two operations.

Group Definition

*Let G be a non-empty set equipped with a binary operation denoted by * i.e., a*b or more precisely as represents the element of G obtained by applying * between a and b of G taken in that order. Then the algebraic structure (G, *), also called mathematical structure, is a group if the binary operation * satisfies the following postulates:*

- **Closure property** i.e., $a*b \in G \ \forall \ a, b \in G$.
- **Associative property** i.e., $(a*b) * c = a * (b*c) \ \forall \ a, b, c \in G$.
- **Existence of Identity** i.e., there exists an element $e \in G$ such that $e*a = a*e = a \ \forall \ a \in G$. The element e is called the identity element of G if it exists.
- **Existence of Inverse** i.e., each element of G possesses inverse. In other word, for $a \in G$ there exists and element b such that $a*b = b*a = e$. The elements b is then called the inverse of a and we write b as a^{-1}. Here a and b are inverse of each other.

Example 1: Show that $(I, +)$ is a group, where I is the set of all integers and '+' is an integer addition operation.

Solution: To prove that $(I, +)$ is a group, we will show that it satisfies all the four postulates for it to be a group.

Closure property: We know that the sum of two integers is also an integer i.e., $a + b \in \boldsymbol{I} \ \forall \ a, b \in \boldsymbol{I}$. So $\boldsymbol{I}$ is closed with respect to addition.

Associative property: We know that addition of integers is an associative operation. Thus, $a + (b + c) = (a + b) + c \ \forall \ a, b, c \in \boldsymbol{I}$.

Existence of Identity: The number $0 \in I$. Also, we have $0 + a = a + 0 = a \ \forall \ a \in I$. Therefore, the integer 0 is the identity element.

Existence of Inverse: If $a \in I$, then $-a \in I$. And, $(-a) + a = a + (-a) = 0$. This shows that every integer possesses additive inverse. Therefore $(I, +)$ is a group.

Proved.

Example 2: Prove that $(I_+, +)$, where I_+, is the set of positive integers is not a group with respect to addition.

Solution: Addition is obviously a binary composition in I_+ i.e., I_+ is closed with respect to addition. Also addition of a positive integers is an associative operation. But, there exists no positive integer $e \in I_+$, such that $e + a = a + e = a \ \forall \ a \in I_+$. For the addition of positive integers, 0 is the identity and $0 \notin I_+$. Therefore, $(I_+, +)$ is not a group.

Proved.

It is superfluous to mention the **closure property** in the group definition as it is implied by the definition of binary operation. However, it is mentioned to the benefit of readers so that one should not forget to show this axiom while showing the group postulates in a problem.

A mathematical structure equipped with one binary operation is also called a **Groupoid** i.e., a mathematical structure $(G, *)$ is said to be a Groupoid if G is closed with respect to the binary operation *. If the binary operation * is associative also, then $(G, *)$ is called a **Semi-group**. Further, if there exists an identity elements in G for *, then $(G, *)$ is called a **Monoid.**

Exercise 5

1. Prove the following
   - The set C of all complex numbers with respect to the operation of addition of complex numbers is a group.
   - The set of all rational numbers with respect to addition is a group.
   - The set of all rational numbers with respect to addition is a group.
   - The set R_0 of all non-zero real numbers with respect to multiplication is a group.
   - The set C_0 of all non-zero complex numbers with respect to multiplication is a group.
2. Define the order of a group. Show that the set of all even integers with zero is an abelian group with respect to addition.
3. Show that the set M of complex numbers z with the condition $|z| = 1$ forms a group with respect to the operation of multiplication of complex numbers.
4. Let Q_+ be the set of all positive rational numbers and * a binary operation on Q_+ defined by $a * b = ab/3$. Determine the identity element in Q_+ and inverse of the element 'a'.
5. Let R be the set of all real numbers and * a binary operation on R defined by $a * b = a + b + ab$. Determine the identity element in R and determine the inverse of a.

6. Show that the set of all rational numbers of the form $2^a\ 3^b$ where a and b are integers, is a group with respect to the multiplication of rational numbers.

7. Consider the binary operation * defined on the set $A = \{a, b, c, d\}$ by the following table.

*	a	b	c	d
a	a	c	b	d
b	d	a	b	c
c	c	d	a	a
d	d	b	a	c

Compute

(a) $c*d$ and $d*c$ (b) $b*d$ and $d*b$

(c) $a*(*c)$ and $(a*b)*c$ (d) Is * commutative; associative?

In exercise 8 and 9, complete the given table so that the binary operation * is associative 8.

8.

*	a	b	c	d
a	a	b	c	d
b	b	a	d	c
c	c	d	a	b
d				

9.

*	a	b	c	d
a	a	b	c	d
b	b	a	c	d
c				
d	d	c	c	d

10. Show that the set of all $m \times n$ matrices with the operation of matrix addition is a group.
11. Determine whether $(P(S), *)$ is a group or not, $P(S)$ is the power set of a non-empty set S and * on $P(S)$ is defined as $A*B = A \oplus B$ for $A, B \in P(S)$.
12. Let $S = \{x \mid x$ is a real number and $x \neq 0, x \neq -1\}$. Consider the following functions $f_i : S \to S$, $i = 1, 2, 3, \ldots, 6$:

    $f_1(x) = x, f_2(x) = 1 - x, f_3(x) = 1/x, f_4(x) = 1/(1 - x), f_5(x) = 1 - 1/x, f_6(x) = x.(x - 1)$.

    Show that $G = \{f_1, f_2, f_3, f_4, f_5, f_6\}$ is a group under the operation of function composition. Find the composition table for $(G, *)$.
13. Prove that the set G of all translations of a fixed point (x, y) in the plane given by $x' = x + a$ and $y' = y + b$, where a and b are real numbers, is a group under the operation of 'composition of translation'.
14. (a) Prove that the set $\{1, -1\}$ is an abelian group with respect to multiplication.

    (*b*) Show that $\{x \mid x$ is cube roots of unity$\}$ from an abelian group under multiplication. Write down the multiplication table.
15. (a) How many elements of the cyclic group of order 7 can be used as generators of the group?

    (*b*) How many elements of the cyclic group of order 8 can be used as generators of the group?

    (c) If $G = \{a\}$ be an infinite cyclic group, prove that a and a^{-1} are the only generator of G.
16. Is the multiplicative group of residue classes {1}, {}, {3}, {4}, {5}, {6} (mod 7) cyclic?
17. Let G is the multiplicative group of all non-singular squares matrices of order n with real elements and S is the set of non-singular matrices of order n with real elements whose determinant is 1. Show whether S is a subgroup of G or not?
18. Show that the set of all elements x of a group G such that $ax = xa$ for every element a of G is a subgroup of G.

19. Find all the subgroups of the cyclic group $G = (a, a^2, a^3, a^4, a^5, a^6 = e)$.
20. Let G be the group of complex numbers under multiplication. Determine whether $S = (1, -1, i, -i)$ is a subgroup of G or not?
21. Prove that the additive group R of real numbers is mapped homomorphically onto the multiplicative group C of complex numbers z such that $|z| = 1$, under function $f: R \rightarrow C$ defined as $f(x) = \cos(x) + i \sin(x)$.
22. Prove that the multiplicative group of $I^*/(5)$ is cyclic and then show that it is isomorphic to the additive group of $I/(4)$.
23. Let (1, *w*, *w*2) be the multiplicative group of the cube roots of unity. Let $G' = \{0, a, b\}$ be the group of rotations of an equilateral triangle about its centroid. Show that G and G' are isomorphic groups.
24. Prove that every abelian group of order 15 is cyclic.
25. Prove that any abelian group of order $p.q$ where p and q are distinct primes, is necessarily cyclic and hence isomorphic to additive group $I/(p.q)$.
26. Let G be the group of all translations of a fixed point (x, y) in the plane and H be the subgroup of all translations parallel to the x-axis. Find the cosets of H in G.
27. Prove that the intersection of two submonoids of a monoid $(S, *)$ is a submonoids of $(S, *)$.
28. Let R^+ be the set of all positive real numbers. Show that the function $f: R^+ \rightarrow R$ defined by $f(x) = \text{In}(x)$ is an isomorphism of the semi-group $(R^+, \times)$ to the semigroup $(R, +)$, where $\times$ and $+$ are ordinary multiplication and addition of real numbers, respectively.
29. Let $(S_1, *)$, $(2_1, *')$ and $(S_3, *'')$ be semi groups, and let $f: S_1 \rightarrow S_3$ and $g: S_2 \rightarrow S_3$ be isomorphism. Show that gof: $S_1 \rightarrow S_3$ is an isomorphism.
30. An element x in a monoid is called an ***idempotent*** if $x^2 = x * x = x$. Show that the set of all idempotents in a commutative monoid S is a submonoid of S.
31. Which of the following tables defines a semi-group?

(a)

*	a	b	c
a	c	b	a
b	b	c	b
c	a	b	c

(b)

*	a	b	c
a	c	b	a
b	b	c	b
c	a	b	c

In exercise 32 to 36, determine whether the relation R on the semigroups S is a congruence relation.

32. $S = 1$, the set of all integers, under the operation of ordinary addition: $a\mathrm{R}b$ iff 2 does not divide $a - b$.
33. S = any semigroup; $a\mathrm{R}b$ iff $a = b$.
34. S = the set of all rational numbers under the operation of addition; $a/b\ R\ c/d$ iff $ad = bc$.
35. $S = I$, the set of all integers, under the operation of integer addition of integers; $a\mathrm{R}b$ iff $a \equiv_3 b$.
36. S = set of all positive integers, under the operations of ordinary multiplication; $a\mathrm{R}b$ iff $|a - b| \leq 2$.
37. Let $A = \{0, 1\}$ and consider the free semigroup A^* generated by A under the operation of catenation. Let N be the semigroup of all nonnegative integers under the operation of ordinary addition.
    (a) Verify that the function f: $A^* \to N$, defined by $f(\alpha)$ = the number of bits in α, is a homomorphism.
    (b) Let R be the relation on A^* defined as $a\mathrm{R}b$ iff $f(\alpha) = f(\beta)$, where f is defined as in (α). Show that R is a congruence relation on A^*.
    (c) Show that A^*/R and N are isomorphic.

38. Let G be a group. Show by mathematical induction that if $ab = ba$, then $(ab)^n = a^n b^n$ for $n \in Z^+$.
39. Let G be a finite group with identity e, and let a be an arbitrary element of G. Prove that there exists a nonnegative integer n such that $a^n = e$.
40. Let H and K be subgroups of a group G. Prove that (a) $H \cap K$ is a subgroup of G and (b) $H \cup K$ need not be a subgroup of G.
41. Let G be an abelian group with identity element e and let $H = \{x \mid x^2 = e\}$. Show that H is a subgroup of G.
42. Show that the group in exercise 12 is isomorphic to S_3.
43. Let G be a group. Show that the function $f: G \to G$ defined by $f(a) = a^{-1}$ is an isomorphism iff G is abelian.
44. Write the multiplication table of the group $Z_2 \times Z_3$.
45. Let G_1 and G_2 be groups. Prove that $G_1 \times G_2$ and $G_2 \times G_1$ are isomorphic.
46. Let $G = S_3$. For each of the following subgroups H of G, determine all the left cosets of H in G.

    (a) $H = \{f_1, g_1\}$ (b) $H = \{f_1, g_3\}$

    (c) $H = \{f_1, f_2, f_3\}$ (d) $H = \{f_1\}$

    (e) $H = \{f_1, f_2, f_3, g_1, g_2, g_3\}$
47. Let $G = Z_8$. For each of the following subgroups H of G, determine all the left cosets of H in G.

    (a) $H = \{[0], [4]\}$ (b) $H = \{[0], [2], [4], [6]\}$
48. Let N be a subgroup of a group G, and let $a \in G$. Let us define $a^{-1}Na = \{a^{-1}na \mid n \in N\}$. Prove that N is a normal subgroup of G iff $a^{-1}Na = N$ for all $a \in G$.
49. Find all the normal subgroups of S_3.
50. Let G be an abelian group and N a subgroup of G. Prove that G/N is and abelian group.
51. Let $(G, *)$ be a mathematical structure such that for all a, b, c, d in G (i) $a*a = a$ and (ii) $(a*b) * (c*d) = (a*c) * (b*d)$. Show that $a*(b*c) = (a*b) * (a*c)$.
52. Let $(G, *)$ be a semigroup. Show that for a, b, c in G if (i) $a*c = c*a$ and (ii) $b*c = c*b$ then $(a*b)*c = c*(a*b)$.

53. Let $(G, *)$ be a mathematical structure such that for all a, b in G (i) $(a * b) * a = a$. (ii) $(a * b) * b = (b * a) * a$ and (iii) $a * b = (a * b) * b$. Show that $*$ is idempotent and commutative.
54. A *central groupoid* is a mathematical structure $(G, *)$ where $*$ is a binary operation such that $(a * b) * (b * c) = b$ for all a, b, c in G. Show that

$$a * ((a * b) * c) = a * b \text{ and}$$

$$(a * (b * c)) * c = b * c$$

in a central groupoid.

6

Partial Ordered Set

Recall the relation on set discussed in the chapter 1. We discussed different types of relation in that chapter. In the chapter five, we have learnt about mathematical (algebraic) structure with one binary operation. A relation between two sets is also a binary operation. Any relation R defined on a non-empty set A is said to be a **Partial Order Relation,** if R is

- Reflexive on A i.e., $xRx \ \forall \ x \in A$
- Anti-symmetric on A i.e., xRy and $yRx \Rightarrow x = y$ and
- Transitive on A i.e., xRy and $yRz \Rightarrow xRz$ for $x, y, z \in A$.

A partial order relation is denoted by the symbol $\leq$. For example, the relation 'less than equal to' is a partial order relation on the set N of natural numbers. Similarly, the relation 'greater than equal to' is also a partial order relation on the set N of natural number. It is for the reader to note that only equivalence relation that is also a partial order, is the *Identity relation or Diagonal relation.* The relation 'less than equal to' is referred as **usual partial order.**

6.1 Poset

Let R be a relation defined on a non-empty set A. The mathematical structure (A, R) is said to be a *Partial order set* or **poset** if the relation R is a partial order relation on A. A general notation for a poset is $(A, \leq)$, where A is any non-empty set and '$\leq$' is any partial order relation defined on the set A. A poset is said to be **finite poset** if A is a **finite set** and **infinite poset** if A is an **infinite set.**

Example 1: Let S be any non-empty set and $P(S)$ be the power set of S. If '$\subseteq$' (a subset of) is a relation defined on $P(S)$, then show that $(P(S), \subseteq)$ is a poset.

Solution: $P(S)$ is obviously a non-empty set. The relation $\subseteq$ on $P(S)$ is reflexive, anti-symmetric and transitive on $P(S)$. So, $\subseteq$ is a partial order relation on $P(S)$, and hence $(P(S), \subseteq)$ is a poset.

Proved.

Example 2: Let A be a non-empty set and R be the set of all equivalence relation defined on A. If '$\subseteq$' (a subset of) is a relation defined on R, then show that the structure $(R, \subseteq)$ is a poset.

Solution: Here, R is a non-empty set, because there always exists some relations (identity, universal etc.) on a set that are equivalence relations. The relation $\subseteq$ on R is reflexive, anti-symmetric and transitive, hence $\subseteq$ is a partial order relation on R. Therefore, $(R, \subseteq)$ is poset.

Proved.

Theorem 1: Let R be a partial order relation on a set A and let R^{-1} be the inverse of relation R, then R^{-1} is also a partial order.

Proof: Since R is a partial order relation on A, R is reflexive, anti-symmetric and transitive.

Reflexivity of $R \Rightarrow \forall x \in A, (x, x) \in R \Rightarrow \forall x \in A, (x, x) \in R^{-1} \Rightarrow$ R^{-1} is reflexive.

Let (x, y) and $(y, x) \in R^{-1}$, then (y, x) and $(x, y) \in R$. Since R is anti-symmetric, it implies that $x = y$. Thus R^{-1} is anti-symmetric.

Let (x, y) and $(y, z) \in R^{-1}$, then (z, y) and $(y, x) \in R$. Since R is transitive, $(z, x) \in R$. This implies that $(x, z) \in R^{-1}$. Hence, R^{-1} is transitive. Therefore R^{-1} is a partial order relation.

Proved.

In the theorem 1, it is proved that if R is a partial order relation on a non-empty set A, then R^{-1} is also a partial order relation on the same set A. Therefore, if (A, R) is a poset then (A, R^{-1}) is also a poset. The poset (A, R^{-1}) is called **Dual** of the poset (A, R) and vice versa. In the example 1, '$\subseteq$' is a partial order relation on $P(S)$. The inverse of this relation is '$\supseteq$' (a superset of) Thus, $P(S), \supseteq)$ is a dual of the poset $(P(S), \subseteq)$.

Example 3: Prove that $(I, \leq)$ is a poset, where I is the set of all integers and $\leq$ is a relation on I defined as $a \leq b$ iff $a = b^k$ for some $k \in I^+$. Find the dual of this poset.

Solution: Since $1 \in I^+$, we have $a = a^1$, $\forall\ a \in I$, i.e., pairs $(a, a) \in \leq\ \forall\ a \in I$. Therefore, $\leq$ is reflexive. Let (a, b) and (b, a) are in $\leq$, then $a = b^k$ and $b = a^j$ for some k and $j \in I^+$. This is possible iff $k = j = 1$, i.e., $a = b$. Thus, $\leq$ is anti-symmetric. Finally, let (a, b) and (b, c) are in $\leq$. Then, by definition of $\leq$, $a = b^k$ and $b = c^j$ for some k and $j \in I^+$. This implies that $a = c^{kj}$, and hence (a, c) is in $\leq$. Thus, $\leq$ is transitive relation. Therefore the structure $(I, \leq)$ is a poset.

In the second part of the given problem, we have to find dual of the poset $(I, \leq)$. The inverse of the relation $\leq$, say $\geq$ can be defined on I as: for any a, b of I, $a \geq b$ iff $a^k = b$ for some $k \in I^+$. This can be easily proved that $\geq$ is a partial order relation on I, and hence, $(I, \geq)$ is dual of the poset $(I, \leq)$ and vice versa.

Proved.

Example 4: Let $A = \{1, 2, 3, 4\}$ and $R = \{(1, 1), (2, 2), (3, 3), (4, 4), (1, 2), (2, 4), (1, 3), (3, 4), (1, 4)\}$ is relation defined on A. Show that (A, R) is a poset and find its dual.

Solution: The relation R is reflexive and transitive. It is anti-symmetric also as the matrix for R is upper triangular. ***If the matrix for a given relation is either upper triangular or lower triangular or diagonal then the relation is anti-symmetric.*** The reverse is not always true. Therefore, R is a partial order relation on A and hence (A, R) is a poset. Now, $R^{-1} = \{(1, 1), (2, 2), (3, 3), (4, 4), (2, 1), (4, 2), (3, 1), (4, 3), (4, 1)\}$. This inverse relation R is also a partial order relation on A, thus (A, R^{-1}) is dual of (A, R).

Proved.

Example 5: Determine whether the relation R defined as aRb iff $a = 2b$, on the set I of all integers, is a partial order relation or not.

Solution: The relation R is not reflexive because no integer, other than zero, is equal to twice of itself. Since R is not reflexive, it is not a partial order relation. Note that there is no need for testing anti-symmetry and transitivity of R.

Ans.

Let $(A, \leq)$ be a poset. Any two elements a and b of A are said to be **comparable,** if either pair (a, b) or pair (b, a) is in $\leq$. That is either $a \leq b$ or $b \leq a$. It is important to mention that in a partial order relation every pair of element need not be comparable. For example, let N be the set of all positive integers and $\leq$ on N is defined as $a \leq b$ iff a divides b. Obviously, $\leq$ is a partial order relation on N and hence, $(N, \leq)$ is a poset. But there exists elements like 3 and 8 which are not **comparable** because neither $3 \leq 8$ nor $8 \leq 3$. *This implies that in a partially ordered set, every pair of elements may not be comparable and that's why it is called partially ordered.* If every pair of elements in a poset $(A, \leq)$ is comparable, then $(A, \leq)$ is called **linearly ordered,** or **totally ordered** or **simple ordered** or **complete ordered** set. A linearly ordered set is also called a **chain.** A set is said to be **anti-chain** if no two distinct elements of the set are related.

Example 6: The set inclusion relation $\subseteq$ on a power set $P(S)$ of a set S, is a partial order relation on $P(S)$ (as proved in the example 1), but not a linear order relation. Let $S = \{1, 2, 3\}$. The sets $\{1\}$ and $\{2\}$ are in $P(S)$ and neither $\{1\} \subseteq \{2\}$ nor $\{2\} \subseteq \{1\}$. Therefore, $(P(S), \subseteq)$ is poset but not a **chain.**

Example 7: The relation $\leq$ (less than equal to) on the set R of all real numbers is a partial order relation on R and hence $(R, \leq)$ is poset. Also for any two real numbers x and y, either $x \leq y$ or $y \leq x$, i.e., every pair of real numbers are comparable with respect to the relation $\leq$. Therefore, $(R, \leq)$ is also a **chain.**

Diagrammatic Representation of Poset

We have learnt in the chapter 1 about the diagrammatic representation of a relation on a finite set. The same concept we can use to represent any finite partial order relation $\leq$ on a set A, in diagrammatic way. Let $A = \{a, b, c\}$ and $R = \{(a, a), (b, b), (c, c), (a, b), (b, c), (a, c)\}$. Obviously, R is a partial order relation on A and hence (A, R) is a poset. The diagram for R is shown in the figure 6.1.1.

Let us simplify this representation for a finite poset. In a poset the reflexivity and transitivity of the relation is implied. The anti-symmetry of the relation we are not considering right now. Since we know that a partial order relation is reflexive, loop on each node in the diagram can be dropped. The arc showing the implied transivity can also be dropped in a poset. Next, all the elements of poset (nodes in diagram) can be arranged in such a way that all edges points upward.

The resulting diagram is a simplified diagram for a given a poset. This simplified diagram is called **Hasse Diagram** of a poset. We shall use this form of representation in this chapter for poset, lattice and for finite Boolean algebra. The above procedure to draw a Hasse diagram can be summarized as below.

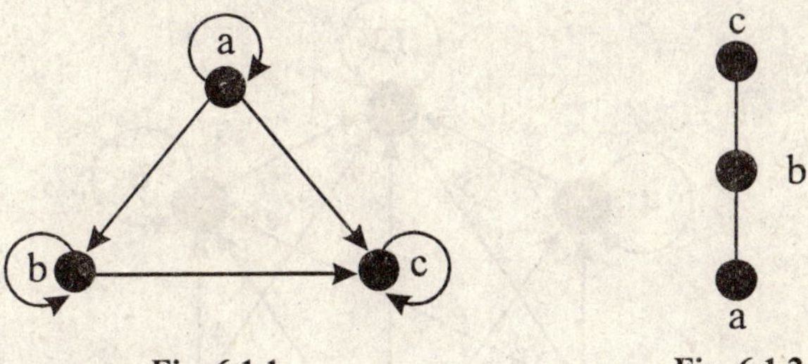

Fig. 6.1.1. **Fig. 6.1.2.**

- Draw a digraph for the partial order relation.
- Drop loops on all nodes.
- Drop all edges implied by the transitivity of the relation.
- Arrange all edges pointing in upward direction and then replace each directed arc with undirected arc.
- Represent each node with **thick** dots (•)

If we apply this procedure on the diagram shown in the figure 6.1.1, we get the diagram for the poset (A, R) as shown in the figure 6.1.2. In the figure 6.1.2 arc (a, c) is implied transitively from (a, b) and (b, c). Thus it is removed from the diagram of R. Other things are obvious.

Example 8: Let us define a set D_n, where n is any positive integer > 1, as the collection of all the positive integers which can evenly divide n. Thus, $D_{12} = \{1, 2, 3, 4, 6, 12\}$, $D_{12} = \{1, 2, 3, 6\}$ etc. Let us define a relation $\leq$ on D_{12} as, for any two integers a and $b \in D_{12}$, $a \leq b$ iff a divides b. Show that (D_{12}, $\leq$) is a poset. Draw a Hasse diagram for this poset.

Solution: Since every integer divides itself, the relation $\leq$ is reflexive on D_{12}. For any two integers a and b, if a divides b and b divides a, then both a and b are equal. Thus $\leq$ is anti-symmetric. Similarly, for any three integers a, b and c if a divides b and b divides c then a divides c. Thus, $\leq$ is transitive. Therefore (D_{12}, $\leq$) is a poset. Since

$|D_{12}| = 6$, the poset is finite. The ordered pairs of the relation $\leq$ are {(1, 1), (1, 2), (1, 3), (1, 4), (1, 6), (1, 12), (2, 2), (2, 4), (2, 6), (2, 12), (3, 3), (3, 6), (3, 12), (4, 4), (4, 12), (6, 6), (6, 12), (12, 12)}. The diagram for this relation is shown in the figure 6.13.

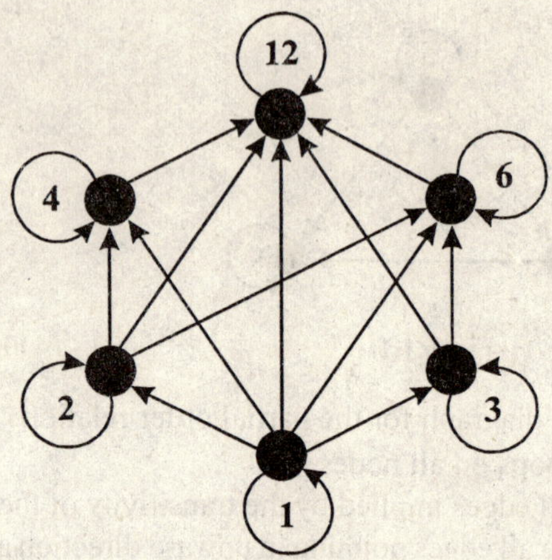

Fig. 6.1.3.

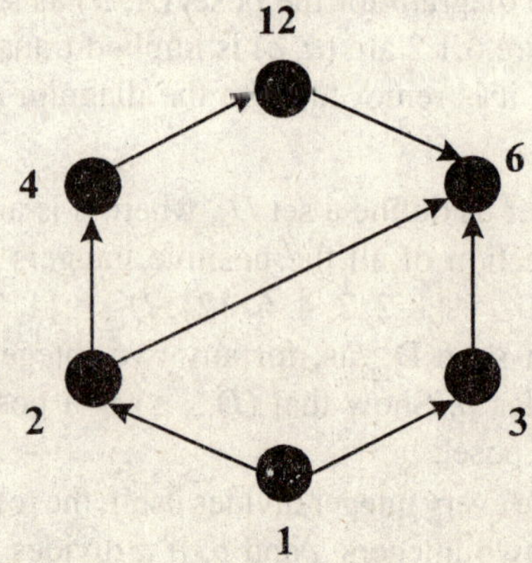

Fig. 6.1.4.

To draw a Hasse diagram for the poset (D_{12}, ≤), first drop all the loops in the figure 6.1.3 and then drop all the arcs that are transitively implied. If we look at the figure 6.1.3, arcs (1, 4), (1, 12), (1, 6), (2, 12), and (3, 12) are transitively implied. Dropping them together with all loops from figure 6.1.3, we get the diagram as shown in the figure 6.1.4. All the arcs are arranged in upward direction, so only thing left is to represent the elements by thick dots and replacement of all the directed arcs with undirected arcs. Applying this on the figure 6.1.4, we get the Hasse diagram as shown in the figure 6.1.5.

Ans.

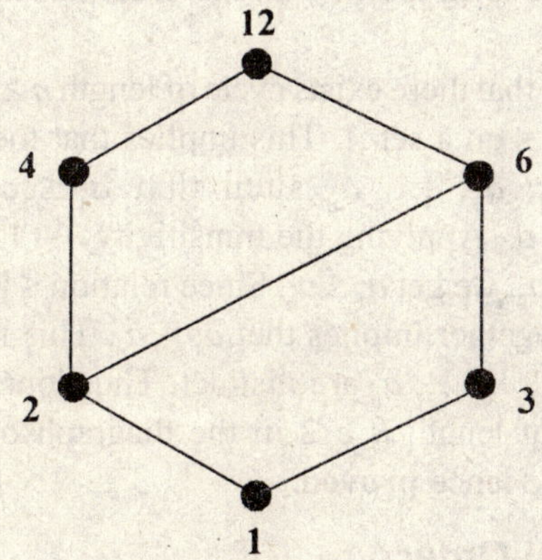

Fig. 6.1.5.

Example 9: From the following Hasse diagram of a poset (A, ≤) find set A and the partial order relation defined on A.

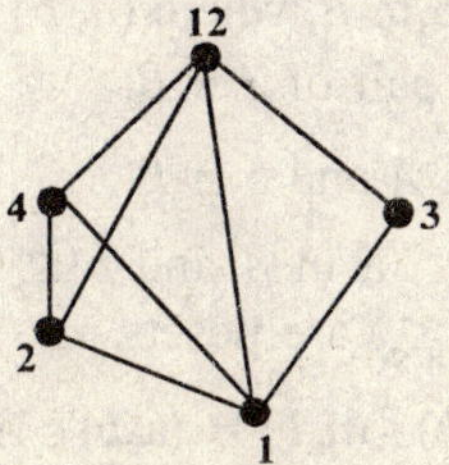

Fig. 6.1.6.

Solution: The set A is the collection of all the nodes i.e. $A = \{1, 2, 3, 4, 12\}$. The relation on A is a partial order. Thus All the reflexive pairs: (1, 1), (2, 2), (3, 3), (4, 4), (12, 12) must be in $\leq$. Next, all the edges when converted to upward directed arcs give the pairs of the relation. These are (1, 2), (1, 3), (2, 4), (4, 12) and (3, 12). The transitively implied arcs are (1, 4), (1, 12) and (2, 12). Therefore, the partial order relation on A as shown in the given Hasse diagram of figure 6.1.6, is {(1, 1), (2, 2), (3, 3), (4, 4), (12, 12), (1, 2), (1, 3), (2, 4), (4, 12), (3, 12), (1, 4), (1, 12), (2, 12)}.

Ans.

Theorem 2: The diagraph of a partial order has no cycle of length greater than 1.

Proof: Suppose that there exists cycle of length $n \geq 2$ in the diagraph of a partial order $\leq$ on a set A. This implies that there are n distinct elements $a_1, a_2, a_3, \ldots, a_n$ such that $a_1 \leq a_2$, $a_2 \leq a_3, \ldots$, $a_{n-1} \leq a_n$ and $a_n \leq a_1$. Applying the transitivity $n - 1$ times on $a_1 \leq a_2$, $a_2 \leq a_3, \ldots, a_{n-1} \leq a_n$, we get $a_1 \leq a_n$. Since relation $\leq$ is anti-symmetric $a_1 \leq a_n$, $a_n \leq a_1$ together implies that $a_1 = a_n$. This is contrary to the fact that all $a_1, a_2, a_3, \ldots, a_n$ are distinct. Thus, our assumption that there is a cycle of length $n \geq 2$ in the diagraph of a partial order relation is wrong. Hence **proved.**

Product Partial Order

Let us see the following theorem before delving into the concept of product partial order.

Theorem 3: If $(A, \leq)$ and $(B, \leq_2)$ are two posets, then $(A \times B, \leq)$ is a poset, where $\leq$ is defined on $A \times B$ as $(a, b) \leq (a_1, b_1)$ if, and only if $a \leq_1 a_1$ in A and $b \leq_2 b_1$ in B, where (a, b) & $(a_1, b_1) \in A \times B$.

Proof: Let (a, b) be a pair of $A \times B$.

Then, $(a, b) \Rightarrow a \in A$ and $b \in B$.

$\Rightarrow$ a, $\leq_1$ a in A and $b \leq_2 b$ in B since $(A, \leq_1)$ & $(B, \leq_2)$ are posets

$\Rightarrow (a, b) \leq (a, b)\ \forall\ (a, b) \in A \times B$ by the definition of $\leq$

$\Rightarrow \leq$ is reflexive on $A \times B$.

Next, if $(a_1, b_1) \leq (a_2, b_2)$ and $(a_2, b_2) \leq (a_1, b_1)$ for any two pairs (a_1, b_1) & (a_2, b_2) of $A \times B$, Then,

$(a_1, b_1) \leq (a_2, b_2) \Rightarrow a_1 \leq_1 a_2$, in A and $b_1 \leq_2 b_2$ in B and

$(a_2, b_2) \leq (a_1, b_1) \Rightarrow a_2 \leq_1 a_1$ in A and $b_2 \leq_2 b_1$ in B

Now, $(a_1 \leq_1 a_2)$ & $(a_2 \leq_1 a_1) \Rightarrow a_1 \Rightarrow a_2$ because $\leq_1$ is a partial order A. Similarly, $(b_1 \leq_2 b_2)$ & $(b_2 \leq_2 b_1) \Rightarrow b_1 = b_2$ therefore, $(a_1, b_1) = (a_2, b_2)$. Thus, $\leq$ is antisymmetric on $A \times B$.

Finally, if $(a_1, b_1) \leq (a_2, b_2)$ and $(a_2, b_2) \leq (a_3, b_3)$ for any three pairs (a_1, b_1), (a_2, b_2) and (a_3, b_3) of $A \times B$, Then,

$(a_1, b_1) \leq (a_2, b_2) \Rightarrow a_1 \leq_1 a_2$, in A and $b_1 \leq_2 b_2$ in B and

$(a_2, b_2) \leq (a_1, b_1) \Rightarrow a_2 \leq_1 a_1$ in A and $b_2 \leq_2 b_3$ in B

Now $(a_1 \leq_1 a_3)$ & $(a_2 \leq_1 a_3)$ because $\leq_1$ is a partial order on A. Similarly, $(b_1 \leq_2 b_2)$ & $(b_2 \leq_2 b_3) \Rightarrow b_1 \leq_2 b_3$. Therefore, $(a_1, b_1) \leq (a_3, b_3)$. Thus, $\leq$ is a transitive relation on $A \times B$. Therefore $\leq$ is a partial order relation on $A \times B$ and hence $(A \times B, \leq)$ is a poset.

Proved.

A partial order as defined in the theorem 3 is called a **product partial order,** a poset $(A \times B, \leq)$ is called **product poset** of posets $(A, \leq_1)$ and $(b_1, \leq_2)$. The same concept can be extended for any finite number of posets. If $(A_1, \leq_1)$, $(A_2, \leq_2)$, $(A_3, \leq_3)$, …, $(A_n, \leq_n)$ are n posets, then $(A_1 \times A_2 \times A_3 \times ... \times A_n), \leq)$ is a poset, where $\leq$ is defined on $A_1 \times A_2 \times A_3 \times \ldots \times A_n)$ as

$(a_1, a_2, a_3, \ldots a_n) \leq (b_1, b_2, b_3, \ldots b_n)$ if, and only if

$a_1 \leq_1 b_1$ in A_1 $a_2 \leq_2 b_2$ in A_2 and so on up to n.

Let x and y be any two real numbers. If $x \leq y$ and $x \neq y$, we say that $x < y$. Let A be a set of natural numbers and let $<$ be a relation "strictly less than" on A. Obviously, $<$ be a reflexive and transitive on A. Any relation that is in reflexive and transitive is called a **Strong order** relation or **Quasi-order relation.** Let us modify the above equation of product post order such that

$(a, b) < (a_1, b_1)$ if and only if

either $a < a_1$ in A or

$a = a_1$ in A and $b_1 \leq_2 b_1$ in B.

An ordering of this type is called **Dictionary Order** or **Lexicographic Order.** If we reason out with some sample words from a dictionary, we can easily to take it out that all words of a dictionary are arranged in lexicographic order. All ring sorting algorithm are based on this concept only.

Example 10: Let $\{a, b, c, \ldots, l\}$ be the set of ordinary alphabets arranged in usual order like $a < c \ldots, l$. Let $\Sigma^n = \Sigma \times \Sigma \times \Sigma \times \ldots n$ times. Obviously Σ^n is set of all words of characters. Let $w_1 = x_1, x_2, x_3, \ldots x_n$ and $w_2 = y_1, y_2, y_3, \ldots y_n$ of two words from Σ^n, where x_1's and y_1's are alphabets from Σ. We define $w_1 < w_2$ and only if $x_1 < y_1$ or $x = y_1$ and $x_2 < y_2$ or $x_1 = y_1; x_2 = y_2$ and $x_3 < y_3; \ldots$ or $x = y_1; x_2 = y_2$ and $x_3 = y_3; x_n \leq y_n$. This relation determines whether word w_1 or word w_2 comes first in a dictionary. For example, 'past' and 'part' $\in \Sigma^4$ and 'part' and past so in a dictionary we get 'part' before we get 'past', if we search for the word from first page without skipping. This is an example of **Lexicographic Order.**

Example 11: Let $(I_o, \leq)$ and $(I_c, \leq)$ be two posets, where I_o and I_c are sets of all odd and even integers respectively and $\leq$ is usual partial order. Show that $(I_o \times I_c, \leq)$ is a product poset of posets $(I_o, \leq)$ and $(I_c, \leq)$.

Solution: To prove that $(I_o \times I_c, \leq^*)$ is the product poset, it is sufficient to prove that $\leq^*$ is a product partial order i.e., the relation $\leq^*$ defined as $(b_1, b_2, \ldots, b_n)$

$(a, b) \times (x, y)$ if, and only if $(a \leq x)$ in I_o and $(b \leq y)$ in Ic is a partial order relation on $I_o \times I_c$. I leave it to the reader to prove this as proved in the theorem 3.

Ans.

Bounding Elements of a Poset

Some elements in a poset are of special importance for understanding certain characteristics of a poset. Whether a subset of a poset is bounded within the poset? Can we designate any element of the poset as **first** one? Can we find a **last** element of the poset? And to get answer to many such queries, it is important to know about these special types of elements of a poset.

Let $(A, \leq)$ be a poset, *An element $x \in A$ is said to be **maximal** element of A if $\exists$ no element $y \in A$ such that $x < y$ i.e., there exists no* *element in A that strictly dominates x. Similarly, An element $x \in A$ is said to be* ***minimal*** *element of A if $\exists$ no element $y \in A$ such that $y < x$ i.e, there is no element in A that strictly precedes x.*

Example 12: Consider the poset of example 9 as shown in the figure 6.1.6. Find the minimal and maximal element of the poset.

Solution: In this poset 1 is the minimal element as there is no element in the poset which precedes 1 i.e., $\exists$ no element y in the poset such that $y < 1$. Similarly, 12 is the **maximal** element as there is no element in the poset which succeeds 12 i.e., $\exists$ no element y in the poset such that $12 < y$.

Ans.

Example 13: Let $(C, \subseteq)$ be a poset, where C is the collection all non-empty subsets of a set $X = \{a, b, c\}$ and $\subseteq$ is 'set inclusion' relation defined on C. Find the maximal and minimal elements of $(C, \subseteq)$.

Solution: Here $C = \{\{a\}, \{b\}, \{c\}, \{a, b\}, \{a, c\}, \{b, c\}, \{a, b, c\}\}$, collection of all non-empty subsets of X. No elements of C are strictly (properly) contained in either $\{a\}$ or $\{b\}$ or in $\{c\}$, so all the singletons are **minimal** elements of poset $(C, \subseteq)$. Similarly, no element of C strictly (properly) contains $\{a, b, c\}$, so $\{a, b, c\}$ is maximal element of $(C, \subseteq)$.

Ans.

Example 14: Let $(1, \leq)$ be a poset, where I is the set of all integers and $\leq$ is the usual partial order. In this poset there is no minimal element because for any integer x we can always find another integer $(x - 1)$ such that $x - 1 < x$. Similarly, there is no maximal element in this poset.

Ans.

Theorem 4: If $(A, \leq)$ is a finite non-empty poset, then A has at least one maximal and at least one minimal element.

Proof: Let $|A| = n$, where n is finite number. Let x_1 be any element of a. If x is not maximal element, then we can find an element x_2, such that $x_1 < x_2$. Again, if x_2 is not maximal, we can find another element x_3 in A such that $x_1 < x_2 < x_3$. If we proceed in this way, the sequence may look like $x_1 < x_2 < x_3, \ldots x_n$, which cannot be extended beyond n. Therefore, the element that is not strictly succeeded by an element, i.e. which appears in the end of the sequence, is the maximal element of poset.

Similarly, we can prove that there exist at least one minimal element in a finite poset.

Proved.

Let (A, ≤) be a poset. *An element $x \in A$ is said to be greatest or largest element of A if $\forall\ y \in A\ y \leq x$, i.e., every element in A precedes x. Similarly, An element $x \in A$ is said to be least or smallest element of A if $\forall\ y \in A\ x \leq y$ i.e., every element in A succeeds x.*

Example 15: Let $(A, \leq)$ be a poset, where A is the set of all non-negative real numbers and ≤ is usual partial order on A. Since $0 \in A$ and $0 \leq x\ \forall\ x \in A$, 0 is the least element of A. There is no greatest element in A, because for any real number x we can always find another real number y such that $x < y$.

Ans.

Example 16: Consider the poset in example 9. Since the element 1 is succeeded by all the elements of the poset, *I* is the least element. Similarly, element 12 succeeds every element of the poset so 12 is the greatest element. Now look at the following posets obtained by removing 1, 12 and both 1 and 12 from the poset of example 9 and shown in figures 6.1.7, 6.1.8 and 6.1.9 respectively.

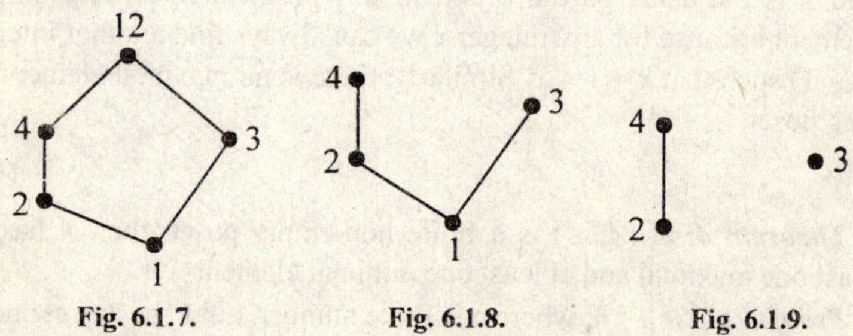

Fig. 6.1.7. **Fig. 6.1.8.** **Fig. 6.1.9.**

In the poset of figure 6.1.7, 12 is the greatest element but there is no least element. Neither 2 succeeds 3 nor 3 succeeds 2 in this poset. However, both 2 and 3 are minimal elements. In the poset of figure 6.1.8, 1 is the least element but there is no greatest element. However, there are two maximal elements 3 and 4 in this poset. Finally in figure 6.1.9, we have a poset in which there are is no least and no greatest element. However, there two minimal elements 2 and 3; two maximal elements 3 and 4. This is an example of a poset in which an element can be both minimal and maximal.

Ans.

Example 17: Let $(P(A), \leq)$ be a poset where A is any non-empty finite set. Find the least and the greatest element of $(P(A), \subseteq)$.

Solution: Since $f \in P(A)$ and every element of $P(A)$ contains ϕ. ϕ is the least element. Similarly $A \in P(A)$ and every element of $P(A)$ are contained in A, A itself is the greatest element of $P(A)$.

Ans.

Theorem 5: A poset has at the most one greatest and at the most one least element.

Proof: Let $(A, \leq)$ be a poset. Suppose the poset A has two least elements x and y. Since x is the least element $x \leq y$. Using the same argument, we can say that $y \leq x$, since y is supposed to be another least element of the same poset. $\leq$ is an anti-symmetric relation, so $x \leq y$ and $y \leq x \Rightarrow x = y$. Thus, there can be at the most one least element. Similarly, there can be at the most one greatest element.

Proved.

It is mentioned that if $(A, \leq)$ is a poset then $(A, \geq)$ is its dual poset. It is easy to infer from this ordering that maximal (minimal) element of a poset is minimal (maximal) element of its dual. Similarly, least (greatest) element of a poset is greatest (least) element of its dual. Further, it is important to note that there may not be any greatest element in a poset, however, if it exists, it is called **Unit** element of the poset and is denoted by 1. The least element of a poset is called **Zero** element and is represented by **0**.

Let us consider the following poset $(A, \leq)$, where $A = \{a, b, c, d, e, f\}$ and $\leq$ is shown by the Hasse diagram in the figure 6.1.10. Let $B = \{a, b\}$ and $C = \{c, d, e\}$ be two subsets of A. In the diagram, there is no element in A, which precedes either a or b of B. However, there are **four** elements c, d, e, f, which succeed both a and b of subset B. This illustration is to convey that there are elements in a poset that forms lower bounds (preceding elements) and there are some that form upper bound (succeeding elements) of a subset. Let us define these terms now.

Let $(A, \leq)$ be a poset. Let B be any non-empty subset of A. *An element $x \in A$ is said to be Lower bound of B if $\forall\ y \in B\ x \leq y$ i.e., every element of B is preceded by x.* Similarly, *An element $x \in A$ is said to be* ***Upper bound*** *of B if $\forall\ y \in B\ y \leq x$* is *i.e., every element of B is succeeded by x.*

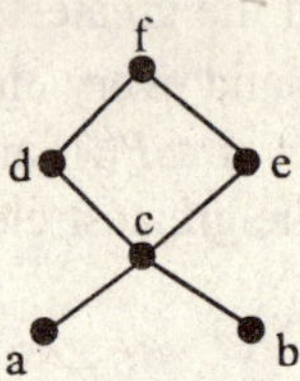

Fig. 6.1.10.

In this example, subset B has no lower bound i.e., **Lower Bound** $(B) = \phi$. **Upper Bound** $(B) = \{c, d, e, f\}$. Let us take subset C now. C has three elements; c, d and e. There are elements a and b in A that precede all these elements of C. Besides a and b, there is one element c that precedes c, d and e. Thus, **Lower Bound** $(C) = \{a, b, c\}$. So far as Upper bound is concerned, there exists only one element $f \in A$, which succeeds every element of C. Thus ϕ **Upper Bound** $(C) = \{f\}$.

In the set of lower bounds of a subset, there may be an element that is the greatest among all of the elements. This element is called ***greatest lower bound (glb)*** *or* ***infimum*** and is defined as:

"An *element* $a \in (A, \leq)$ is **glb** of a non-empty sunset B of A if, and only if

- *a is a lower bound of B and*
- *If there exists any b which is lower bound of B, then $b \leq a$.*"

In the continuing example based on the figure 6.1.10, the **glb**(B) = NIL, since there is no lower bound so there is no question of any greatest among them. However, **glb**$(C) = c$, because c is the greatest element in $\{a, b, c\}$. Likewise, in the set of upper bounds of a subset, there may be an element that is least among all of the elements. This element is called ***least upper bound (lub) or supremum*** and is defined as:

"An *element* $a \in (A, \leq)$ is **lub** of a non-empty subset B of A if, and only if

- *a is upper bound of B and*
- *If there exists any b which is upper bound of B, then* $a \leq b$."

In the poset of the figure 6.1.10 **lub**(B) = c and **lub**(C) = f.

Example 18: Let $(R, \leq)$ be a poset, where R is the set of real numbers and $\leq$ is the usual partial order. Let $B = \{x \mid x$ is a real number such that $1 < x < 2\}$. Find the **lub** and **glb** of B in R.

Solution: Here Lower bound $(B) = \{x \mid x$ is a real such that $x \leq 1\}$ and Upper bound $(B) = \{x \mid x$ is real such that $x \geq 2\}$. The greatest element in Lower bound (B) is 1 and least element in Upper Bound (B) is 2. Therefore, **glb** (B) = 1 and **lub** (B) = 2.

Ans.

Theorem 6: If $(A, \leq)$ is a poset, then any subset B at A has at the most one **lub** and at the most one **glb**.

Proof: Let us first take the case of **lub**. A subset B may not have any upper bounds and hence no **lub**, If it has only one upper bound, then that will be the **lub**. In case of many upper bounds, we have to show that if there is any greatest then it is all the most one only. The proof is similar to that in theorem 5.

Proved.

Isomorphic Posets

Let $(X, \leq)$ and $(Y, \leq)$ be two posets. A function $f: X \rightarrow Y$ is said to be isomorphism from X to Y, if

- f is one to one onto function, and
- For any $a, b \in A$, $a \leq_x b \Leftrightarrow f(\text{a}) \leq_y f(b)$, i.e. images preserve the order of pre-images.

If f is an isomorphism, we say that $(X, \leq)$ and $(Y, \leq)$ are **isomorphism** posets.

Example 19: Let $(X, \leq)$ and $(Y, \leq)$ be two posets, where X is the set of all integers and Y be the set of all even integers. The partial order $\leq$ is the usual partial order. Let $f: X \rightarrow Y$ be a function defined by $f(x) = 2x$. Show that $(X, \leq)$ and $(Y, \leq)$ are isomorphic posets.

Solution: To prove that the two posets $(X, \leq)$ and $(Y, \leq)$ are isomorphic, it is sufficient to prove that $f: X \to Y$ is an isomorphism. Let x and y be any two elements of X, then $x \neq y \Leftrightarrow 2x \neq 2y \Leftrightarrow f(x) \neq f(y)$. Thus, f is **one to one.** Let z be any element of Y, then there exists an integer $x \in X$ such that $z = 2x$, since Y is the set of all even integers, i.e. for every even integer z in Y, $\exists$ an integer x in X such that $f(x) = 2x = z$. Hence f is onto also

Next, $x \leq y \Leftrightarrow 2x \leq 2y \Leftrightarrow f(x) \leq f(y)$. This implies that images preserve the order. Therefore, $f: X \to Y$ is an isomorphism and hence $(X, \leq)$ and $(Y, \leq)$ are isomorphic posets.

Proved.

Theorem 7: Let $((X, \leq)$ and $(Y, \leq)$ are isomorphic posets under the isomorphism $f: X \to Y$, then show that

- If x is a minimal (maximal) element of the poset $(X, \leq_x)$, then $f(x)$ is a minimal (maximal) element of $(Y, \leq_y)$.
- If $f(x)$ is the greatest (least) element of $(X, \leq_x)$, then $f(x)$ is the greatest (least) element of $(Y, \leq_y)$.
- If x is an upper bound (lower bound, lub, glb) of a subset A of $(X, \leq_x)$, then $f(x)$ is an upper bound (lower bound, lub, glb) of subset $f(A)$ of $(Y, \leq_y)$.
- If every subset of $(X, \leq_x)$ has a lub (glb), then every subset of $(Y, \leq_y)$ has a lub(glb).

Proof: Since $(X, \leq_x)$ and $(Y, \leq_y)$ are isomorphic posets under the isomorphism $f: X \to Y$, all elements of X and theirs images in Y preserves the order under the respective partial order relation. If $x \in X$ is a minimal element of $(X, \leq_x)$, then according to the definition of 'minimal element', $\exists$ no y in X such that $y < x$.

Now, $x \in X \Rightarrow f(x) \in Y$. Let z be any element of Y, then $\exists$ an element y in X such $f(y) = z$ because f is onto. And $x \leq_x y \Leftrightarrow f(x) \leq_y f(y)$. Since there exists no y in X such that $y < x \Leftrightarrow \exists$ no $f(y)$ in Y such that $f(y) < f(x)$. This implies that $f(x)$ is a minimal element of $(Y, \leq_y)$.

Similarly other parts of the theorem can be proved. It is left as an exercise to the reader.

Proved.

Topological Sorting

Let $(A, \leq)$ be a poset. Sometimes it is required to extend the given partial order to a linear order. The *process of constructing linear order from a given partial order is called* ***Topological Sorting.*** Let us consider a poset as shown in the figure 6.1.11 (a). This poset is not a linear ordered set (since, 2 and 3 are not related, find the other pairs that are not related). The extension of partial order to a linear order simply means, an element must be designated as **first** one, a **second** element must follow it, then **third**, then **fourth**, and so on. The algorithm for topological sorting is as below.

- **Step 0:** Initialize A with the poset and SORT = ϕ.
- **Step 1:** Select a minimal element from A, say it is a. If there is only one minimal element then that will be selected, otherwise, select any one of the minimal elements.
- **Step 2:** Make $A \leftarrow A - \{a\}$ and SORT $\leftarrow$ SORT $\cup$ $\{a\}$
- **Step 3:** Repeat step 1 and step 2 until $A = \phi$

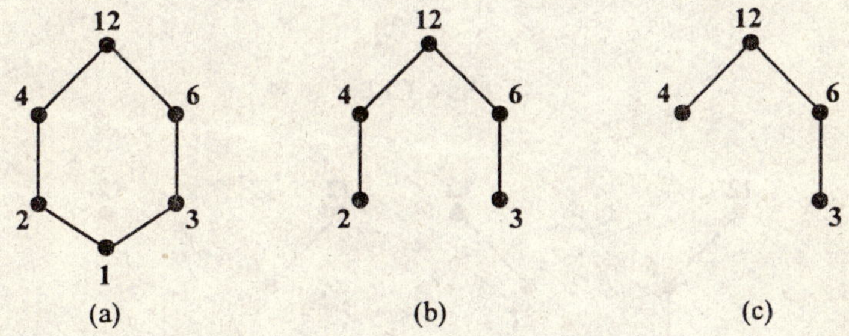

Fig. 6.1.11.

Let us consider poset of figure 6.1.11 (a) and see how this algorithm works. Step 0 initializes SORT to {} and A to {1, 2, 3, 4, 6, 12}. Step 1 picks 1 as the minimal element from A, since it is the only minimal element. Step 2 now sets A to {2, 3, 4, 6, 12} and SORT = {1}. The resulting Hasse diagram for A is shown in the figure 6.1.11 (b). Now there are two minimal elements: 2 and 3. We can pick up any one, say 2. Then Sort = {1, 2} and A = {3, 4, 6, 12}. The resulting Hasse diagram for A, now, is shown in the figure 6.1.11(c). Had we selected 3, the resulting Hasse diagram would have been as shown in the figure 6.1.12(a). Again, there are two minimal elements:

3 and 4. We select 4. It makes Sort = {1, 2, 4} and A = {3, 6, 12}. See diagram 6.1.13(a). Now there is only one minimal element 3. Selecting 3 makes SORT = {1, 2, 4, 3} and A = {6, 12}. Processing in the same way, we get SORT = {1, 2, 4, 3, 6, 12} after two iterations from now.

Diagrams for all possible cases of selection of minimal elements in different steps are given. We have obtained one linear order $1 < 2 < 4 < 3 < 6 < 12$ for the given partial order. There is many more such linear order for the same poset. Reader may try to find those.

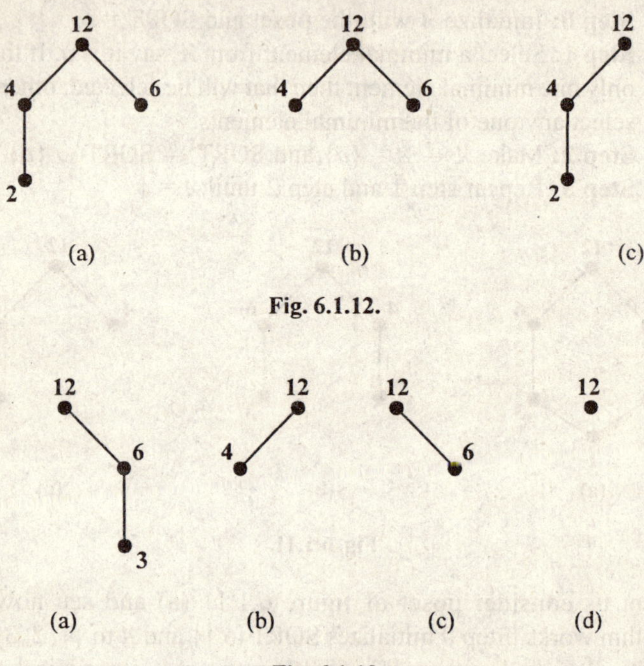

Fig. 6.1.12.

Fig. 6.1.13.

A poset $(A, \leq)$ is said to be **well ordered set** if every non-empty subset of A contains a least element. For example, $(N, \leq)$ is a well ordered set, where N is the set of all natural numbers and $\leq$ is usual partial order. The poset of figure 6.1.11(a) is **not well ordered set,** because $\exists$ a subset {2, 3} such that it has no least element. However, posets of figures 6.1.12(c), 6.1.13(a), 6.1.13(b) and 6.1.13(c) are all **well ordered.** In the next section, we will learn another type of ordered set called **Lattice** in which each pair of element will have a **glb** and a **lub.**

6.2 Lattice

Let $(L, \leq)$ be a poset. If ever, subset $\{x, y\}$ containing any two elements of L, has a **glb** (Infimum) and **lub** (Supremum), then the poset $(L, \leq)$ is called a **lattice.** A **glb** $(\{x, y\})$ is represented by $x \wedge y$ and it is called ***meet*** of x and y. Similarly, **lub** $(\{x, y\})$ is represented by $x \vee y$ and it is called ***join*** of x and y. Therefore, **lattice** is a mathematical structure equipped with two binary operations **meet** and **join**. If set L is a finite set, lattice $(L, \leq)$ is called ***finite*** lattice otherwise it is called ***infinite*** lattice.

A lattice $(L, \leq)$ is said to be a complete lattice if, and only if every non-empty subset S of L has a **glb** and a **lub**.

A non-empty subset S is called a **sublattice** of a lattice $(L, \leq)$ if $x \wedge y, \in S$ and $x \vee y \in S$ whenever $x, y \in S$ i.e., S is closed with respect to the binary operations of meet and join.

We know that join (meet) of a subset $\{x, y\}$ in a poset is a meet (join) of the same subset $\{x, y\}$ in the corresponding dual poset. If a poset $(L, \leq)$ is a lattice, then its dual $(L, \geq)$ is also a lattice called **dual lattice** of $(L, \leq)$. The direct implication of this properties is that whenever a statement is valid for a lattice $(L, \leq)$, the same statement can be made valid for its **dual lattice** by replacing **join operations** with **meet operations** and **meet operations** with **join operations.** This is referred as the **principle of duality for lattice**.

Example 1: Consider the set R of all real numbers. This set is partially ordered by the usual partial order relation $\leq$. Thus $(R, \leq)$ is a poset. If x and y be any elements of R, then $x \vee y = \max(x, y)$ and $x \wedge y = \min(x, y)$. For any subset $\{x, y\}$ containing two elements x and y of R, the join and meet are real numbers. Therefore, $(R, \leq)$ is closed with respect to join and meet and hence it is a lattice. However, $(R, \leq)$ is not a **complete** lattice since $A = \{x \in R \mid x \geq 1\}$ a subset of R has no LUB. In fact it has no upper bound. Here, $(A, \leq)$ is sub-lattice of $(R, \leq)$.

Ans.

Example 2: Consider the set I_+ of all positive integers. This set is partially ordered with relation $\leq$ defined by setting $x \leq y$ if, and only if x divides y. Thus, $(I_+, \leq)$ is a poset. If x and y are any two elements of

I_+ then $x \wedge y$ is the highest common factor (HCF) of x & y and $x \vee y$ is the least common multiple (LCM) of x & y. Both HCF and LCM of two positive integers are positive integers. Therefore, $(I_+, \leq)$ is closed with respect to join and meet and hence it is a lattice. It is however not a complete lattice.

Ans.

Example 3: Let S be a set and $L = P(S)$. The mathematical structure $(L, \subseteq)$ is a poset as it is shown in the previous section. Let A and B be any two subsets of S i.e. A and B be any two elements of $P(S)$. We can define $A \vee B = A \cup B$ and $A \wedge B = A \cap B$. The poset $(L, \subseteq)$ is closed with respect to join and meet operation i.*e*., for any subset $\{A, B\}$ of $P(S)$ there is a join and meet in $P(S)$. Thus $(L, \subseteq)$is a lattice. A Hasse diagram is shown in the figure 6.2.1 for the lattice $(L, \subseteq)$ where $I = P(S)$ and $S = \{a, b, c\}$.

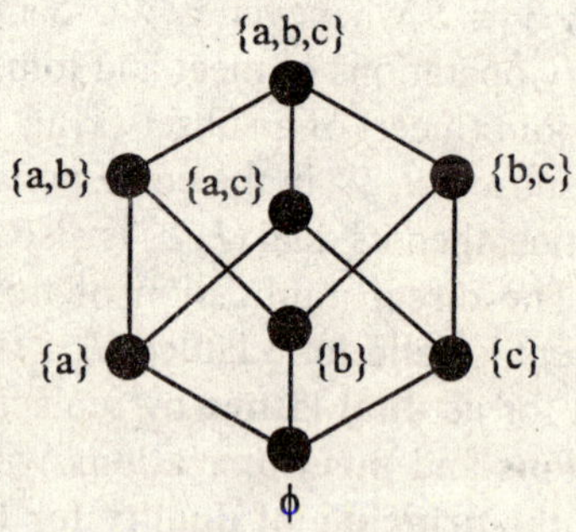

Fig. 6.2.1.

From this diagram, it is obvious than every pair of elements of $P(S)$, there is a lub (join) and a glb (meet) in the $P(S)$. A Hasse diagram is, sometimes, very handy to test whether a poset is lattice or not.

Ans.

Example 4: Let n be a positive integer and D_n be the set of all positive divisors of n. Let $\leq$ be a relation of divisibility defined on D_n. As proved in the previous section 6.1, $(D_n, \leq)$ is a poset. For any x and $y \in D_n$, the join and meet are defined as $x \vee y = \text{LCM}(x, y)$ and $x \wedge y = \text{HCF}(x, y)$. The poset $(D_n, \leq)$ is closed with respect to join and meet operations and hence (D_n, ≤) is a lattice. The figures 6.2.2(a) and 6.2.2(b) are Hasse diagrams for the lattices (D_{20}, ≤) and (D_{105}, ≤) respectively. Here D_{20} = {1, 2, 4, 5, 10, 20} and D_{105} = {1, 3, 5, 7, 15, 21, 35, 105}. A lattice (D_n, ≤) is a **sublattice** of lattice of example 2.

Ans.

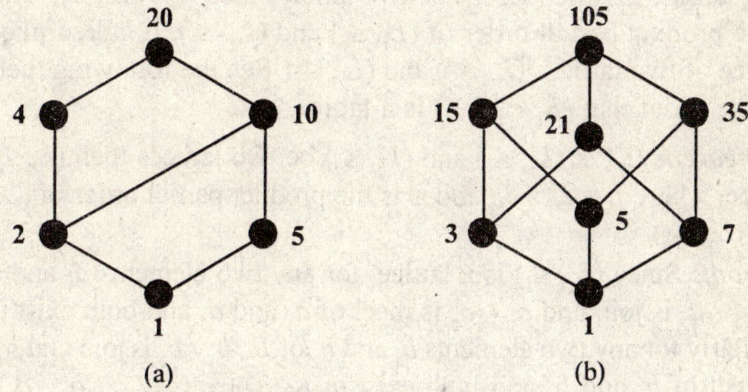

Fig. 6.2.2.

Example 5: Which of the Hasse diagrams in figure 6.2.3 represent lattices?

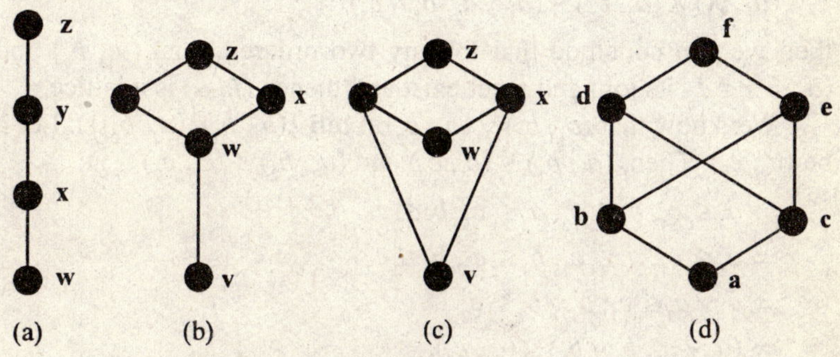

Fig. 6.2.3.

Solution: In this example, (a) and (b) are lattices, whereas (c) and (d) are not lattice. In (c), {x, y} has lower bounds {v, w}. Since v and w are not related, we cannot find greatest of the two, and hence

glb (meet) does not exist for the element x and y. In (d), $\{d, c\}$ has lower bounds $\{a, b, c\}$ and no **glb (meet)** can be determined.

Ans.

Product Lattice

Let $(L_1, \leq_1)$ and $(L_2, \leq_2)$ be two lattices then $(L_1 \times L_2, \leq)$, where $\leq$ is the product partial order of $(L_1, \leq_1)$ and $(L_2, \leq_2)$, is called **product lattice** of two lattices $(L_1, \leq_1)$ and $(L_2, \leq_2)$. See the following theorem for the proof that $(L_1 \times L_2, \leq)$ is a lattice.

Theorem 1: Let $(L_1, \leq_1)$ and $(L_2, \leq_2)$ be two lattices then $(L, \leq)$ is a lattice, where $L = L_1 \times L_2$ and $\leq$ is the product partial order of $(L_1, \leq_1)$ and $(L_2, \leq_2)$.

Proof: Since $(L_1, \leq_1)$ is a lattice, for any two elements a_1 and a_2 of L_1, $a_1 \vee a_2$ is join and $a_1 \wedge a_2$ is meet of a_1 and a_2 and both exist in L_1. Similarly for any two elements b_1 and b_2 of L_2, $b_1 \vee b_2$ is join and $b_1 \wedge b_2$ is meet of b_1 and b_2 and both exist in L_2. Thus, $(a_1 \vee a_2, b_1 \vee b_2)$ and $(a_1 \wedge a_2, b_1 \wedge b_2) \in L$. Also (a_1, b_1) and $(a_2, b_2) \in L$ for any $a_1, a_2 \in L_1$ and any $b_1, b_2 \in L_2$.

Now if we prove that

$$(a_1, b_1) \vee (a_2, b_2) = (a_1 \vee a_2, b_1 \vee b_2) \text{ and}$$
$$(a_1, b_1) \wedge (a_2, b_2) = (a_1 \wedge a_2, b_1 \wedge b_2)$$

then we can conclude that for any two ordered pairs (a_1, b_1) and $(a_2, b_2) \in L$, its join and meet exist and thence $(L, \leq)$ is a lattice.

We know that $(a_1, b_1) \vee (a_2, b_2) =$ **lub** $\{(a_1, b_1), (a_2, b_2)\})$. Let it be (c_1, c_2). Then, $(a_1, b_1) \leq (c_1, c_2)$ and $(a_2, b_2) \leq (c_1, c_2)$

$$\Rightarrow a_1 \leq_1 c_1, b_1 \leq_2 c_2; a_2 \leq_1 c_1, b_2 \leq_2 c_2$$
$$\Rightarrow a_1 \leq_1 c_1, a_2 \leq_1 c_1, b_1 \leq_2 c_2, b_2 \leq_2 c_2$$
$$\Rightarrow a_1 \vee a_2 \leq_1 c_1; b_1 \vee b_2 \leq_2 c_2$$
$$\Rightarrow (a_1 \vee a_2, b_1 \vee b_2) \leq (c_1, c_2)$$

This shows that $(a_1 \vee a_2, b_1 \vee b_2)$ is an upper bound of $\{(a_1, b_1), (a_2, b_2)\}$ and if there is any other upper bound (c_1, c_2) then $(a_1 \vee a_2, b_1 \vee b_2) \leq (c_1, c_2)$. Therefore, **lub** $(\{a_1, b_1), (a_2, b_2)\}) = (a_1 \vee a_2, b_1 \vee b_2)$ i.e.

$$(a_1, b_1) \vee (a_2, b_2) = (a_1 \vee a_2), (b_1 \vee b_2)$$

Similarly, we can show for the meet.

Example 6: Let $L_1 = \{a, b\}$ and $L_2 = \{x, y, z, w\}$ and lattices $(L_1, \leq_1)$ and $(L_2, \leq_2)$ are given by the Hasse diagrams of the figure 6.2.4(a) and 6.2.4(b) respectively. Find the lattice $((L_1 \times L_2, \leq)$ where $\leq$ is the product partial order.

Solution: Here $L_1 \times L_2 = \{((a, x), (a, y), (a, z), (a, w), (b, x), (b, y), (b, z), (b, w)\}$. The Partial order relation $\leq$ on $L_1 \times L_2$ is defined as product partial order. The resulting product poset has 8 elements (ordered pairs) and the order relationships among them is shown as in the Hasse diagram of the figure 6.2.4(c). We have $(a, x) \leq (b, x) \mid$ since $a \leq_1 b$ and $x \leq_2 x \mid$; $(a, x) \leq (a, y) \mid$ since $a \leq_1 a$ and $x \leq_2 y \mid$; $(a, x) \leq (b, z) \mid$ since $a \leq_1 b$ and $x \leq_2 z \mid$; and so on. An arc for $(a, x) \leq (b, z)$ is not shown in the diagram as it is transitively implied by the presence of relationships $(a, x) \leq (a, z)$ and $(a, z) \leq (b, z)$.

Ans.

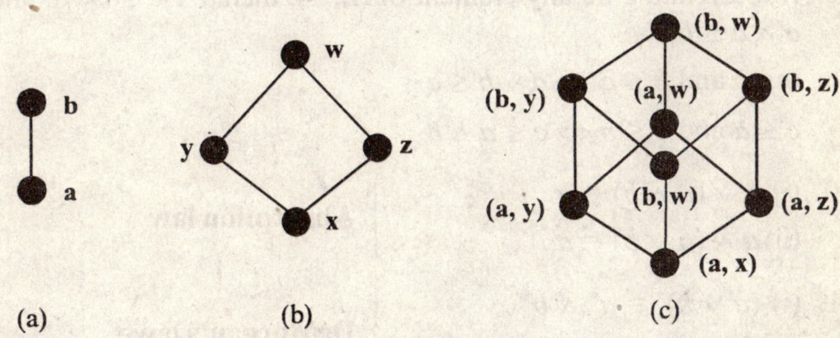

Fig. 6.2.4.

Properties of Lattice

Let $(L, \leq)$ be a lattice and a, b, c and d be any elements of L. The following properties hold good in a lattice. The proof of some of the properties are given at the end. Rest are left as an exercise to the reader.

1. $a \leq a \vee b$ and $b \leq a \vee b$ i.e. $a \vee b$ is an upper bound of a and b. Also for any element c, if $a \leq c$ and $b \leq c$ then $a \vee b \leq c$. This implies that $a \vee b$ is least upper bound of a and b. Recall the definition of join of two elements a and b.
2. $a \wedge b \leq a$ and $a \wedge b \leq b$ i.e. $a \wedge b$ is a lower bound of a and b. Also for any element c, if $c \leq a$ and $c \leq b$ then $c \leq a \wedge b$. This

implies that $a \wedge b$ is greatest lower bound of a and b. Recall the definition of meet of two elements a and b.

3. (i) $a \vee b = b$ if, and only if $a \leq b$
   (ii) $a \wedge b = a$ if, and only if $a \leq b$
   $\Rightarrow$ **a ∧ b = a if, and only if a ∨ b = b**

4. (i) $a \vee a = a$
   (ii) $a \wedge a = a$
   Idempotent law

5. (i) $a \vee b = b \vee a$
   (ii) $a \wedge b = b \wedge a$
   Commutative law

6. (i) $a \vee (b \vee c) = (a \vee b) \vee c$
   (ii) $a \wedge (b \wedge c) = (a \wedge b) \wedge c$
   Associative law

7. If $a \leq b$ and c be any element of $(L, \leq)$, then $a \vee c \leq b \vee c$ and $a \wedge c \leq b \wedge c$.

8. $a \leq c$ and $b \leq c \Leftrightarrow a \vee b \leq c$

9. $c \leq a$ and $c \leq b \Leftrightarrow c \leq a \wedge b$

10. (i) $a \vee (a \wedge b) = a$
    (ii) $a \wedge (a \vee b) = a$
    Absorption law

11. (i) $(a \vee b)' = a' \wedge b'$
    (ii) $(a \wedge b)' = a' \vee b'$
    DeMorgan's law

12. If $a \leq b$ and $c \leq d$ then $c \leq a \wedge b$, then $a \vee c \leq b \vee d$ and $a \wedge c \leq b \wedge d$.

Proof: Here, properties 3(i), 3(ii), 6(i) and 11(i) have been proved. Other properties can be proved easily using almost similar arguments, so is left as exercise to the reader.

3 (i) Let $a \vee b = b$. Since $a \vee b$ is an upper bound of a, $a \leq a \vee b$. Therefore, $a \leq b$.

Next, let $a \leq b$. Since $\leq$ is a partial order relation, $b \leq b$. Thus, $a \leq b$ and $b \leq b$ together implies that b is an upper bound of a and b. We know that $a \vee b$ is least upper bound of a and b, so $a \vee b \leq b$. Also $b \leq a \vee b$ because $a \vee b$ is an upper bound of b. Therefore, $a \vee b \leq b$ and $b \leq a \vee b \Leftrightarrow a \vee b = b$ by the anti-symmetry property of partial order relation $\leq$. Hence, it is proved that $a \vee b = b$ if and only if $a \leq b$.

3 (ii) Let $a \wedge b = a$. Since $a \wedge b$ is a lower bound of b, $a \wedge b \leq b$. Therefore, $a \leq b$.

Next, let $a \leq b$. Since $\leq$ is a partial order relation, $a \leq a$. Thus, $a \leq b$ and $a \leq a$ together implies that a is a lower bound of a and b. We know that $a \wedge b$ is greatest lower bound of a and b, so $a \leq a \wedge b$. Also $a \wedge b \leq a$ because $a \wedge b$ is a lower bound of a. Therefore, $a \wedge b \leq a$ and $a \leq a \wedge b \Leftrightarrow a \wedge b = a$ by the anti-symmetry property of partial order relation $\leq$. Hence, it is proved that $a \wedge b = a$ if and only if $a \leq b$.

6 (i) Here we have to prove that $a \vee (b \vee c) = (a \vee b) \vee c$.
Let $a \vee (b \vee c) = x$ and $(a \vee b) \vee c = y$. Since $a \vee (b \vee c)$ is join (least upper bound) of a and $(b \vee c)$, we have $a \leq x$ and $b \vee c \leq x$. Also $b \vee c$ is join of b and c, so $b \leq b \vee c \leq x$ and $c \leq b \vee c \leq x$. Therefore, $a \leq x$, $b \leq x$ and $c \leq x$. Since $a \vee b$ is join of a and b, we have $a \vee b \leq x$. Similarly, $(a \vee b) \vee c$, so $(a \vee b) \vee c \leq x$. We have assumed that $(a \vee b) \vee c = y$, thus $y \leq x$.

Using the same argument, we can show that $x \leq y$. Since $\leq$ is an anti-symmetric relation, $y \leq x$ and $x \leq y$ together implies that $x = y$. Hence proved.

11 (i) We have to prove that $(a \vee b)' =$ '$a \wedge b$' where a' and b' are complements of a and b respectively. To prove that $(a \vee b)' = a' \wedge b'$, we simply prove that $(a' \wedge b')$ is complement of $(a \vee b)$ i.e.,

$$(a \vee b) \vee (a' \wedge b') = I \text{ and}$$

$$(a \vee b) \wedge (a' \wedge b') = 0$$

We have
$$\begin{aligned}(a \vee b) \vee (a' \wedge b') &= [(a \vee b) \vee a'] \wedge [(a \vee b) \vee b'] \\ &= [(a \vee a') \vee b] \wedge [(a \vee (b \vee b')] \\ &= [I \vee b] \wedge [a \vee I] \\ &= I\end{aligned}$$

and
$$\begin{aligned}(a \vee b) \wedge (a' \wedge b') &= [(a \wedge (a' \wedge b')] \vee [b \wedge (a' \wedge b')] \\ &= [(a \wedge a') \wedge b'] \vee [a' \wedge (b \wedge b')] \\ &= [0 \wedge b'] \vee [a' \wedge 0] \\ &= 0\end{aligned}$$

This shows that $(a' \wedge b')$ is complement of $(a \vee b)$. Hence proved.

This completes the answer.

Types of Lattice

There are some more structured classes of mathematical structures that are very useful in computer science. This gradual structuring of a mathematical structure, called lattice, leads us to the concept of **Finite Boolean Algebra**, which we shall learn about in the next section of this chapter.

A lattice $(L, \leq)$ is said to be a **bounded** if it has a greatest element **1** and a least element **0**.

Example 7: Let $(I_+, 1)$ be a lattice, where I_+, is a set of all positive integers and 'I' is the relation of divisibility i.e., $a \mid b$ if and only if a divides b. This lattice is not bounded since it has no greatest element ***I***. Reader may note that this lattice has a least element (*zero element*), the number 1.

Ans.

Example 8: Let $(P(S), \subseteq$ be a lattice where S is any non-empty set. This lattice is bounded since its greatest element is S and its least element is 0 and both are in $P0$ $S0$.

In a bounded lattice $(L \leq)$ for any element x in L we have

(i) $0 \leq x \leq \mathrm{I}$

(ii) $x \vee 0 = x$ and $x \wedge 0 = 0$

(iii) $x \vee \mathrm{I} = \mathrm{I}$ and $x \wedge \mathrm{I} = \mathrm{I}$

A finite lattice is always bounded. Its greatest element is the **join** of all of its elements and its least element is given by the **meet** of all its elements. A Hasse diagram can be drawn for a finite poset. If the Hasse diagram corresponds to a lattice the lattice is bounded. In other words, if there is a Hasse diagram for a lattice, then the lattice is bounded.

A lattice $(L, \leq)$ is said be **distributive** if the **meet** operation distributes over join operation and join operation distributes over **meet** operation. That is, for any x, y, and of I.

1. $x \wedge (y \vee z) = (x \wedge y) \vee (x \wedge z)$
2. $x \vee (y \wedge z) = (x \vee y) \wedge (x \vee z)$

If $(L, \leq)$ is not distributive then L, is called a **non-distributive** lattice.

Example 9: The lattice $(L, \leq)$ is a distributive lattice, where $\leq$ is the usual partial order on the set of all integers 1. This lattice is a distributive lattice. It is important to note however, that it is not bounded because neither the least element nor the greatest element exists in this lattice.

Ans.

Example 10: Let $(P(S), \subseteq)$ be a lattice where S is any non-empty set. Then this lattice is a distributive lattice since union is distributive over intersection and intersection is distributive over union

Ans.

Example 11: The lattice $(D_{12}, |)$ is distributive. The lattices $(A, |)$ and $(B, |)$ where $A = \{1, 2, 3, 4, 12\}$ $B = \{1, 2, 3, 5, 30\}$ and is the partial order of divisibility are not distributive lattice as shown below. The Hasse diagrams are given in the figures 6.2.5.

Solution: The Hasse diagram of figures 6.2.5(a) is for the lattice $(D_{12}, |)$.

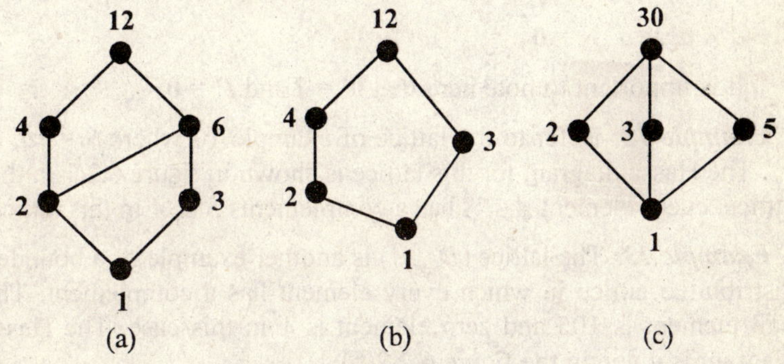

Fig. 6.2.5.

It can be easily verified that for every three elements of $(D_{12}, |)$, the distributive properties are satisfied. However, in the case of $(A, |)$, as shown in the diagram 6.2.5(b), we have

$4 \wedge (2 \vee 3) = 4 \wedge 12 = 4$ and

$(4 \wedge 2) \vee (4 \wedge 3) = 2 \vee 1 = 2$

Since $4 \neq 2$, the meet is not distributive over join and hence, the lattice $(A, |)$ is a non-distributive lattice. This lattice is bounded with zero element 1 and Unit element 12. in the case of $(B, |)$ as shown in the diagram 6.2.5 (c), we have

$2 \wedge (3 \vee 5) = 2 \wedge 30 = 2$ and

$(2 \wedge 3) \vee (2 \wedge 5) = 1 \vee | = 1$

Since $1 \neq 2$, the meet is not distributive over join and hence, the lattice $(B, |)$ is a non-distributive lattice. This lattice, like $(A, |)$, is bounded with zero element 1 and Unit element 30.

Ans.

Let $(L, \leq)$ be a bounded lattice having unit element $\boldsymbol{I}$ and zero element $\mathbf{0}$. Sometimes a zero element of a lattice is also called *universal lower bound* and unit element is called *universal upper bound* of the lattice. Let $a \in L$, be any element. An element a of L is said to be complement of a in L, if

$a \vee a' = a' \vee = I$ and

$a \wedge a' = a' \wedge = \mathbf{0}$

It is important to note here that $\mathbf{0'} = \boldsymbol{I}$ and $\boldsymbol{I'} = \mathbf{0}$.

Example 12: Refer to the lattice of example 10 where $S = \{a, b, c\}$. The Hasse diagram for this lattice is shown in figure 6.2.1.In this lattice, every element $A \subseteq S$ has a complements $S \subseteq A$ in this lattice.

Example 13: The lattice $(D_{105}, |)$ is another example of a bounded distributive lattice in which every element has a complement. The unit element is 105 and zero element is 1 in this case. The Hasse diagram is given in the figure 6.2.2(b).

Example 14: Refer to the lattice of figure 6.2.5(a). It has an element 6 such that it has no complement. It can be tested that $6 \wedge 2 = 2$, $6 \vee 2 = 6$; $6 \wedge 4 = 2$, $6 \vee 4 = 12$; $6 \wedge 3 = 3$, $6 \vee = 6$, . But for any element b to be complement of 6 we must have $6 \wedge b = 1$, $6 \vee b = 12$. There exists no such b in this lattice. Similarly, element 2 does not have a complement in this lattice. In the case of the lattice of figure 6.2.5 (b), every element has a complement. In fact the element 3 has two complements: 2 and 4. Finally, see the lattice of figure 6.2.5 (c). Here, every element except unit element (12) and zero element (1), has two complements, 3 has two complements; 5 and 7; 5 has two complements: 3 and 7 and 7 has two complements: 3 and 5.

It is interesting to note from the above examples that in a bounded distributive lattice if a complement exists it is unique whereas in a non-distributive lattice an element may have more than one complements. This is true not only in a particular case but is essentially valid for all bounded and distributive lattice as stated in the following theorem.

Theorem 2: Let $(L, \leq)$ be a bounded distributive lattice. Prove that if a complement for an element $a \in L$ exists, then it is unique.

Proof: Let I and 0 are the unit and zero elements of L repectively. Let b and c be two complements of element $a \in L$. Then from the definition, we have

$a \wedge b = 0 = a \wedge c$ and

$a \vee b = I = a \vee c$

We can write $b = b \vee 0 = b\ (a \wedge c)$

$= (b \vee a) \wedge (b \vee c)$ [Since lattice is distributive]

$= I \wedge (b \vee c))$

$= (b \vee c)$

Similarly, $c = c \vee 0 = c \vee (a \wedge b)$

$= (c \vee a) \wedge (c \vee b)$ [Since lattice is distributive]

$= I \wedge (b \vee c)$ [Since $\vee$ is a commutative operation]

$= (b \vee c)$

The above two results show that $b = c$.

Proved.

A lattice $(L, \leq)$ is said to be ***complemented*** lattice if it is bounded and if every element of L has a complement in L. Lattices of example

12 and 13 are examples of complemented lattice as they are bounded and every element has a complement in the respective lattice. Both these lattices are distributive. A non-distributive lattice may also be complemented. Lattices of figures 6.2.5(b) and 6.2.5(c) are the examples of bounded non-distributive and complemented lattice. A bounded, distributive and complemented lattice is of special concern in computer science and is known as ***finite Boolean algebra***. Before taking up this topic, let us know about isomorphic lattice.

Let $(L_1, \leq_1)$ and $(L_2, \leq_2)$ be two lattices. They are said to be *isomorphic* to each other if they are isomorphic as posets i.e. if there exists a function f: $L_1 \rightarrow L_2$ such that

- f is one to one onto and
- It preserves the partial order i.e., for $a \leq_1$ b $\Leftrightarrow f((a) \leq_2 f(b)$

If $(L_1, \leq_1)$ is isomorphic to a lattice $(L_2, \leq_2)$, then $(L_1, \leq_1)$ possesses all the properties of $(L_2, \leq_2)$. For example L_2 is bounded iff L_1 is bounded, L_2 is distributive iff L_1 is distributive, L_2 is complemented iff L_1 is complemented and so on. In general, any formula that is valid in L_1 can be made valid in L_2 by replacing elements of L_1 in the formula by its corresponding images in L_2. Isomorphic lattices have identical Hasse Diagrams.

6.3 Finite Boolean Algebra

Let $B = \{0, 1\}$. This is the set of Boolean symbols: 1 for **true** and 0 for **false**. Then, $B \times B = \{(0, 0), (0, 1), (1, 0) (1, 1)\}$ if we write (0, 0) as 00, (0, 1) as 01 and so on and $B \times B$ as B_2 then, we h a v e $B_1 = \{0, 1\}$; $\{B_2 = 00, 01, 10, 11\}$;... and $B_n = \{x \mid x$ is a binary string of length $n\}$. Also $|B_n| = 2^n$. Let $\leq$ be a partial order relation on B_n defined as, if $x = a_1\ a_2\ a_3\ ....a_n$ and $y = b_1\ b_2 \cdot b_3 ...\ b_n$, where a_1's and b_i's are either 0 or 1.

1. $x \leq y$ if and only if $a_k \leq b_k$ for $k = 1, 2, 3, ..., n$. For example, $0101 \leq 0111$ because for $k = 1$: $0 \leq 1$; for $k = 2$: $1 \leq 1$; for $k = 3$; $0 \leq 1$; and for $k = 4$: $1 \leq 1$. The two strings 0101 and 1000 are not related since neither $0101 \leq 1000$ nor $1000 \leq 0101$. Also every n-bit long string is related to itself i.e., for all x, $x \leq x$.

2. $x \wedge y = c_1\, c_2\, c_3 \ldots\, c_n$ where $c_k = \min(a_k, b_k)$, e.g., 0101 $\wedge$ 0110 = 0100 and 1001 $\wedge$ 1101 = 1001. Notice that here meet ($\wedge$) is a Boolean AND operation. .
3. $x \vee y = c_1\, c_2\, c_3 \ldots\, c_n$ where $c_k = \max(a_k - b_k)$, e.g, 0101 $\wedge$ 0110 = 0111 and 1001 $\vee$ 1101 = 1101. Notice here too that join ($\vee$) is a Boolean OR operation.
4. x has a complement $x = c_1\, c_2\, c_3 \ldots\, c_n$ where $c_k = 0$ if $a_k = 1$ and $c_k = 1$ if $a_k = 0$, e.g., (0101)' =1010; (1001) = 0110 and so on. In this case, the complement is just one's complement of the binary string.

Obviously, this relation $\leq$ is a partial order relation on B_n. Thus $(B_n, \leq)$ is a poset. Since every pair of elements has a join and a meet as defined above, $(B_n, \leq)$ is a lattice. This is bounded because it is a finite lattice. Since Boolean AND) is distributive over Boolean OR and Boolean OR is distributive over Boolean AND, the join distributes over meet and the meet distributes over join in this lattice. (Refer to the definition of join and meet above). Therefore, $(B_n, \leq)$ is a distributive lattice. The one's complement of n bit long string is again a n-bit long string, every element in $(B_n, \leq)$ has a complement. Hence, $(B_n, \leq)$ is a ***finite Boolean algebra*** since it is a bounded, distributive and complemented lattice. The figure 6.3.1 shows the Hasse diagrams of $(B_n, \leq)$ for $n = 1$, 2 and 3.

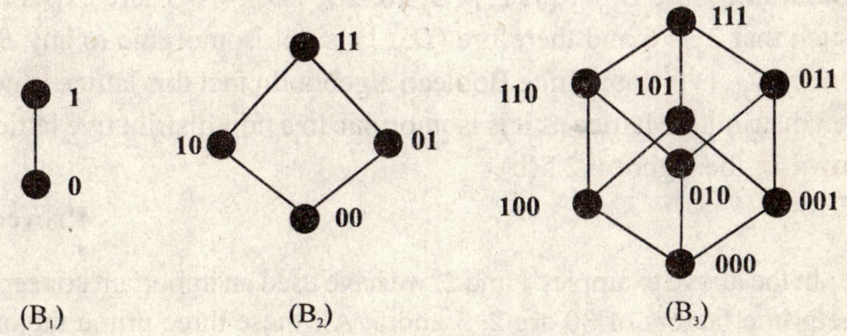

Fig. 6.3.1.

Let S be any set with $|S| = n$ then $|P(S)| = 2^n$. Thus we can always find a one to one onto function from $P(S)$ to B_n. We know that $(P(S), \subseteq)$ is a bounded, distributive and complemented lattice and hence $(P(S), \subseteq)$ is a ***finite Boolean algebra.*** See the Hasse diagram in the figure

6.2.1. If we replace each subset by its binary representation, we get the Hasse diagram of B_3 as shown in the figure 6.3.1. Therefore, for any set S such that $|S| = 3$, $(P(S), \subseteq)$ is isomorphic B_3. In general, lattice $(P(S), \subseteq)$ is isomorphic to $B_{|S|}$. Since isomorphic lattices possess same properties, if one is a finite Boolean algebra then other is also a finite Boolean algebra. A finite Boolean algebra may be defined in terms of B_n as "*any finite lattice is said to be a finite Boolean algebra if and only if it is isomorphic to some* B_n."

Example 1: Show that D_{30} is a finite Boolean algebra with the partial order of divisibility.

Solution: Here $D_{30} = \{1, 2, 3, 4, 5, 6, 10, 15, 30\}$. $|D_{30}| = 8 = |B_3|$. Let us define a function $f: D_{30} \rightarrow B_3$ as $f(1) = 000, f(2)\ 100$, $f(3) = 010, f(5) = 001, f(6) = 110, f(10) = 101, f(15) = 011$ and $f(30) = 111$. Now a Hasse diagram for $(D_{30}, |)$ can be obtained by just replacing the corresponding pre-images in the Hasse diagram of B_3 in figure 6.3.1. This shows that $(D_{30}, |)$ is isomorphic to B_3. Hence $(D_{30}, |)$ is a finite Boolean algebra.

Proved.

Example 2: Show that D_{20} is not a finite Boolean algebra with the partial order of divisibility.

Solution: Here $D_{20} = \{1, 2, 4, 5, 10, 20\}$ $|D_{20}| = 6$ There exists no n such that $2_n = 6$ and therefore $(D_{20}, |)$ is not isomorphic to any B_n. Hence $(D_{20}, |)$ is not a finite Boolean algebra. In fact this lattice is not even distributive lattice as it is isomorphic to a non-distributive lattice shown in the figure 6.2.5(b).

Proved.

In the above examples 1 and 2, we have used an important concept. The prime factors of 30 are 2, 3 and 5. All these three prime factors are distinct. All divisors of 30 are product of one or more of these primes. For example, 6 is the product of 2 and 3, 10 is the product of 2 and 5, 15 is the product of 3 and 5 and 30 itself is the product of 2, 3 and 5. Whereas 20 can be written as $2 \times 2 \times 5$. All these prime factors are not distinct. The concept has been summarized in the following theorems.

Theorem 1: Let $n = p_1 \times p_2 \times p_3 \times \ldots \times p_m$; where $p_1, p_2, p_3, \ldots p_m$ are all distinct primes. Prove that D_n is a Boolean algebra.

Proof: Let $S = \{p_1, p_2, p_3, \ldots p_m\}$. $P(S)$ is the collection of all subsets of S including Null set. The product of all primes in a subset of S is a divisor of n. This is true for all subsets if we take 1 corresponding to the $\phi \subseteq S$. There is no divisor for n other than those obtained from the subsets of S. Since there are 2^m subsets of S, there will be 2^m distinct divisions of n. Therefore.

$$\begin{aligned} D_n &= \{x \mid x \text{ is a divisor of } n\} \\ &= \{x \mid x \text{ is the product of elements of subsets of } S \text{ and } x = 1 \text{ for } \phi\} \end{aligned}$$

Here, $|D_n| = 2^m$ thus $(D_n, |)$ is isomorphic to B_m where m is a positive integer. Therefore, $(D_n, |)$ is finite Boolean algebra.

Proved.

Theorem 2: Let n be a positive integer greater than 1. If in the prime factorization of n. all the factors are not distinct i.e. p^2 divided n where p is a prime number, then prove that D_n is not a finite Boolean algebra.

Proof: Let us suppose that D_n is a finite Boolean algebra. Then it is bounded and complemented lattice according to the definition of a finite Boolean algebra. Since p^2 divides n, we can write $n = p^2 q$ for some integer q. Also, $P \in D_n$ because p is a divisor of n. This implies that p has a complement p' in D_n such that LCM$(p, p') = n$ and HCF $(p, p') = 1$. Both p and p' are divisors of n and p is a prime number, we have

$$\begin{aligned} \text{LCM } (p, p') = n &\Rightarrow pp' = n \\ &\Rightarrow pp = p^2 \Rightarrow p' = pq \\ &\Rightarrow \text{LCM}(p, p') = pq \neq n \text{ and HCF}(p, p') = p \neq 1 \end{aligned}$$

This is obviously a contradiction which we arrive at because our assumption that D_n is a finite Boolean algebra. Therefore D_n is **not** a finite Boolean algebra.

Proved.

Example 3: Determine whether D_n is a finite Boolean algebra where (a) $n = 12$ (b) $n = 40$ (c) $n = 75$ (*d*) $n = 21$ (e) $n = 70$?

Solution: From the theorems 1 and 2, it is obvious that if n can be written as product of distinct primes then corresponding D_n is a finite

Boolean algebra otherwise it is not. (a) Since all the prime factors of 12 (2×2×5) are not distinct, D_{12} is not a finite Boolean algebra. Similarly, D_{40} (40 = 2×2×2×5), D_{75} (75 = 3×5×5) are not finite Boolean algebra. However, D_{21} (21 = 3×7)and D_{70} (70 = 2×5×7) are finite Boolean algebra as both 21 and 70 have distinct prime factors.

Ans.

Since a Boolean algebra is essentially a lattice, a finite Boolean algebra possesses all properties possessed by a lattice. Thus, all the properties listed in 6.2 under heading **"properties of Lattice"** are valid for any finite Boolean algebra.

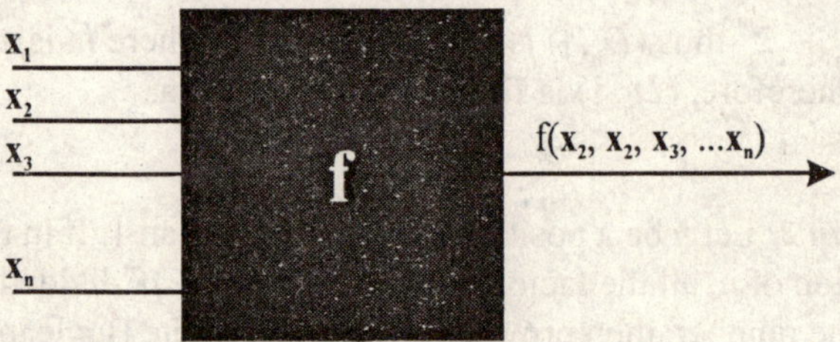

Fig. 6.3.2.

Let us consider function f: $B_n \rightarrow B$ where B_n and B are sets as explained before. Obviously this is a **Boolean function** in n Boolean variables. The schematic diagram is shown is the figure 6.3.2. If we construct a table showing every n-tuple (n-bit long binary string) of B_n and its image in B under f then the table will contain 2^n rows. Each row contains an n-tuple and its image in B under f. A table so obtained is called **truth table** for the Boolean function f.

A typical function f: $B_3 \rightarrow B$ in three Boolean variables x, y and z is given in the table 6.3.1.

Here, if we write x' as complement of x i.e. whenever x is 1 x' is 0 and whenever x is 0 x' is 1, then $f(x, y, z) = x' \wedge y' \wedge z'$. Consider a two variable function f: $B_2 \rightarrow$ as given in table 6.3.2. We can write $f(x, y) = (x' \wedge y) \vee (x' \wedge y') = x'$. Similarly, we can write $f(x, y) = (x \wedge y) \vee (x' \wedge y) = y$ for the function f: $B_2 \rightarrow B$ as given in table 6.3.3. The function f: $B_2 \rightarrow B$ as given in table 6.3.4 below, can be written as $f(x, y, z)$ $(x' \wedge y \wedge z) \vee (x \wedge y \wedge z)$.

x	y	z	$f(x, y, z)$
0	0	0	0
0	0	1	0
0	1	0	1
0	1	1	0
1	0	0	0
1	0	1	0
1	1	0	0
1	1	1	0

Table 6.3.1.

x	y	$f(x, y)$
0	0	1
0	1	1
1	0	0
1	1	0

Table 6.3.2.

x	y	$f(x, y)$
0	0	0
0	1	1
1	0	0
1	1	1

Table 6.3.3.

x	y	z	$f(x, y, z)$
0	0	0	0
0	0	1	0
0	1	0	0
0	1	1	1
1	0	0	0
1	0	1	0
1	1	0	0
1	1	1	1

Table 6.3.4.

This shows that a Boolean function can be expressed in terms of Boolean variables. Any expression that involves only Boolean variables. Boolean constants and Boolean operators is called a **Boolean expression** or **Boolean polynomial.** The Boolean operators are $\wedge$ (AND), $\vee$ (OR) and $\neg$ (NOT). $\neg$ is an unary operator whereas $\wedge$ & $\vee$ are binary operators. The operator $\neg$ is also called complement or negation.

Boolean Polynomial and Functions

A Boolean variable is defined over the set $B = \{0, 1\}$. If a Boolean function is of n variables. We can say that f is a function from B_n to B. Any Boolean function can be written as a Boolean polynomial as shown in the above examples. This we will prove as a theorem here. Before that let us define what is a Boolean polynomial (Boolean expression).

Let $x_1, x_2, x_3, \ldots, x_n$ be a set of n Boolean variables. A **Boolean polynomial** $p(x_1, x_2, x_3, \ldots, x_n)$ in a Boolean variables x_k's is defined recursively as follows:

1. $x_1, x_2, x_3, \ldots, x_n$ are all Boolean polynomial.
2. The symbols 0 and 1 (Boolean constants) are Boolean polynomial.
3. If p and q are two Boolean polynomials then $p \wedge q$ and $p \vee q$ are Boolean polynomial.
4. If p is a Boolean polynomials then **negation of** p ($\neg p$ or p') is a Boolean polynomial.
5. A Boolean polynomial can also be obtained by repeated use of any or more or all of the above four rules 1 to 4. There exists no other way to define a Boolean polynomial.

Again look at the table 6.3.2. There are four entries in the table under columns x and y. These are {(0 and 0) or $(0\wedge 0)$}, {(0 and 1) or $(0 \wedge 1)$}, {(1 and 0) or $(1 \wedge 0)$} and {(1 and 1) or $(1 \wedge 1)$}. The same entries can be written involving variable names as $(x' \wedge y')$, $(x' \wedge y)$, $(x \wedge y')$ and $(x \wedge y)$. Each of these Boolean expressions corresponding to entries in a truth table is called a **minterm**. There are eight minterms in table 6.3.1 and either in tables 6.3.4. In general, there are 2^n minterms in a truth table for a function f from B_n to B. If b is any n-tuple of B_n, then $b = (x_1, x_2, x_3, \ldots, x_n)$, where each x_i is either 0 or 1. Generally, when $x_i = 0$, we say that the variable is complemented in b and is denoted by x_i (complement of x_i) and when $x_i = 1$. We say that the variable is not complemented in b and is denoted by x_i. The minterm corresponding to n-tuple b is then given by the Boolean expression

$$z_1 \wedge z_2 \wedge\ z_3 \wedge \ldots \wedge z_n$$

where $z_i = x_i$ if $x_i = 1$ and $z_i = x_i$' if $x = 0$. Notice that a m8interm is the meet of all the variables in present form (complemented or non-complemented) in an n-tuple. Similarly, a maxterm is defined as the join of all the variables in present form in an n-tuple. The maxterm corresponding to n-tuple b is then given by the Boolean expression.

$$z_1 \vee z_2 \vee z_3 \vee \ldots \vee z_n$$

where $z_i = x_i$ if $x_i = 0$ and $z_i = x_1$' if $x = 1$.

The meet operator $\wedge$ (referred as AND) is also called **conjunction**. Therefore a minterm is also called **conjunctive**. Similarly, the join operator $\vee$ (referred as OR) is called **disjunction**. Thus, a maxterm is also called **disjunctive**. A Boolean function may be expressed in **Conjunctive Normal Form (CNF)** of Boolean polynomial or in **Disjunctive Normal Form (DNF)** of a Boolean polynomial. A Boolean polynomial is said to be in CNF. If it is expressed as the meet of those maxterms that produce output 1 for the function. Notice that a minterm that produces 0 for a function corresponding maxterm produces 1 for the function. **(minterm)'** = **(maxterm)**. For example, function $f(x, y)$ corresponding to the truth table 6.32.can be written in CNF as

$$f(x, y) = (x' \vee y) \wedge (x' \vee y')$$

And $f(x, y, z)$ corresponding to the truth table 6.3.4 can be written in CNF as

$$f(x, y, z) = (x \vee y \vee z) \wedge (x \vee y \vee z)\ (x \vee y' \vee z) \wedge (x' \vee y \vee z)$$
$$\wedge (x' \vee y \vee z') \wedge (x' \vee y' \vee z).$$

Likewise, a Boolean polynomial is said to be in DNF, if it is expressed as the join of those minterms that produce output 1 for the function. For example, function $f(x, y)$ corresponding to the truth table 6.3.2 can be written in DNF as

$$f(x, y) = (x' \vee y') \vee (x' \vee y)$$

And $f(x, y, z)$ corresponding to the truth table 6.3.4 can be written in DNF as

$$f(x, y, z) = (x' \wedge y \wedge z) \vee (x \wedge y \wedge z)$$

Example 4: Let $f: B_3 \to B$ is given as $f(x, y, z) = x' \wedge (y \vee z) \wedge (x \vee y \vee z)$. Find truth table for this function and then reduce this into CNF and DNF representation.

Solution: The truth table for the given function is shown in the table 6.3.5. This table contains three minterms producing 1 for the function and five minterms that produce 0 for the function. To represent the given function in DNF, we have to take only those minterms that produce 1 for the function. The join of these minterms is the DNF representation for f. Thus, f can be written in DNF as

$$f(x, y, z) = (x' \wedge y' \wedge z) \vee (x' \wedge y \wedge z') \vee (x' \wedge y \wedge z)$$

For the **CNF** representation, we take only those minterms that produces 0 for f. Convert the minterm to maxterm and then meet of these maxterm is the CNF representation for f. Thus, f can be written in CNF as

$$f(x, y, z) = (x \vee y \vee z) \wedge (x' \vee y \vee z) \wedge (x' \vee y \vee z) \wedge (x' \vee y' \vee z) \wedge (x' \vee y' \vee z')$$

Ans.

Let us define a set $M_l(f) = \{b \in B_n \mid f(b) = 1\}$ of all minterms that produce a 1 for a function $f: B_n \to B$. If $|M_l(f)| = 1$ then f equal to the minterm $b \in B_n$. This implies that f can be produced by a Boolean expression. In general, this is true for any Boolean function.

Theorem 3: Let f, g and h are three functions from B_n to B and $M_l(f)$, $M_l(g)$ and $M_l(h)$ are set of minterms for f, g and h respectively. Prove that

(a) If $M_l(f) = M_l(g) \cup M_l(h)$ then $f(b) = g(b) \vee h(b)\ \forall\, b \in B_n$

(b) If $M_l(f) = M_l(g) \cap M_l(h)$ then $f(b) = g(b) \wedge h(b)\ \forall\, b \in B_n$

Proof: (a) Let b be any n-tuple of B_n. If $b \in M_l(f)$ then $f(b) = 1$.

$$b \in M_l(f) \Rightarrow b \in M_l(g) \cup M_l(h)$$

$$\Rightarrow b \in M_l(g) \text{ or } b \in M_l(h)$$
$$\Rightarrow g(b) = 1 \text{ or } h(b) = 1$$
$$\Rightarrow g(b) \vee h(b) = 1$$

Now, if $b \notin M_l(f)$ then $f(b) = 0$.

$$b \notin M_l(f) \Rightarrow b \notin M_l(g) \cup M_l(h)$$
$$\Rightarrow b \notin M_l(g) \text{ and } b \notin M_l(h)$$
$$\Rightarrow g(b) = 0 \text{ and } h(b) = 0$$
$$\Rightarrow g(b) \vee h(b) = 0$$

This shows that $f(b) = g(b) \vee h(b)$

(b) Let b be any n-tuple of B_n. If $b \in M_l(f)$ then $f(b) = 1$.

$$b \in M_l(f) \Rightarrow b \in M_l(g) \cap M_l(h)$$
$$\Rightarrow b \in M_l(g) \text{ and } b \in M_l(h)$$
$$\Rightarrow g(b) = 0 \text{ and } h(b) = 0$$
$$\Rightarrow g(b) \wedge h(b) = 1$$

Now, if $b \notin M_l(f)$ then $f(b) = 0$.

$$b \notin M_l(f) \Rightarrow b \notin M_l(g) \cap M_l(h)$$
$$\Rightarrow b \notin M_l(g) \Rightarrow \text{ or } b \notin M_l(h)$$
$$\Rightarrow g(b) = 0 \text{ or } h(b) = 0$$
$$\Rightarrow g(b) \wedge h(b) = 0$$

This shows that $f(b) = g(b) \wedge h(b)$

Proved.

Theorem 4: Let $f: B_n \rightarrow B$ be a Boolean function then prove that it is produced by a Boolean expression.

Proof: Let $M_l(f) = \{b_1, b_2, b_3, \ldots b_m\}$. We can define a function f_i for each of b_i where $1 \leq i \leq$ m such that $f_i\ (b_i) = 1$ and $f_i\ (b_i) = 0$ if $b_i \neq b$. This definition implies that $M_l(f_i) = \{b_i\}$ and so

$$M_l(f) = M_l(f_1) \cup M_l(f_2) \cup M_l(f_3) \cup \ldots \cup M_l(f_m)$$

Therefore from the theorem 3 above, we have

$$f = f_1 \vee f_2 \vee f_3 \vee \ldots \vee f_m$$

If M_{b_i}'s are the corresponding minterms for each f_i, then f can be written as join of these minterms as below.

$$f = M_{b_1} \vee M_{b_2} \vee M_{b_3} \vee \ldots M_{b_m}$$

Proved.

In the table 6.3.4 above. $f(x, y, z) = (x' \wedge y \wedge z) \vee (x \wedge y \wedge z)$. This function can also be written, in simple form as $f(x, y, z) = y \wedge z$. The process of expressing a Boolean expression as join of minterms and then simplifying the resulting expression can be formalized. There are different procedures to simplify a Boolean expression so that minimum number of **gates** (fundamental circuits for AND, OR and NOT) can be used to achieve the same output as desired by the function. In this text, we shall discuss Boolean Expression simplification by Karnaugh Map and Quine McChiskey Method.

6.4 Simplification of Boolean Expression

The information contained in a truth table may be expressed in compact form by listing decimal equivalent of those minterms that produce a **1** for the function. For example, for the truth table 6.3.5, we can write

$$f(x, y, z) = \Sigma(1, 2, 3)$$

Because in the truth table 6.3.5, the minterms $x' \wedge y' \wedge z$, $x' \wedge y \wedge z'$ and $x' \wedge y \wedge z$ produce a **1** for the function, the decimal equivalent for these minterms are 1, 2 and 3 respectively.

Sometimes a minterm is irrelevant for a function. Consider a situation of a circuit having 10 input lines numbered 1 to 10. It is designed to work in such a way that it takes only first 8 input lines and ignores the last two input lines. Thus, out of 2^{10} total possible minterms, only 2^8 are relevant and others are of no use. There are many such situations in which one or more minterms are not taken into consideration while determining the nature of a function. *A minterm that is irrelevant for a function output belongs to a set of* ***don't care conditions of the function***. If there is any minterm belonging to the set of ***don't care conditions,*** it is mentioned together with the function definition. For example, if the output of a Boolean function in four variables: x, y, z and w, is 1 for minterms 1, 3, 5, 6, 12 & 15 and 2, 7, 9 & 11 belong to *don't' care conditions,* then function is represented as

$$f(x, y, z, w) = \Sigma\,(1, 3, 5, 6, 12, 15)$$

$$d\,(x, y, z, w) = \Sigma\,(2, 7, 9, 11)$$

Simplification of a Boolean Expression means finding of essential ***prime implicants.*** What is this prime implicant and how to find it? Let $A = \{a_1, a_2, a_3, \ldots a_n\}$ be a set of Boolean expressions. We say that the Boolean expression b covers or subsumes A, precisely, when

$$\boldsymbol{b} + x = \boldsymbol{b} \;\forall\; x \in A$$

i.e., x is a product of b with some other terms such that $x = \boldsymbol{bc}$,

$$\therefore\; \boldsymbol{b} + x = \boldsymbol{b} + \boldsymbol{bc} = \boldsymbol{b}(1 + c) = b$$

In general, covering term contains fewer variables than the terms that is covers. For example, both ***ab*** and ***ab'*** are covered by ***a***, so ***a*** is the covering term for the set of Boolean expressions {***ab***, ***ab***'}. A ***prime implicant*** is a minimal covering term i.e. one that covers a maximal set of terms. There are different ways to find essential prime implicants. In this text, we shall discuss **Karnaugh** Map and **Quine McCluskey** method.

Karnaugh Map (K-Map)

The K-Map is a visual tool to minimize a Boolean expression involving a few Boolean variables. It consists of a rectangle divided into squares. Each square corresponds to a minterm. These squares are arranges in a way that the contiguous squares differ in exactly one variable being complemented or uncomplemented. The K-Map must be thought of as covering the surface of a sphere, so that, squares on opposite edges of the map are contiguous and all four corner squares of the map have a point in common. K-Map for 2, 3 and 4 variables are shown in the figure 6.4.0(a), 6.4.0(b) and 6.4.0(c) respectively. Minterm and its decimal equivalent that corresponds to a square are also shown in the respective figure.

This K-Map is for two variables p and q. A square in right hand side figure contains decimal equivalent of a minterm.

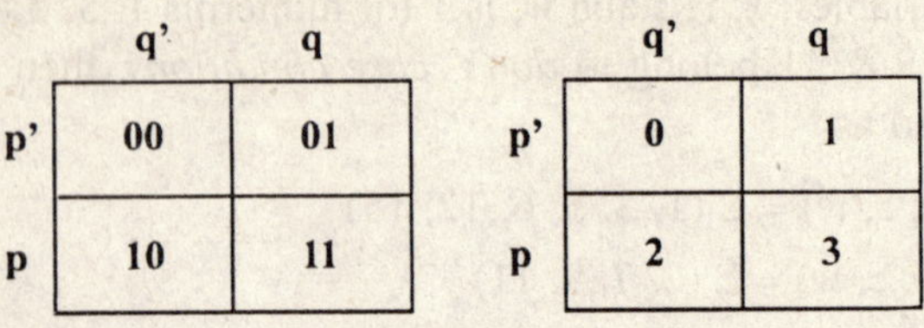

Fig. 6.4.0(a).

The following K-Map is for three variables *p*, *q* and *r*. A square in right hand side figure contains decimal equivalent of a minterm of corresponding square of left hand side.

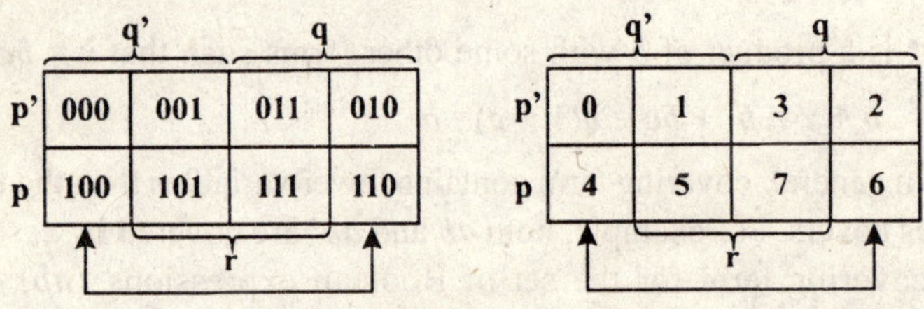

Fig. 6.4.0(b).

The K-Map shown in figure 6.4.0(c_1) and 6.4.0(c_2) are for four variables *p*, *q*, *r* and *s*. A square in figure 6.4.0(c_2) contains decimal equivalent of a minterm of corresponding square of figure 6.4.0(c_1).

A square in a K-Map for a function contains a **1** if the corresponding minterm produces 1 for the function, if a minterm produces **0** for the function then the corresponding square contains **0**. If the minterm belongs to ***don't care conditions,*** the corresponding square contains an '*X*'. This is the way to fill K-Map for a function.

The next step involves finding of one or more **duet** or **quad** or **octet** and so on. A **duet** is contiguous two squares either horizontally or vertically, but not diagonally in any case, having 1 as entry. A **quad** is contiguous four squares either horizontally or vertically or in square or rolling having **1** as entry. An **octet** is contiguous eight squares either horizontally or vertically or in rectangle or rolling having **1** as entry. Each duct, quad and octet can be expressed in terms of product of variables. The sum of these products is the simplified expression for the given function. The procedure is explained through the following examples.

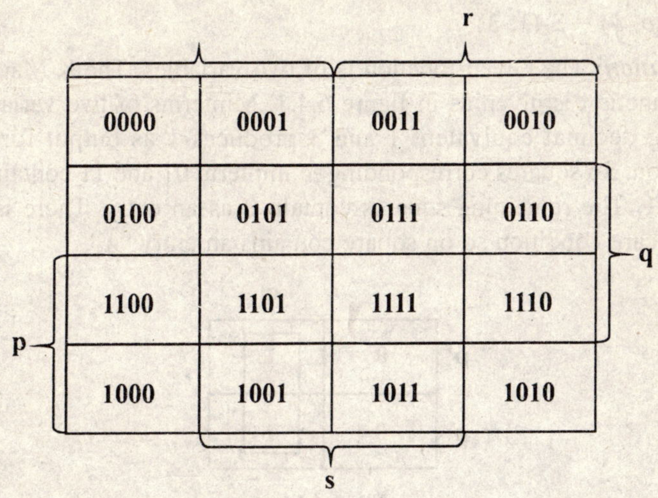

Fig. 6.4.0(c_1).

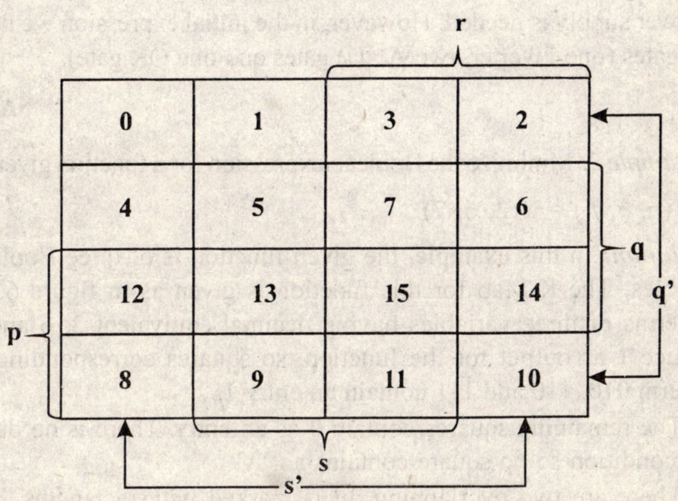

Fig. 6.4.0(c_2).

Example 1: Minimize the Boolean expression for a function given as

$f(p, q) = \Sigma\ (1, 3)$

Solution: The Given function is of two variables. The K-Map for this function is given as in figure 6.4.1. Minterms of two variables having decimal equivalent 1 and 3 produces 1 as output for the function. So squares corresponding to minterm 01 and 11 contain an entry **1**. The remaining squares contain 0 as an entry. There is no don't care condition so on square contains an entry "*X*".

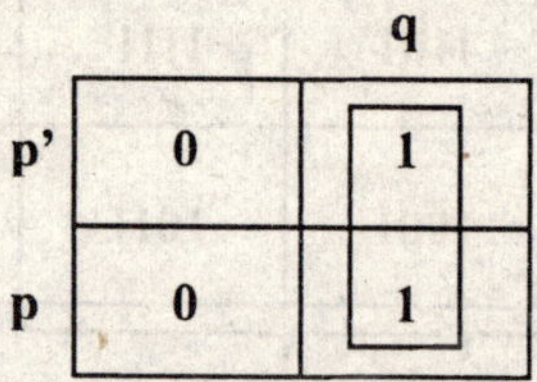

Fig. 6.4.1.

The duet corresponds to column q, so the given function on $f(p, q) = p'q + pq$ can be expressed in **simplified** form as $f(p, q) = q$. No gate is required to implement this function. Only a simple wire connected to power supply is needed. However, in the initial expression we need four gates (one inverter, two AND) gates and one OR gate).

Ans.

Example 2: Minimize the Boolean expression for a function given as

$f(p, q, f) = \Sigma(2, 6, 7)$

Solution: In this example, the given function is of three Boolean variables. The K-Map for this function is given as in figure 6.4.2 Minterms of three variables having decimal equivalent 2, 6 and 7 produce **1** as output for the function, so squares corresponding to minterm 010, 110 and 111 contain an entry 1.

The remaining squares contain 0 as an entry. There is no don't care condition so no square contains an '*X*'.

There are two overlapping duets marked with rectangles. The vertical rectangle corresponds to the intersection of q and r'. the horizontal rectangle corresponds to the intersection of q. Therefore the given function

$f(p, q, r) = p'qr' + pqr' + pqr$

can be expressed in **simplified** form as

$f(p, q, r) = qr' + pq$

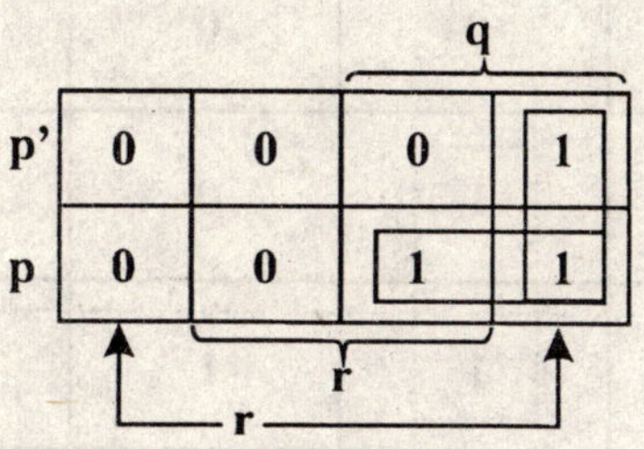

Fig. 6.4.2.

Now number of gates required to implement this function is four (one inverter, two AND gates). Compare this with the original requirement of ten gates (two inverters, six AND) gates and two OR gates).

Ans.

Example 3: Minimize the Boolean expression given by the function

$f(p, q, r, s) = \Sigma(5, 8, 9, 10, 11, 12, 13, 14, 15)$

Solution: The function is of four Boolean variables. The K-Map for this function is given as in figure 6.4.3. Minterms of four variables having decimal equivalent 5, 8, 9, 10, 11, 12, 13, 14, and 15 produce 1 as output for the function, so squares corresponding to minterm 0101, 1000, 1001, 1010, 1011, 1100, 1101, 1110 and 1111 contain an entry 1. The remaining squares contain 0 as an entry. There is no don't care condition so no square contains an '*X*'.

There is one duet and one octet marked with rectangles. Duet is overlapping with octet. If overlapping duet (or quad or octet) reduces the number of variables in an expression, it is worth considering. The octet corresponds to *p* and duet corresponds to *qr*'s. Therefore the given function can be expressed in **simplified** form as

$$f(p, q, r, s) = p + qr's.$$

Compare the requirement of gates in this simplified expression with that in the original expression

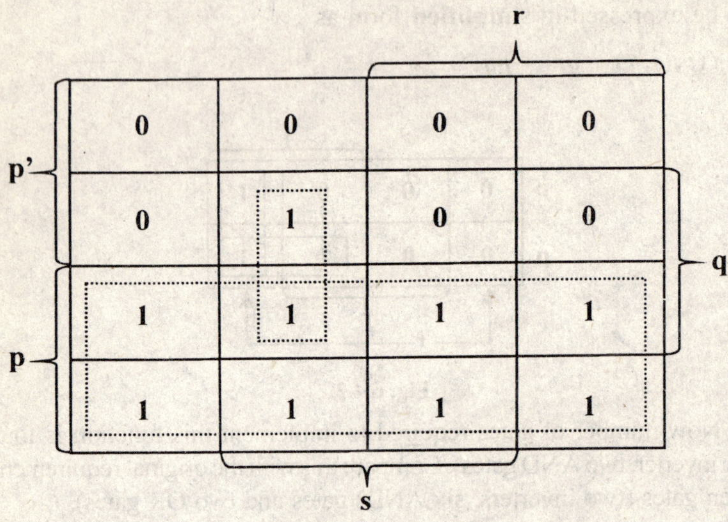

Fig. 6.4.3.

Ans.

Example 4: Minimize the Boolean Expression given by the function

$f(p, q, r, s) = \Sigma\ (1, 2, 3, 6, 8, 9, 10, 12, 13, 14)$

Solution: The function is of four Boolean variables. The K-Map for this function is given as in figure 6.4.4. Minterms of four variables having decimal equivalent 1, 2, 3, 6, 8, 9, 10, 12, 13 and 14 produce **1** as output for the function, so squares corresponding to minterm 0001, 0010, 0011, 0110, 1000, 1001, 1010, 1100, 1101 and 1110 contain an entry 1. The remaining squares contain 0 as an entry. There is no don't care condition so no square contains an '*X*'.

There is one duet and two quads marked with rectangles. The duet corresponds to $p'q's$. The vertical quad corresponds to rs'. The other quad corresponds to pr'. Therefore the given function can be expressed in **simplified** form as

$$f(p, q, r, s) = p'\, q's + rs' + pr'.$$

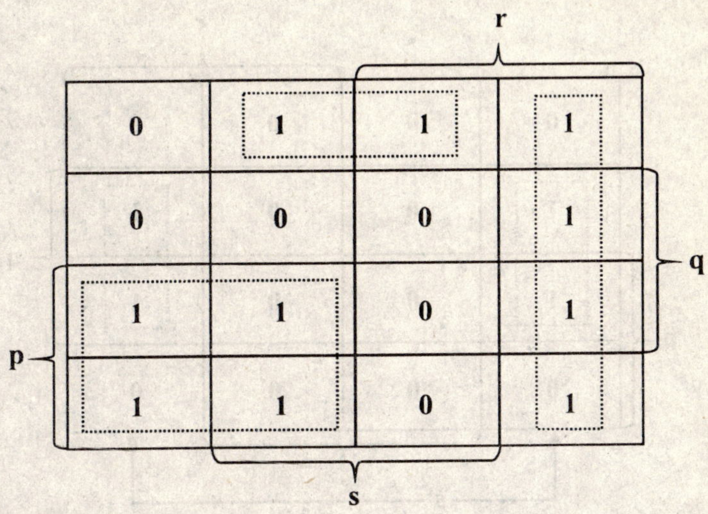

Fig. 6.4.4.

Ans.

Example 5: Minimize the Boolean Expression given by the function

$f(p, q, r, s) = \Sigma(4, 6, 12, 14)$

Solution: The function is of four Boolean variables. The K-Map for this function is given as in figure 6.4.5. Minterms of four variables having decimal equivalent 4, 6, 12 and 14 produce **1** as output for the function, so squares corresponding to minterm 0100, 0110, 1100, and 1110 contain an entry 1. The remaining squares contain 0 as an entry. There is no don't care condition so no square contains an '*X*'.

There is one **rolling** quad marked with half-open half-closed rectangles at two ends. This rectangle gets completed if we conceptually consider the K-Map as a surface of a cylinder. This quad corresponds to *qs*'. Therefore the given function can be expressed in **simplified** form as

$f(p, q, r, s) = qs'$

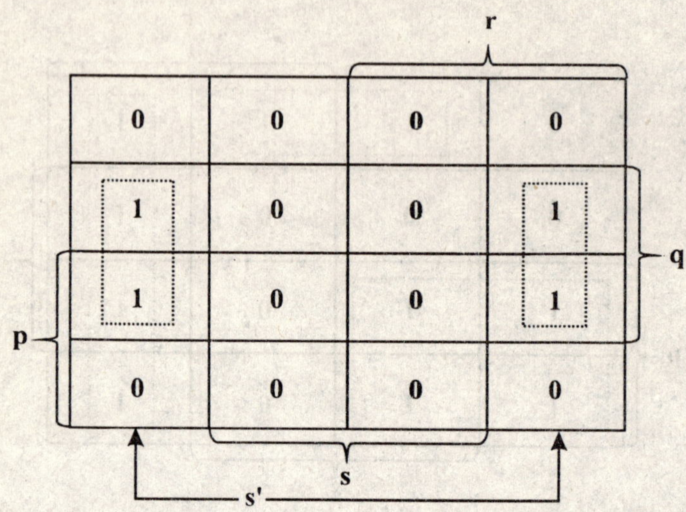

Fig. 6.4.5.

Example 6: Minimize the Boolean Expression given by the function

$$f(p, q, r, s) = \Sigma(5, 11, 13, 15)$$

Solution: The function is of four Boolean variables. The K-Map for this function is given as in figure 6.4.6. Minterms of four variables having decimal equivalent 5, 11, 13 and 15 produce 1 as output for the function, so squares corresponding to minterm 0101, 1011, 1101, and 1111 contain an entry 1. The remaining squares contain 0 as an entry. There is no don't care condition so no square contains an '*X*'.

There are three duets marked with rectangles. Here one duet marked with dashed rectangle is **redundant**. We do not take into consideration the product term corresponding to a redundant duet (or quad or octet) in the simplified expression of the function. The non-redundant duets corresponds to qr's and prs. Therefore the given function can be expressed in simplified form as

$$f(p, q, r, s) = qr's + prs$$

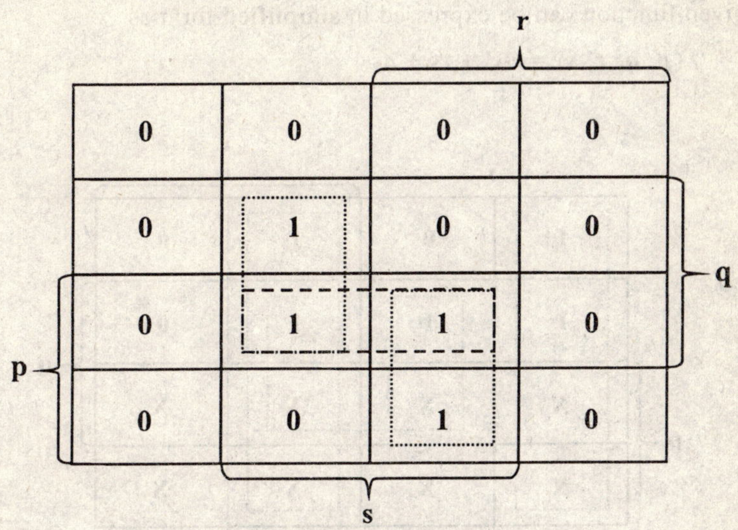

Fig. 6.4.6.

Ans.

Example 7: Minimize the Boolean Expression given by the function

$f(p, q, r, s) = \Sigma\ (0, 3, 4, 5, 7)$
$d(p, q, r, s) = \Sigma\ (8, 9, 10, 11, 12, 13, 14\ 15)$

Solution: The function is of four Boolean variables. The K-Map for this function is given as in figure 6.4.7. Minterms of four variables having decimal equivalent 0, 3, 4, 5 and 7 produce 1 as output for the function, so squares corresponding to minterm 0101, 1011, 1101, and 1111 contain an entry 1. There are minterms having decimal equivalent of 8, 9, 10, 11, 12, 13, 14, and 15 that belong to don't care condition. Thus squares corresponding to minterm 1000, 1001, 1010, 1011, 1100, 1101, 1110 and 1111 contain an entry *X*. The remaining squares contain 0 as an entry.

Whenever, there is a square with entry *X* in a K-Map, we can consider that square as having entry either 1 to 0 depending upon our requirement to form duet, quad or octet. Thus we have three quads and corresponding product terms are $r's'$, rs and qs. Therefore the given function can be expressed in **simplified** form as

$$f(p, q, r, s) = r's' + rs + qs$$

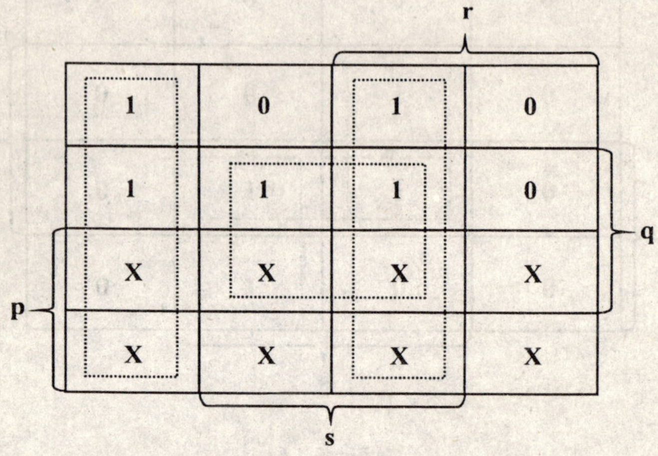

Fig. 6.4.7.

Ans.

K-Map method may be used for the simplification of a Boolean expression involving five or more, variables but the visualization of the result becomes increasingly difficult. To deal with such problems, we use Quine-McCluskey method.

Quine-McCluskey Method

This programmable deterministic algorithm based method is used to automate the technique of circuit minimization. This method is very robust and best suitable in the situation when number of Boolean variables involved are four or more. This method has two phases:

Phase 1: Distributive law i.e., $ax + ax' = a(x + x') = a$. This is used repetitively to get prime implicants.

Phase 2: Selection Phase i.e., only those prime implicants are selected which are necessary to represent Boolean function.

The sum of these prime implicants is the result. The result so obtained, at the end, is the simplest sum of the products of Boolean variables for the given function. The step by step working procedure is explained below with an example.

Example 1: Find the minimal expression for the Boolean expression

wx + *xy* + *yz* + *wz* + *w'x'yz'* + *w'x'y'z*

Solution: There are four variables: *w*, *x*, *y*, and *z*, used in the given Boolean expression. The initial is to find all the prime implicants. This is what is done in the phase 1 and this involves a number of steps as outlined below.

Step 1: Represent each term in the given expression by a string of 1, 0 and – where '*1*' stands for a variable, '0' for complemented variable and '–' for absent variable in the term. Thus, the given terms can be written as follows:

***wx* = 11 – – ;** ***xy* = – 11 – ;** ***yz* = – – 11;**

***wz* = 1 – – 1;** ***w'x'yz'* = 0010** and ***w'x'y'z* = 0001**

Step 2: An absent variable in a Boolean term implies that it is ignored while finding the output. Expand all terms containing '–' in the **Step 1**, with all possible combinations of '1' and '0'. Expanding the above terms, we get

***wx* = 11 – –;**	**1111**	**1110**	**1101**	**1100**
***xy* = –1 1–;**	**1111**	**1110**	**0111**	**0110**
***yz* = – – 11;**	**1111**	**1011**	**0111**	**0011**
***wz* = 1– –1;**	**1111**	**1101**	**1011**	**1001**
***w'x'yz'* =**	**0010**			
***w'x'y'z* =**	**0001**			

Step 3: Select the distinct terms from the exhaustive list of terms as obtained in the Step 2. Arrange them in the decreasing order of the number of 1's in the term. If there are many terms having same number of 1's then arrange them in decreasing order of magnitude

within the same group. Applying this on the terms obtained in the step 2, we get

$$1111\}\quad \textit{Group}\quad 1\ \textit{containing}\quad 1\ 1\text{'s}$$

$$\left.\begin{matrix}1110\\1101\\1011\\0111\end{matrix}\right\}\quad \textit{Group}\quad 2\ \textit{containing}\quad 3\ 1\text{'s}$$

$$\left.\begin{matrix}1100\\1001\\0110\\0011\end{matrix}\right\}\quad \textit{Group}\quad 3\ \textit{containing}\quad 2\ 1\text{'s}$$

$$\left.\begin{matrix}0010\\0001\end{matrix}\right\}\quad \textit{Group}\quad 4\ \textit{containing}\quad 1\ 1\text{'s}$$

Step 4: Combine those terms that differ by a 0 or 1 in exactly one position *e.g.* 1111 and 1110 are combined in yield 111– i.e. $wxyz + wxyz' = wxy$. It is important to note that a term can combine with a term in the group either immediately above or immediately below its group. The result of this step applied on the arranged terms in the step 3 is given below.

$$\left.\begin{matrix}111- & (1111, & 1110)\\11-1 & (1111, & 1101)\\1-11 & (1111, & 1011)\\-111 & (1111, & 0111)\end{matrix}\right\}A_1\quad \textit{Group } 1 \textit{ combined with Group } 2$$

$$\left.\begin{array}{ll} 11-0 & (1110,\ \ 1100) \\ -110 & (1110,\ \ 0110) \\ 110- & (1101,\ \ 1100) \\ 1-01 & (1011,\ \ 1001) \\ 10-1 & (1011,\ \ 1001) \\ -011 & (1011,\ \ 0011) \\ 011- & (0011,\ \ 0010) \\ 0-11 & (0111,\ \ 0011) \end{array}\right\} A_2 \quad \textit{Group 2 combined with Group 3}$$

$$\left.\begin{array}{ll} -001 & (1001,\ \ 0001) \\ 0-10 & (0110,\ \ 0010) \\ 001- & (0011,\ \ 0010) \\ 00-1 & (0011,\ \ 0001) \end{array}\right\} A_3 \quad \textit{Group 3 combined with Group 4}$$

Step 5: Repeat step 4 on the terms currently obtained until no more terms can be combined. It is important to mention here that number of groups of terms containing different number of 1's decreases by one, after every such repetition. During the process, we may get a term that cannot combine with any other terms. Those terms that do not combine further are *prime implicants*. Keep on collecting all such terms, if any, from every repetition. Repeating the step 4, we get

$$\left.\begin{array}{ll} 11-- & (111-,\ \ 110-) \\ -11- & (111-,\ \ 011-) \\ 11-- & (11-1,\ \ 11-0) \\ 1--1 & (11-1,\ \ 10-1) \\ 1--1 & (1-11,\ \ 1-01) \\ --11 & (1-11,\ \ 0-11) \\ -11- & (-111,\ \ -110) \\ --11 & (-111,\ \ -011) \end{array}\right\} B_1 \quad \textit{By combining } A_1 \textit{ and } A_2$$

$$\left.\begin{array}{lll} -0-1 & (10-1, & 00-1) \\ 0-1- & (011-, & 001-) \\ -0-1 & (-011, & -001) \\ 0-1- & (0-11, & 0-01) \end{array}\right\} B_2 \quad \textit{By combining } A_2 \textit{ and } A_3$$

Step 6: Steps 4 and 5 give all the prime implicants for the given Boolean expressions. In this step, collect all the unique prime implicants. These are, in this case, 11 – –, – 11 –, 1 – – 1, – – 11, –0 –1 and 0 –1–. In terms of the variable, these prime implicants can be written as *wx*, *xy*, *wz*, *yz*, *x'z* and *w'y* respectively. All these prime implicants may not be essential for representing the given Boolean expression in minimized form. Therefore, the selection phase follows.

Step 7: This is **selection phase**. Create a table of $m + 1$ rows and $n + 1$ columns, where m is the number of expanded terms uniquely identified in the step 3 and n is the number of prime implicants uniquely identified in the step 6. One extra row and column is taken for headings. Each cell in the table is uniquely determined by ordered pairs (Expanded term, prime implicants). We put an '*X*' in a cell (Expanded term, Prime implicant), when the prime implicant covers the expanded term. For example, 11 – – covers 1111, so there is an '*X*' in the cell (1111, 11 – –), i.e., first row and first column. Since 1111 is not covered by – 0 –1 , so there is no '*X*' in the cell (1111, – 0 –1), i.e., fifth column of first row. We have the following table.

Expanded Term	11—	–11–	1—1	—11	–0–1	0–1–
1. 1111	*X*	*X*	*X*	*X*		
2. 1110	*X*	*X*				
3. 1101	*X*		*X*			
4. 1011			*X*	*X*	*X*	
5. 0111		*X*		*X*		*X*
6. 1100	**X**					

7. 1001			*X*		*X*	
8. 0110		*X*				*X*
9. 0011				*X*	*X*	*X*
10. 0010						**X**
11. 0001					**X**	

Table 6.4.1.

From the table 6.4.1, prime implicants are selected as follows:

Step 7.1 Look at the rows from top to bottom and select the row, if any, having single entry '*X*'. The prime implicant corersponding to the column, in which '*X*' is present, is essential prime implicants.

Step 7.2 Mark all the rows in which there is a '*X*' in the column corresponding to the essential prime implicants determined in the step 7.1. This prime implicant covers all the expanded terms for which there is a '*X*' in this column.

Step 7.3 Repeat steps 7.1 and 7.2 on the table created in step 7 until we get a covering prime implicants for all expanded terms (rows).

In the table 6.4.1, we have selected row 6. Thus the prime implicant 11 – – is essential. Also there are '*X*' in the column corresponding to 11 – – for rows 1, 2 and 3 besides for row 6. Thus, 11 – – covers the expanded terms 1111, 1110, 1101 and 1100 corresponding to the rows 1, 2, 3 and 6 respectively. Next, rows 10 and 11 having single entry '*X*'. The corresponding prime implicants are 0 – 1– and –0 –1 respectively. It can be verified that 0– 1– covers expanded terms 0111, 0110, 0011 and 0010 corresponding to the rows 5, 8, 9 and 10 respectively. And, 0– 1– covers expanded terms 1011, 1001, 0011 and 0001 corresponding to the rows 4, 7, 9 and 11 respectively. The expanded term 0011 is covered by both 0– 1– and –0 –1. This does not create any problem as long as both are essential. Thus out of the six prime implicants, only three are essentially required to represent the given Boolean expression. These are 11—, 0– 1– and –0 –1 i.e., wx, $w'y$ and $y'z$. Therefore, the minimized Boolean expression is

$$wx + w'y + y'z$$

Ans.

It is important to mention that, in most of the cases, the answer does not come so easily as above. There may not be any row having single entry '*X*'. How to proceed then? A tie has to be broken. Select a row, if any, having two '*X*'s otherwise look for three '*X*'s and so on. After selecting such row, one prime implicant has to be selected. We select the prime implicant that covers maximum number of expanded terms. In case of tie, select the prime implicant having minimum number of variables. If there is a tie, again, select having minimum number of complemented variables. See the following example.

Example 2: Find the minimal expression for the Boolean expression

$xyz + x'z + y'z$

Solution: There are three variables *x*, *y* and *z*, used in the given Boolean expression.

Step 1: Representing each term in the given expression by a string of 1, 0 and we get

xyz = 111; $x'z$ = 0–1; $y'z'$ = –00;

Step 3: Expanding all terms containing '–' in the Step 1, with all possible combinations of '1' and '0', we get

xyz = 111

$x'y$ = 0–1: 011 001

$y'z'$ = – 00: 100 000

Step 3: Arranging the distinct terms in the decreasing order of the number of 1's in the term, we get

111} *Group* 1 *containing* 3 1's

011} *Group* 2 *containing* 2 1's

100, 001} *Group* 3 *containing* 1 1's

000} *Group* 4 *containing* *no* 1's

Step 4: Combining the terms that differ by a 0 or 1 in exactly one position, we get

$-11 \quad (111, \quad 011)\}A_1$ *Group* 1 *combined with Group* 2

$0-1 \quad (011, \quad 001)\}A_2$ *Group* 2 *combined with Group* 3

$\left.\begin{array}{ll} 0-1 & (011, \quad 001) \\ 00- & (001, \quad 000) \end{array}\right\} A_3$ *Group* 3 *combined with Group* 4

Step 5: None of the terms of the step 4 can combine with the other terms, so the repetition of the step 4 is not required in this case.

Step 6: The Unique prime implicants are –11, 0–1, –00 and 00–. In terms of variables, these prime implicants can be written as *yz*, *x*'*z*, *y*'*z*' and *x*'*y*' respectively.

Step 7: In the **selection phase,** we now create a table of 5 + 1 rows and 4 + 1 columns for 5 expanded terms uniquely identified in the step 3 and prime implicants uniquely in the step 6.

Expanded Terms	**–11**	**0–1**	**–00**	**00–**
1. 111	<u>*X*</u>			
2. 011	*X*	*X*		
3. 100			<u>*X*</u>	
4. 001		*X*		*X*
5. 000			*X*	*X*

Table 6.4.2.

In the table 6.4.2, we select row 1. Thus the prime implicants –11 is essential. Also there is '*X*' in the column corresponding to –11 for row 2 besides for row 1. Thus –11 covers the expanded terms **111** and **011.** Next, row 3 has single entry '*X*' in it. The corresponding prime implicant is –00. It covers expanded terms **100** and **000**. Thus expanded terms corresponding to rows 1, 2, 3 and 5 have been covered by the prime implicants –11 and –00. The expanded term **001** is covered either by 0–1 or 00–. We have to select one of them. Both the terms have two variables. 0–1 has one complemented variables whereas 00– has two. According to the selection criteria mentioned above, the term having minimum number of complemented variable is selected. So we, select here 0–1. Thus the essential prime implicants are –11, –00 and 0–1. When expressed in terms of variables, these prime implicants are *yz*, *y*'*z*' and *x*'*z* respectively. Therefore, the minimized Boolean expression is

$$yz + y'z' + x'z$$

Ans.

6.5 Propositional Calculus

Consider statements like:

1. Today is Monday.
2. It will rain tomorrow.
3. For any positive integer x, $x + 3 > 3$.
4. You are at home.

These statements are either true or false at a given point of time. None of them can be both true and false at the same time. Also, the answer cannot be like 'may be' or 'probably' etc. in any of the above case, i.e., there is no ambiguity in the answer. Statements like above are known as declarative sentences.

A ***proposition*** is any declarative sentences that is either true or false but never both at the same time. The two valued (T, F) Boolean algebra is very closely related to the system of basic logic called ***propositional calculus***. This logical system is concerned with propositions. It is important to note that this system does not allow for such ideas as "possibly true" or "indeterminate". A proposition that corresponds to a single declarative sentence is called ***atomic proposition*** and is generally represented by lower case alphabets **a,** ***b*, ...*q*, *r*, ..., *x*, *y*, *z*.** The truth values are denoted by *T*(true) and *F*(False). Value *T* & *F* directly correspond to Boolean constants **1 and 0** respectively. The concept of propositional calculus can be extended to **predicate calculus** that allows us to use quantifiers ($\forall$, $\exists$) and propositional functions in a Boolean expression. In this text, our discussion is restricted to propositional calculus.

An atomic proposition in itself is not sufficient to deal with many situations of a logical system. Generally atomic propositions are combined to give a logical meaning to a compound sentence. For example, let p be the proposition "you are at home" and let q be the proposition "you are in office". The compound sentence "you are either at home or in office" can be represented as (**p or q**). *A proposition that is obtained from the combination of other propositions is called* ***compound proposition***. In the propositional calculus we use connectives to combine propositions. Sometimes we need to use negative of a proposition also in compound proposition. The connectives are shown in the table 6.5.1 with its corresponding meaning in Boolean Algebra.

Let us correlate atomic proposition with Boolean variables and compound proposition with Boolean function. Like, every Boolean function has a truth table, every compound proposition has a truth table showing the truth values of the compound proposition in terms of its component part (atomic proposition). Truth table for the negation of a proposition p is shown in the table 6.5.2. The possible values (truth) for the compound proposition $p \vee q$ and $p \wedge q$ are listed in the table 6.5.3. Similarly, table 6.5.4 shows possible values (truth) for proposition $p \rightarrow q$, $q \rightarrow p$ and $p \equiv q$ under the respective columns of the table.

Sl. No.	*Connective*	*Called as*	*Boolean equivalent*	*Set equivalent*
1.	$\vee$	**Disjunction**	**Inclusive OR**	**Union**
2.	$\wedge$	**Conjunction**	**AND**	**Intersection**
3.	$\rightarrow$	**Implies that**	**If...Then..**	$\Rightarrow$
4.	$\equiv$	**Equivalence**	**Equality**	**Equality** $\Leftrightarrow$
5.	~(or $\neg$)	**Negation**	**1's complement**	**Complement**

Table 6.5.1.

p	-p
F	T
T	F

Table 6.5.2.

p	q	p∨q	p∧q
F	F	F	F
F	T	T	F
T	F	T	F
T	T	T	T

Table 6.5.3.

p	q	p→q	q→p	(p→q)∧(q→p) or (p≡q)
F	F	T	T	T
F	T	T	F	F
T	F	F	T	F
T	T	T	T	T

Table 6.5.4.

If a compound proposition has two atomic propositions as components then the truth table for the compound proposition contains four entries. These four entries may be all *T*, may be all *F*, may be one *T* and three *F* and so on. There are in total 16 (2^4) possibilities as shown in the table 6.5.5 in the 16 different columns. Column 1 contains all *T*. This implies that if the column corresponding for a compound proposition in a truth table contains all *T*, then the proposition is always true. This is called **tautology**. Now look at column 16. It contains all *F*. This implies that if the column corresponding to a compound proposition in a truth table contains all F, then the proposition is **never true**. This situation is referred as **contraction**. If the column corresponding to the compound proposition in a truth table contains both *T* and *F*, then the proposition is true for some contradiction of its component and false for some other combination. This is called **contingency.**

Out of the 16 columns, columns from 9 to 16 are just negation of the columns from 1 to 18 in reverse order i.e. column $I\,(1 \le I \le 8)$ is just negation of the column 17–1. Two propositions p and q are said to be equal if and only if we have a **tautology** from the truth table $p = q$ (or $p \equiv q$)

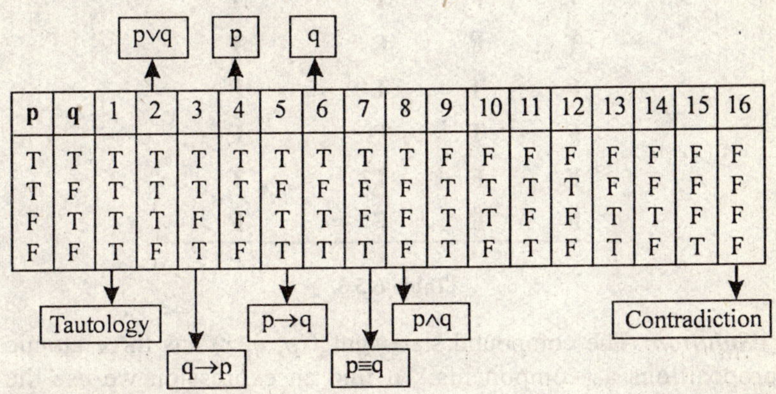

p	q	1	2	3	4	5	6	7	8	9	10	11	12	13	14	15	16
T	T	T	T	T	T	T	T	T	T	F	F	F	F	F	F	F	F
T	F	T	T	T	T	F	F	F	F	T	T	T	T	F	F	F	F
F	T	T	T	F	F	T	T	F	F	T	T	F	F	T	T	F	F
F	F	T	F	T	F	T	T	T	F	T	F	T	F	T	F	T	F

Table 6.5.5.

Example 1: Find the truth table for the proposition $p \to (q \wedge \sim p)$

Solution: The given compound proposition has two atomic propositions p and q as its components. The truth table will have four entries as shown in the table 6.5.6.

$p \to q$	p	q	$p \to (q \vee \sim p)$
T	F	T	T
T	F	T	F
F	T	F	T
T	T	T	T

Table 6.5.6.

Ans.

Example 2: Determine the compound proposition (logical expression $f(p, q, r)$ having truth values as shown in the following table 6.5.7.

p	q	r	$f(p, q, r)$
T	T	T	T
T	T	F	F
T	F	T	T
T	F	F	T
F	T	T	F
F	T	F	F
F	F	T	T
F	F	F	F

Table 6.5.6.

Solution: The compound statement $f(p, q, r)$ has three atomic propositions as components. To find an expression we use the concept of minterms discussed in the section 6.4 of this chapter. Look for those combinations of propositions p, q and r that produce a truth value T for f. The minterms are $p \wedge q \wedge r$, $p \wedge \sim q \wedge r$, $p \wedge \sim q \wedge r$ and $\sim p \wedge \sim q \wedge r$. These are called conjunctive. A conjunctive is defined as conjunction of atomic proposition or negation of an atomic proposition. The disjunction of these conjunctives is the required expression. Thus,

$$f(p, q, r) = (p \wedge q \wedge r) \vee (p \wedge \sim q \wedge r) \vee (p \wedge \sim q \wedge \sim r) \wedge (\sim p \wedge \sim q \wedge r)$$

Ans.

Properties of Propositional Calculus

Let p, q and r be any three propositions then following are true.

1. Conjunction and disjunction are commutative i.e.

$$p \wedge q = q \wedge p$$

$$p \vee q = q \vee p$$

2. Conjunction and disjunction are associative i.e.

$$p \wedge (q \wedge r) = (p \wedge q) \wedge r$$

$$p \vee (q \vee r) = (p \vee q) \vee r$$

3. Conjunction distributes over disjunction and vice versa i.e.

$$p \wedge (q \vee r) = (p \wedge q) \vee (p \wedge r)$$

$$p \vee (q \wedge r) = (p \vee q) \wedge (p \vee r)$$

4. Idempotent law is obeyed by proposition with respect to conjunction and disjunction i.e.

$$p \wedge p = p$$

$$p \vee p = p$$

5. De Morgan's law i.e. complement (negation) of conjunction is disjunction of complements (negations) and complement of disjunction is conjunction of complements.

$$\sim(p \wedge q) = \sim p \vee \sim q$$

$$\sim(p \vee q) = \sim p \wedge \sim q$$

6. Involution i.e. negation of a negation of a proposition p is p itself

$$\sim(\sim p) = p$$

7. $p \rightarrow q = (\sim p) \vee q$
   $p \rightarrow q = ((\sim q) \rightarrow \sim p)$
   $p \rightarrow q = (p \rightarrow q) \wedge (p \rightarrow p)$ [$\leftrightarrow$ is also called equivalence i.e. $\equiv$]
   $\sim(p \rightarrow q) = p \wedge \sim q$
   $\sim(p \leftrightarrow q) = (p \wedge \sim q \vee (q \wedge \sim p)$

All these properties can be proved using truth table. To prove any of these properties and other compound propositions like this, we simply find truth table for that proposition. If the truth table is a tautology then the given proposition is true and hence treated as proved. The proof of these properties is left as exercise to the reader.

We have used atomic propositions and different connectives to form compound propositions. There is a rule to form any compound proposition from atomic proposition. Any arbitrary combination may not be a valid proposition. Thus, any proposition to be a valid proposition, it must be a **well-formed formula (wff)**. Any proposition is said to be a well-formed formula if it is formed using the following rule

1. **<Atomic proposition>::=** $a|b|c...|y|z$
2. **<Connectives>::=** $\rightarrow|\wedge|\vee|\equiv$
3. **<wff> ::= <Atomic proposition>|~(<wff>)|(<wff>) <Connectives>(<wff>).**

The above rule is interpreted as

- An atomic proposition is a well-formed formula.
- Negation of a well-formed formula is a well-formed formula.
- If p and q are any well-formed formula then $p \wedge q$, $p \vee q$, $p \rightarrow q$ and $p \equiv$ are well-formed formula.

Connective	Precedence	Polish Symbol
~	1	N
∧	2	K
∨	3	A
→	4	C
≡	5	E

Table 6.5.7

To determine that whether a given proposition is well-formed or not, we use Polish notations. First we convert the given proposition in infix form into **polish prefix notation (operator operand (operand)) or into Polish, postfix notation (operand (operand) operator).** Then we apply a partial and total sum rule to validity of proposition. This is explained in the following example. The following notations, as shown in table 6.5.7, are used to denote different connectives in a polish notation of a proposition. The connectives have the precedence as shown in precedence column. Lower number means higher precedence. Parentheses are taken into account on top priority. Deeper is the parentheses, higher is its precedence.

Example 3: Convert the proposition

$$(p \rightarrow (q \rightarrow r)) \equiv (\sim(p \vee r) \wedge \sim q)$$

into a polish prefix notation.

Solution: Recall the concept you have studied in your fundamental courses of data structure for conversion of expression from infix to prefix, infix to postfix prefix to postfix and so on. See the steps followed here.

Step 1: Here we first take the $(q \rightarrow r)$ and $(p \vee r)$ as they have higher precedence. Converting them to Polish notation, we have

$$(p \rightarrow Cqr) \equiv (\sim Apr \wedge \sim q)$$

Step 2: Now we take the $\sim Apr$ and $\sim q$ as they have now higher precedence. Converting them to Polish notation, we have

$$(p \rightarrow Cqr) \equiv (NApr \wedge Nq)$$

Step 3: Next we take $p \rightarrow Cqr$ and $NApr \wedge Nq$ as they have now higher precedence. Converting them to polish notation, we have

$$CpCqr \equiv KNAprNq$$

Step 4: Now, only connective left is $\equiv$, converting them, we have

$$ECpCqrKNAprNq$$

Ans.

Example 4: Convert the proposition

$$(\sim(p \equiv \sim(q \rightarrow r))) \equiv ((p \vee r) \vee \sim(\sim p \rightarrow \sim q)) \text{ and}$$
$$((p \vee \sim q) \rightarrow (p \wedge \sim r)) \equiv (\sim p \rightarrow (q \vee r))$$

into polish prefix and postfix notation.

Solution: It is left as an exercise to the reader.

Now, once we have a Polish prefix or Polish postfix notation we can apply the partial and total sum rule to determine the validity of the given expression. This is a three steps method.

Step 1: Write the expression in Polish prefix (postfix) notation in a line and in the next line can apply next line assign a value +1 to all atomic proposition or constant, 0 to all unary connective (there is only one unary connective ~) and –1 to all binary connectives.

Step 2: Find the partial sum for all symbols (atomic proposition, constants and connectives) from **RHS** is the notation is in Polish prefix and from **LHS** if the notation is in Polish postfix.

Step 3: The sum corresponding to left most symbol is called a **total** sum and all other are called **partial sums**. If all the partial sums are **positive** and total sum is equal to **1,** then the given expression is valid otherwise the expression is not valid i.e., not a well-formed formula.

Example 5: Determine whether the proposition given below in Polish prefix notation is valid

$$CECpEpNqCprNq$$

Solution: It is mentioned here that the notation is in prefix form. Even if it is not mentioned it is not difficult to know that whether the given expression is in prefix of postfix notation. (*Look at first symbol, if it is an operator the expression is in prefix, if the very last symbol is an operator the expression is in postfix otherwise the expression is in infix.*)

In the figure 6.5.1, we have arranged the symbol of the given expression in first line. In the second line each symbol has been assigned a value according to the rule described in the step 2. Since the notation is in Polish prefix, we have calculated the partial sums from RHS. Thus, the partial sum for the first symbol from RHS is 1, for second symbol it is 1 (previous partial + its own assigned value = 1 + 0 = 1), for third it is 2 and so on. The last symbol (left most side) has partial sum 1. **We call this partial; sum as total sum.** From the table it is obvious that all the partial sums are positive and total sum is equal to 1. Therefore, the given expression is a valid well-formed formula.

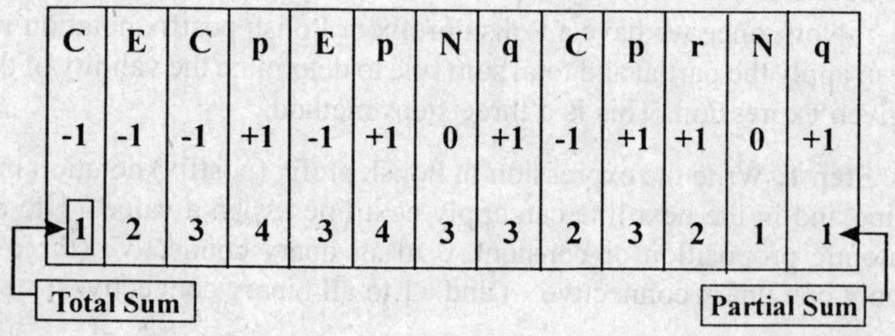

C	E	C	p	E	p	N	q	C	p	r	N	q
-1	-1	-1	+1	-1	+1	0	+1	-1	+1	+1	0	+1
1	2	3	4	3	4	3	3	2	3	2	1	1

Fig. 6.5.1.

Ans.

Example 6: Determine whether the proposition given below is valid

KpACpqNqp

Solution: Look at first symbol, if it is an operator so the expression is in Polish prefix notation.

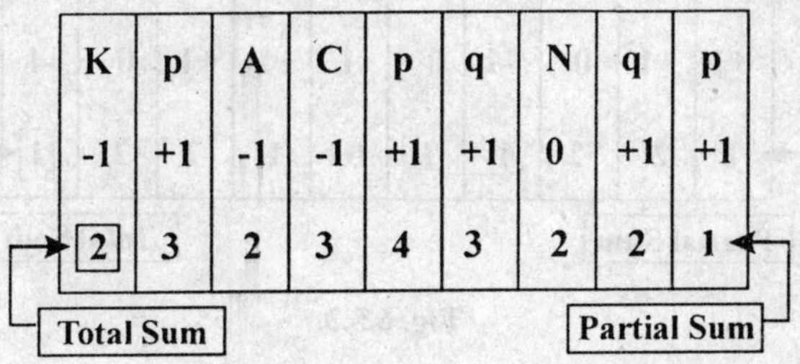

Fig. 6.5.2.

In the figure 6.5.2, we have arranged the symbols of the given expression in first line. In the second line each symbols have been assigned a value according to the rule described in the step 2. Since the notation is in Polish prefix, We have calculated the partial sums from RHS. Thus, the partial sum for the first symbol from RHS is 1, for second symbol it is 2 (previous partial + its own assigned value = 1+1= 1), for third it is 2 and so on. The total sum is 2. From the table it is obvious that all the partial sums are positive, but the total sum is equal to 2. Therefore, the given expression is **not** a valid well-formed formula.

Ans.

Example 7: Determine whether the proposition given below is valid.

pqNCNAprNA

Solution: Look at first symbol if it is not an operator so the expression is not in Polish prefix notation. The last symbol is an operator so it is in Polish postfix notation.

In the figure 6.5.3, we have arranged the symbol of the given expression in first line. In the second line each symbols have been assigned a value according to the rule described in the step 2. Since the notation is in Polish postfix, we have calculated the partial sums from LHS. Thus, the partial sum for the first symbol from LHS is 1, for second symbol it is 2 (previous partial + its own assigned value =

1 + 1 = 1), for third it is 2 and so on. The total sum is 1. From the table it is obvious that all the partial sums are **not** positive. One partial sum is **zero.**

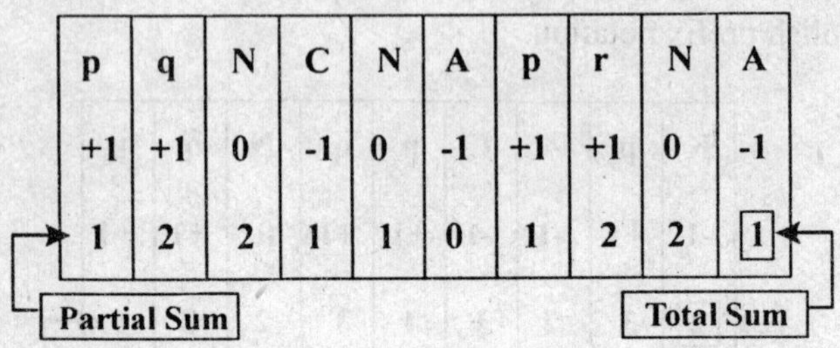

p	q	N	C	N	A	p	r	N	A
+1	+1	0	-1	0	-1	+1	+1	0	-1
1	2	2	1	1	0	1	2	2	1

Fig. 6.5.3.

Though the total sum is equal no 1, the given expression is **not** a valid well-formed formula.

It is worth mentioning, as a conclusion, that sets, lattices, finite Boolean algebra and proposition have many properties in common. These similarities have been outlined in this chapter to make the reader aware that how the concepts of hardware circuit design has evolved over the period of time. This shows that how a basic mathematics theory can bring a revolutionary change in industry when supported by an appropriate technology e.g. integration techniques.

Exercise

1. Determine whether the relation R is a linear order on the set A.
   (a) $A = R$ –the set of real numbers, and aRb iff a is less than or equal to b ($a \leq b$).
   (b) $A = 1$ –the set of all integers, and aRb iff a divides b^2.
   (c) $A = R$ –the set of real numbers, and aRb iff $a \geq b$.
   (d) $A = R \times R$, where R is the set of real numbers, and (a, b) R (a', b') iff $a \leq b$ and $b \leq b'$, where $\leq$ is usual partial order.
2. What can you say about the relation R on a set A if R is a partial order and an equivalence relation?
3. If $(A, \leq)$ is a poset and A' is a subset of A, show that $(A', \leq')$ is also a poset, where $\leq'$ is the restriction of $\leq$ to A'.

4. Show that if R is a linear order on A then its inverse R^{-1} is also a linear order on A.
5. Let $A = \{x \mid x \text{ is a real number and } -5 \leq x \leq 20\}$. Show that the usual relation < is a quasi-order.
6. Let $A = \{1, 2, 3, 5, 6, 10, 15, 30\}$ and consider the partial order $\leq$ of divisibility on A. That is $a \leq b$ iff a divides b $(a \mid b)$. Let $A' = P(S)$, where $S = \{x, y, z\}$, be the poset with partial order $\subseteq$. Show that $(A, \leq)$ and $(A', \subseteq)$ are isomorphic.
7. Show that if R is a quasi-order on A then its inverse R^{-1} is also a quasi-order on A.
8. Determine the Hasse diagram of the relation on $A = \{1, 2, 3, 4, 5\}$, the matrix of which is given below.

(a) $\begin{bmatrix} 1 & 1 & 1 & 1 & 1 \\ 0 & 1 & 1 & 1 & 1 \\ 0 & 0 & 1 & 1 & 1 \\ 0 & 0 & 0 & 1 & 1 \\ 0 & 0 & 0 & 0 & 1 \end{bmatrix}$ (b) $\begin{bmatrix} 1 & 0 & 1 & 1 & 1 \\ 0 & 1 & 1 & 1 & 1 \\ 0 & 0 & 1 & 1 & 1 \\ 0 & 0 & 0 & 1 & 0 \\ 0 & 0 & 0 & 0 & 1 \end{bmatrix}$

9. Determine the matrix of the partial order relation of which Hasse diagram is given by the following figures.

(a)

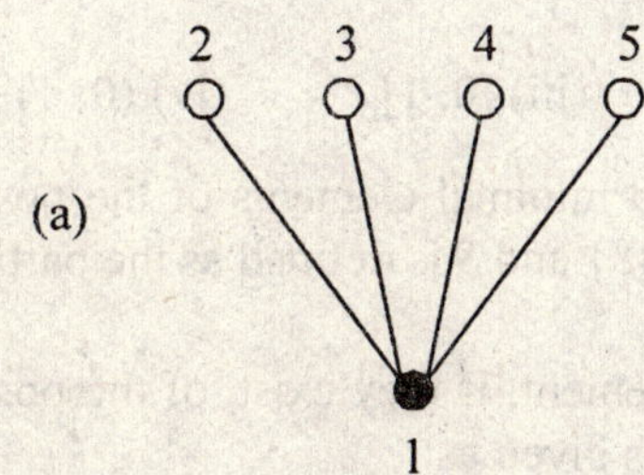

(b)

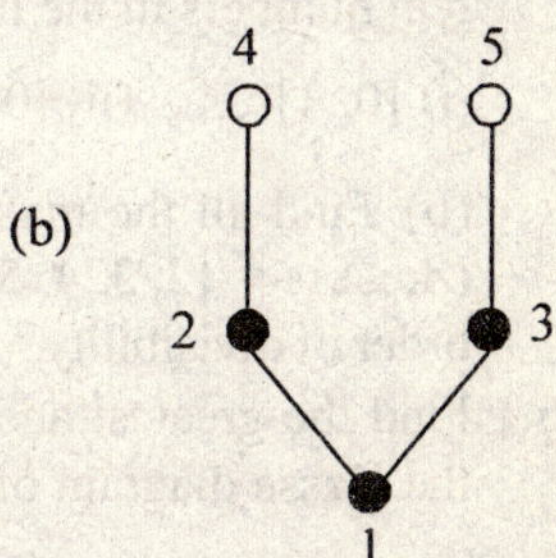

10. Let $A = I^+ \times I^+$ have lexicographic order. Mark each of the follwoing as true or false.

(a) $(2, 12) < (5, 3)$ (b) $(3, 6) < (3, 24)$

(b) $(4, 8) < (4, 6)$ (c) $(4, 8) < (4, 6)$

(d) $(15, 92) < (12, 3)$

11. Let $R = \{(a, a), (b, b), (c, c), (d, d), (e, e), (c, a), (c, b), (d, a), (d, b), (d, e), (b, a), (e, a)\}$ be a relation defined on set $A = \{a, b, c, d, e\}$. Draw the Hasse diagram for R.
12. Let $A = \{1, 2, 3, 5, 6, 10, 15, 30\}$ be a set and R be a relation of divisibility on A. Show that R is a partial order on A and draw a Hasse diagram of R. Is R a linear order relation? What about if $A = \{2, 4, 8, 16, 32\}$?
13. Find all the maximal and minimal elements of the poset shown by the following Hasse diagrams.

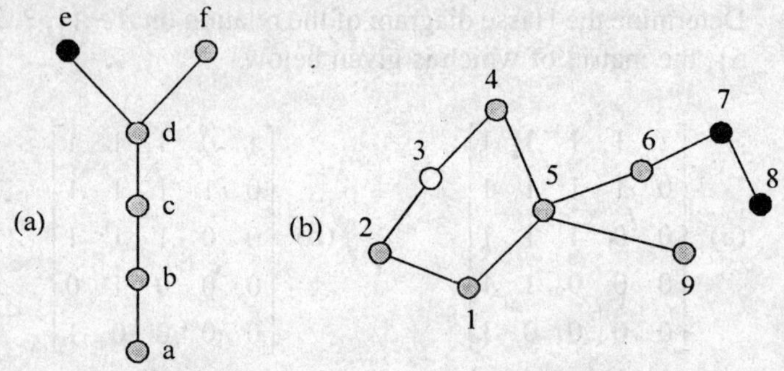

14. (a) Find all the maximai and minimal elements of the poset $(A, \leq)$, where $\leq$ is the usual partial order and A is the set of all real numbers in the intervals

    (i) [0, 1] (ii) [0, 1] (iii) [0, 1] (iv) (0, 1)

    (b) Find all the maximal and minimal elements of the poset $(A, \leq)$, $A = \{2, 3, 4, 5, 8, 24, 48\}$ and S is defined as the partial order of divisibility.
15. Find the greatest and least element, if they exist, of the poset the Hasse diagram of which is given as:

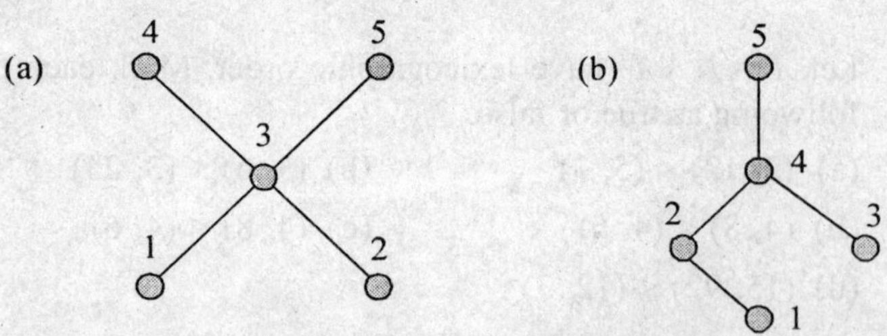

16. Find the greatest and least elements in poset given in exercise 14.
17. Find all the upper bounds and lower bounds of $B = \{x \mid x$ is a real number in $[1, 2]$, in the poset $(A, \leq)$, where $A = \{x \mid x$ is any real number$\}$ and $\leq$ is usual partial order.
18. Construct the Hasse diagram of a topological sorting of the poset whose Hasse diagram is shown in 15(a).
19. Let $A = \{2, 3, 4, 6, 8, 12, 24, 48\}$ and $\leq$ denotes the relation of divisibility i.e., $a \leq b$ iff a divides b. Show that $(A, \leq)$ is a poset and find all the lower bounds, upper bound, glb and lub if any, of subset $B = \{4, 6, 12\}$.
20. Let $A = R$ –the set of all real numbers, and $\leq$ is the usual partial order on A. Find all the lower bounds, upper bounds, glb and lub, if any, of subset $B = \{x \mid x$ is a real number in $[1, 2]\}$.
21. Let $A = R$ the set of all real numbers, and $\leq$ is the usual partial order on A. Find all the lower bounds, upper bounds, glb and lub, if any, of subset $B = \{x \mid x$ is a real number in $(1, 2)\}$.
22. Let L and M are two lattices shown by the following Hasse diagram. Draw the Hasse diagram of $L \times$ M with product partial order.

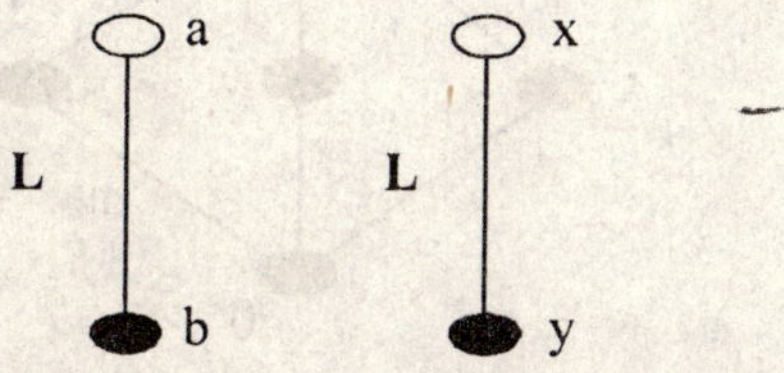

23. Let $L = P(S)$ be the lattice of all subsets of a set S under the relation of set containment. If T is a subset of S then show that $P(T)$ is a sub-lattice of 1.
24. Let $S = \{a, b, c\}$ and $L = P(S)$. Prove that $(L, \subseteq)$ is isomorphic to D_{42}.
25. Let S is a set. Let $\mathfrak{R}$ be the set of all equivalence relations on S and $\prod$ be the set of all equivalence partitions of S. We know that every equivalence relation induces a partition, and there is an equivalence relation corresponding to every equivalence

partition, we can define a function f from $\Re$ to $\prod$ which is one to one and onto. We know that $(\Re, \subseteq)$ is a poset. Answer the following:

(a) Show that $(\Re, \subseteq)$ is a lattice.

(b) Let us define a relation $\leq$ on $\prod$ as for any P_1 and P_2 of $\prod$, $P_1 \leq P_2$ iff $R_1 \subseteq R_2$, where R_1 and R_2 are equivalence relations corresponding to partition P_1 and P_2 respectively. Show that $(\prod, \leq)$ is a lattice.

(c) Let $P_1 = \{A_1, A_2, \ldots\}$ and $P_2 = \{B_1, B_2 \ldots\}$ be two partitions of S. Show that $P_1 \leq P_2$ iff each A_m is contained in some B_1.

26. Let L be a bounded lattice with at least two elements. Show that no elements of L is its own complement.

27. A lattice is said to be **modular** if, for all a, b, and c, $a \leq c \Rightarrow a \vee (b \wedge c) = (a \vee b) \wedge c$. Show that a distributive lattice is modular. Further show that the lattice shown in the following Hasse diagram is a non-distributive lattice and is modular.

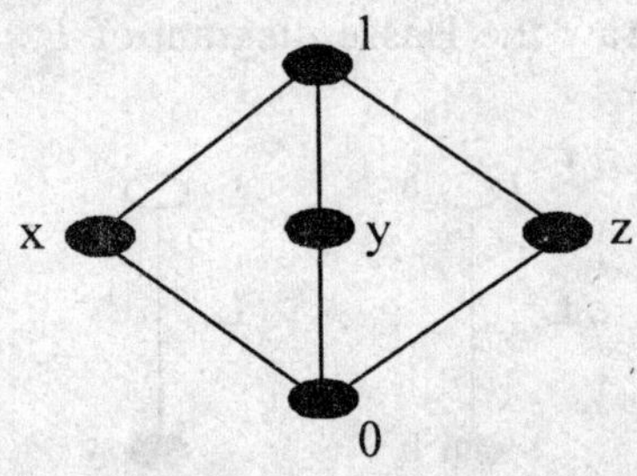

28. Prove that if a and b are elements of a bounded distributive lattice and if a has a complement a', then

(a) $a \vee (a' \wedge b) = a \vee b$ (b) $a \wedge (a' \vee b) = a \wedge b$

30. Show that if L_1 and L_2 are two distributive lattices then $L = L_1 \times L_2$ is also distributive, where partial order in L is the product of partial orders in L_1 and L_2.

31. Find the complement of each element in D_{42}.

32. Find the complement of each element in D_{105}.

33. Show that set of all positive integers together with the usual partial order 'Less than or equal to' forms a distributive lattice.
34. Show that the lattice D_n is distributive for any n.
35. Let L be a lattice and let a and b be elements of L such that $a \leq b$. The interval $[a, b]$ is defined as the set of all $x \in L$ such that $a \leq x \leq b$. Prove that $[a, b]$ is a sub-lattice of L.
36. Find all sub-lattice of D_{24} that contains at least five elements.
37. Show that a subset of a linearly ordered poset is a sub-lattice.
38. Let L is a distributive lattice. Show that if there exists an a with $a \wedge x = a \wedge y$ and $a \vee x = a \vee y$ then $x = y$.
39. Let a, b and c be elements of a lattice $(A, \leq)$. Show that

    (a) $a \wedge b = b$ if and only if $a \vee b = b$.

    (b) If $a \leq b$, then $a \vee (b \wedge c) \leq b \wedge (a \vee c)$.

    (c) $a \vee (b \wedge c) \leq (a \vee b) \wedge (a \vee c)$.

40. Prove that a relation corresponding to a triangular matrix is anti-symmetric relation. Show by the method of counter example that the reverse of the above statement is not true.
41. Simplify the following Boolean expressions.

    (a) $(a \wedge b) \vee (a \wedge b \wedge c) \vee (b \wedge c)$.

    (b) $(a \wedge b) \vee (a' \wedge b \wedge c') \vee (b \wedge c)$.

    (c) $((a \wedge b') \vee c) \wedge (a \vee b') \wedge c$.

42. Show that in a finite Boolean algebra the following statements are equivalent for any a and b.

    (a) $a \vee b = b$ (b) $a \wedge b = a$

    (c) $a' \vee b = 1$ (d) $a \wedge b' = 0$ and

    (e) $a \leq b$

43. Show that in a finite Boolean algebra, for any a, b and c $((a \vee c) \wedge (b' \vee c))' = (a' \vee b) \wedge c'$.
44. Show that in a finite Boolean algebra, for any a, b and c, if $a \leq b$, then $a \vee (b \wedge c) = b \wedge (a \vee c)$.
45. Consider the Boolean polynomial $p(x, y, z) = (x \vee y) \wedge (z \vee x')$. If $B = [0, 1]$, compute the truth table of the function $f: B_3 \rightarrow B$ defined by p.

46. Consider the Boolean polynomial $p(x, y, z) = (x \wedge y) \vee (x' \wedge (y \wedge z'))$. If $B = \{0, 1\}$, compute the truth table of the function $f: B_3 \to B$ defined by p.

47. Show that the following Boolean polynomials are equivalent

    (a) $(x \vee y) \wedge (x' \vee y)$ and y
    (b) $x \wedge (y \vee (y' (y \vee y')))$ and x
    (c) $(z' \vee x) \wedge ((x \wedge y) \vee z) \wedge (z' \vee y)$ and $x \wedge y$.

48. Reduce the following Boolean expressions, defined over the two-valued Boolean algebra, to Conjunctive Normal Form and Disjunctive Normal Form.

    (a) $f(x, y, z) = (x \wedge y) \vee (x \wedge z) \vee (y' \wedge z)$
    (b) $f(x, y, z) = \{(x \vee y)' \vee (x' \wedge z)\}'$
    (c) $f(w, x, y, z) = (w \wedge x \wedge y') \vee (w \wedge x' \wedge z) \vee (x \wedge y' \wedge z')$

49. Using Karnaugh map minimize the following Boolean expressions:

    (a) $xyz + x'z + y'z'$
    (b) $x (y'z + y(xz + y')) + x'(z + y (z + x'))$
    (c) $w (x + y (z + x') + y') + w'x'y'z'$
    (d) $wx'y + wx'z + w(y' + z) + w'x'(y' + z')$
    (e) $v'wxy + vw'xy + vwx'y + vwxy' + vxz + x'y'z + w'xy$.

50. Using Quine McCluskey method minimize the following Boolean expressions:

    (a) $x (y'z + y(xz + y')) + x'(z + y(z + x'))$
    (b) $wx'y + wx'z + w(y' + z) + w'x(y' + z')$
    (c) $v'wxy + vw'xy + vwx'y + vwxy' + vxz + x'y'z + w'xy$
    (d) $u'xy' + v'wz' + u'xz' + v'wy + x'y'z + uvw$
    (e) $u (v + w (x + y (z + x) + v) + w') + u'vx'z + u'wx'y' + v'wz$.

51. Convert the following formula into Polish prefix and Polish postfix notation:

    (a) $((p \vee \neg q) \to (p \wedge \neg r)) \equiv (\neg p \to (q \vee r))$.
    (b) $\neg(\neg p \equiv \neg(q \to \neg r)) \wedge (p \to (q \vee r) \equiv (\neg r \vee \neg p))$.
    (c) $((p \to \neg r) \wedge ((q \vee s) \equiv (p \wedge \neg s))) \to (p \vee \neg q \vee r \vee s)$.

(d) $((p \vee \neg q \vee s) \wedge (q \vee r \vee \neg s)) \rightarrow (((\neg p \vee q) \rightarrow (\neg r \vee s)) \equiv (p \equiv s))$.

52. Test the formula to determine whether it is well-formed or not.

(a) ***ACpEpNqAKprCNrNEpq.***

(b) ***pqrAqNpNsKCNpNCrpNqACE.***

(c) ***EACpqrNpsNAKpqNsKACpNCqs.***

(d) ***spEpNqAKNprCApNprsAKspCqKApKCE.***

(e) ***pqKrNpAqsNrNKApNqAK.***

(f) ***CAsEpACqNpAKrpqs.***

53. Convert the following formula to infix notation:

(a) ***CAKEpqNrAKpNrNqp.***

(b) ***KAEpEqrENpApqKANprNq.***

(c) ***pqrKANpNqrEEprKApNCE.***

(d) ***CpCqKNqNCAprKANpNAqrp.***

(e) ***pNqANpqNprpKANACpEqAK.***

(f) ***rqNANpCqNEprqNENAC.***

7

Finite Automata

Since the days of realization that an electronic device can compute any complex mathematical expression, substantial effort has been made to define a computing device that will be general enough to compute every "computable" function. In 1936 ***Turing*** suggested the use of a machine. Since then it is known as the ***Turing machine***, which is considered to be the most general computing device. The study of the Turing machine is beyond the scope of this course. It is however, important to mention here that any computation that can be described by means of ***Turing machine*** can be mechanically carried out. Conversely, any computation that can be performed on a modern day digital computer can be described by means of a ***Turing machine***. This is according to the definition of a ***Turing machine.***

The ***Turing machine is the most general possible computing device.*** There are different classes of ***Turing machines***. We devote this chapter to discuss the simplest possible class of computing device called ***finite automata*** or ***finite states machine***. We shall study a finite state machine as

- an acceptor of a *language*
- a generator of a language
- an algorithm to test whether a sentence is valid?

Any language is suitable for communication provided the syntax and semantic of the language is known to the participating sides. It is made possible by forcing a standard on the way to make sentences from words of that language. This standard is forced through a set of rules. This set of rules is called ***grammar*** of the language. We shall study in this text the most general type of grammars called **Chomsky grammar** and the languages defined by it.

7.1 Grammar and Languages

Let A be a set of symbols. Let A^* be a set of all finite sequences of symbols of A. The set A is referred as **alphabets** and a sequence in A^* is called **word**. The number of symbols in a word is called **length of the word**. A word of length zero is called **empty string.** A^* may contains an empty string. An empty string is also called **NULL** string and is denoted by $\wedge$ or by ε (pronounced as epsilon). For example, let

$A \;= \{a, b, c, \ldots, z\}$ then
$A^* = \{x \mid x$ is an ordinary word of dictionary in case insensitive term including $\varepsilon\}$

Grammar is probably the most important class or generator of languages. A grammar is a mathematical system for defining a language. As well, it is a device for giving the sentences in the languages a useful structure.

A grammar for a language L uses two finite disjoint sets of symbols. These are the set of **non-terminal** symbols, denoted by N in this text, and the set of **terminal** symbols, denoted by Σ. The set of terminal symbols is the set of alphabets over which the language is defined. Non-terminal symbols are used in the generation of words in the language. The heart of a grammar is a finite set P – production rules, which describe how the sentences of the language are to be generated. A production rule is of the form

$\alpha \to \beta$

where $\alpha \in (N \cup \Sigma)\, N\, (N \cup \Sigma)$ and $\beta \in (N \cup \Sigma)$. That is, α is any string containing at least one non-terminal and β is any string including a null string. The union $N \cup \Sigma$ of sets of non-terminal and of terminal symbols is also called **total vocabulary** and is denoted by V. Now, we can give a formal definition of a grammar as below.

A grammar is defined as a 4-tuple mathematical structure $G\,(N, \Sigma, P, S)$ where

- N is a non-empty, finite set of non-terminal symbols (sometimes called variables or syntactic categories).

- Σ is a non-empty, finite set of terminal symbols. The set N is disjoint from Σ i.e. *a symbol cannot be both terminal and non-terminal in the same grammar.*
- P is an non-empty, finite set of production rules of the form $\alpha \to \beta$. Notice that P is an non-empty finite subset of Cartesian product of sets $(N \cup \Sigma)\ N(N \cup \Sigma)$ and $(N \cup \Sigma)$.
- S is the distinguished symbol in N called the ***sentence*** (or **start** symbol).

Example 1: An example of a grammar is $G(\{A, S\}, \{1, 0\}, P, S)$ where, P consists of

1. $S \to 0A1$ 2. $0A \to 00A1$ 3. $A \to \varepsilon$

The non-terminal symbols are A, S and terminal symbols are 1, 0. The symbol ε is used to denote an empty string.

A grammar defines a language in a recursive manner. A ***sentential form*** of a grammar $G(N, \Sigma, P, S)$, is defined recursively as

- S is a ***sentential form***
- If $\alpha\ \beta\ \gamma$ is a sentential form and $\beta \to \delta$ is in P, then $\alpha\ \delta\ \gamma$ is also a sentential form.

A sentential form of G containing no non-terminal symbols is called a ***sentence*** generated by G. A ***language*** generated by a grammar G is the set of sentences generated by G and is denoted by $L(G)$.

Example 2: Find the language of the grammar G given in the example 1 of this section 7.1.

Solution: We apply the production rules one by one. From rule 1 we have

$S \to 0A1$ [From rule 1]

$\to 00A11$ [From rule 2 $0A \to 00$ 1]

$\to 000A111$ [Applying rule 2 2nd time]

$\to 0000A1111$ [Applying rule 2 3rd time]

...

...

$\to 0^nA\ 1^n$ [Applying rule 2 $(n-1)$ times]

$\to 0^n\ 1^n$ [Applying rule 3 $A \to \varepsilon$]

Optionally, when rule 3 is applied just after rule 1, it gives $S \rightarrow 01$. This shows that rule 2 can be applied 0 or more times giving the string 0^n1^n for $n > 0$. Therefore, $L(G) = \{0^n1^n \mid n > 0\}$.

Ans.

A sentence in a language can be derived from its grammar. Before proceeding further, let us get acquainted with some terminology. Let (N, Σ, P, S) be a grammar G. We define a relation $\Rightarrow$ (read as *directly derives*) on V^* as:

If $\alpha\beta\gamma$ is a string in V^* and $\beta \rightarrow \delta$ is a production rule in P, then

$\alpha\beta\gamma \Rightarrow \alpha\, \delta\, \gamma$

Then n-steps transitive closure of $\Rightarrow$ is denoted as $\Rightarrow^n$. That is to say $\alpha \Rightarrow^n \beta$ if there is a sequence $a_0, a_1, a_2, \ldots, a_k, \ldots, a_n$ of $n + 1$ strings (not necessarily distinct) such that

$\alpha = \alpha_0$;
$\alpha_{k-1} \rightarrow \alpha_k$ **for** $\mathbf{1 \leq k \leq n}$ and
$\alpha_n = \beta$.

This sequence of strings is called a derivation of length n of β from α in G. In general, the transitive closure of $\Rightarrow$ is denoted as $\Rightarrow^+$ (read as *derives in a nontrivial way*). It implies that $\alpha \Rightarrow^+ \beta$ if and only if β has been derived from α by applying at least one production from P. Similarly, the reflexive and transitive closure of $\Rightarrow$ is denoted as $\Rightarrow^*$ (read as *derives*). It implies that $\alpha \Rightarrow^* \beta$, if and only if β has been derived from α by applying zero or more productions from P.

Example 3: Let $G = (N, \Sigma, P, S)$ be a grammar, where

$N = \{S, L, D, W\}$,

$\Sigma = \{a, b, c, 0, 1, 2, 3, 4, 5, 6, 7, 8, 9\}$ and

P is given by:

1. $S \rightarrow L$ 2. $S \rightarrow LW$ 3. $W \rightarrow LW$ 4. $W \rightarrow DW$
5. $W \rightarrow L$ 6. $W \rightarrow D$ 7. $L \rightarrow a$ 8. $L \rightarrow b$
9. $L \rightarrow c$ 10. $D \rightarrow 0$
11. $D \rightarrow 1$ ....19. $D \rightarrow 9$

Which of the following statements are true for this grammar.

(a) $ab092 \in L(G)$ (b) $2a3b \in L(G)$

(c) $aaaa \in L(G)$ (d) $S \Rightarrow a$

(e) $S \Rightarrow^* ab$ (f) $DW \Rightarrow 2$

(g) $DW \Rightarrow^* 2$ (h) $W \Rightarrow^* 2abc$

(i) $W \Rightarrow^* ba2c$

Solution: (a) $ab092 \in L(G)$ is true if $ab092$ is a valid sentence i.e. it is derived from starting symbol. S using one or more productions in P. We have the following derivation.

$S \to LW$	[From rule 2]
$\to LLW$	[From rule 3, $W \to LW$]
$\to LLDW$	[From rule 4, $W \to DW$]
$\to LLDDW$	[From rule 4, $W \to DW$]
$\to LLDDD$	[From rule 6, $W \to D$]
$\to LLDD2$	[From rule 12, $D \to 2$]
$\to LLD92$	[From rule 19, $D \to 9$]
$\to LL092$	[From rule 10, $D \to 0$]
$\to Lb092$	[From rule 8, $L \to b$]
$\to ab092$	[From rule 7, $L \to a$]

As shown above, $ab092$ is a sentence in $L(G)$. Therefore, this statement is true. The above derivation can be better shown as the following **derivation tree.** (on next page)

In the above derivation tree, the starting symbol S is at the root, all the intermediate nodes are non-terminal symbols and leaves are terminal symbols from G. *For a string to be a valid sentence of a language L(G), all the leaves in the derivation tree must be terminal symbols. Otherwise the sentence does not belong to the language.* We read symbols at the leaves from left to right to get the sentence derived and in the case it is **ab092.**

(b) Here again, $2a3b \in L(G)$ is true if $2a3b$ is a valid sentence. See the 19 productions in P. As usual, we have to begin with S to derive any sentence or to test validity of any string. While proceeding from S, we can have either L or LW in the next step. This means, the left most non-terminal will always be an L. The non-terminal L can be replaced either by ***a*** or ***b*** or *c*. Therefore, any string acceptable (derivable) by this grammar must have either ***a*** or ***b*** or ***c*** as its left most symbol. The given string $2a3b$ does not satisfy this condition. Hence, $2a3b$ does not belong to $L(G)$.

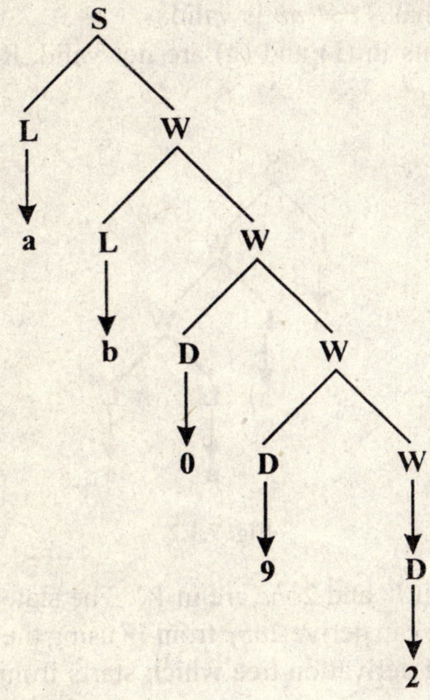

Fig. 7.1.1.

(c) The derivation of the strings is shown in the derivation tree on page 284.

From the derivation tree, it is obvious that string ***aaaa*** is a valid sentence of the language $L(G)$. Hence the statement is true.

(d) As $\Rightarrow$ stands for directly derives. "$S \Rightarrow a$" will be valid if there is any production in P of the form $S \rightarrow a$ which directly provides for substituting S by a. Since there is no production of the form $S \rightarrow a$ in P, the statement "$S \Rightarrow a$" is not true.

(e) As $\Rightarrow^*$ stands for derives in zero or more steps (reflexive and transitive closure), "$S \Rightarrow^* ab$" will be valid if the string ***ab*** belongs to $L(G)$. Let us derive this as follows:

$S \Rightarrow^1 LW$	[From rule 2 $S \to LW$]
$\Rightarrow^2 aW$	[From rule 7 $L \to a$]
$\Rightarrow^3 aL$	[From rule 5 $W \to L$]
$\Rightarrow^4 ab$	[From rule 8 $L \to b$]

This shows that $S \Rightarrow^* ab$ is valid.

The statements in (f) and (g) are not valid. Reader may verify this.

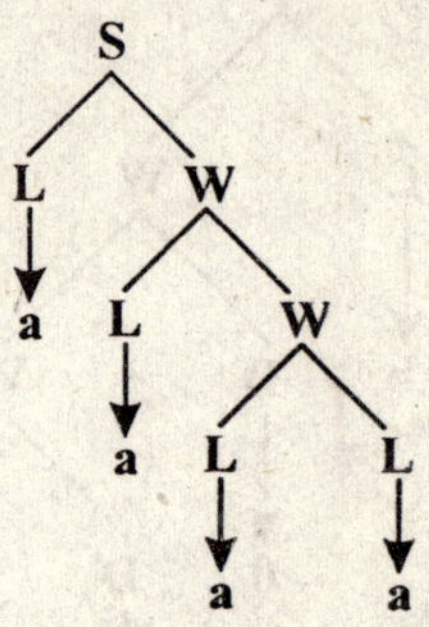

Fig. 7.1.2.

(h) Here, both W and $2abc$ are in V^*. The statement $W \Rightarrow^*$ 2abc will be valid if we can derive $2abc$ from W using the productions in P. See the following derivation tree which starts from W as it's root.

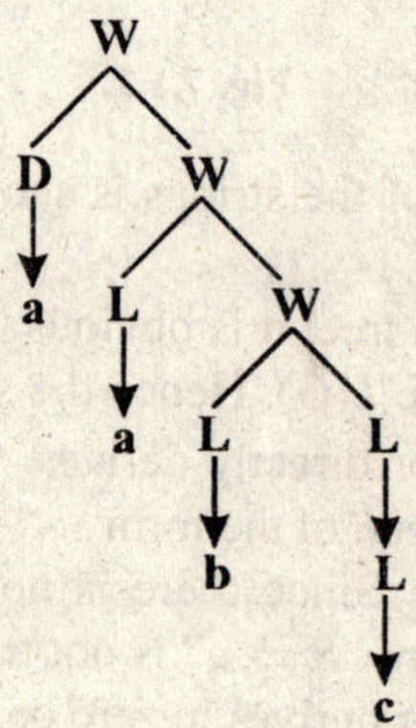

Fig. 7.1.3.

In the above derivation tree all the leaves are terminal symbols. This shows that starting from W we can derive $2abc$ using productions of P. Therefore, it is a valid statement.

The statement in (i) is also true. Reader may verify this as an exercise.

This completes the Answer.

For examples 2 and 3, it should be clear, how to find language of a grammar and test whether a given sentence is valid according to the grammar. The process of testing whether a sentence is valid begins from sentence itself. Derivation tree is an important tool to test the validity of a sentence under some types of grammar. Let us now summarize the definition of a ***derivation tree.***

Derivation Tree

A derivation tree is also called a *parse tree*. A tree is a derivation tree of a valid sentence in a grammar $G = (N, \Sigma, P, S)$ if

- Every vertex is a symbol of the set $N \cup \Sigma \cup \{\varepsilon\}$.
- The root is the start symbol S.
- If A is an interior vertex in the tree, then A must be in N.
- If a vertex A has n Siblings $X_1, X_2, X_3, \ldots, X_n$ in order from left to right, then $A \rightarrow X_1X_2X_3 \ldots X_n$ must be a production in P.
- If ε is a vertex, then ε must be a leaf and it must be the only offspring of its parent.

There are certain grammars in which derivations of sentences cannot be expressed as trees. Construction of a derivation tree works only if the left hand side of every production in G contains a single, non-terminal symbol. See the grammar of ***example 1*** above. In the grammar of this example, the left hand side does not have this simple form. Although it is possible to construct a graphical representation of derivations (***example 2***) under this grammar, the resulting digraph would not be a tree. Many other problems arise in an unrestricted grammar. To simplify the situation, some restrictions are imposed on α in the format of production $\alpha \rightarrow \beta$ of a general grammar. The subsequent restrictions yield different types of grammar. The general format of the grammar we have been discussing is also known as ***phrase structure grammar, or Chomsky grammar.*** A grammar $G = (N, \Sigma, P, S)$ is said to be of

Type 0: if there is no restriction on the production rules i.e., in $\alpha \to \beta$, where $\alpha, \beta \in (N \cup \Sigma)^*$. This type of grammar is also called an ***unrestricted grammar.***

Type 1: if in every production $\alpha \to \beta$ of P, $\alpha, \beta \in (N \cup \Sigma)^*$ and $|\alpha| \leq |\beta|$. Here $|\alpha|$ and $|\beta|$ represent number of symbols in string α and β respectively. This type of grammar is also called a ***context sensitive*** grammar (or ***CSG***).

Type 2: if in every production $\alpha \to \beta$ of P, $\alpha \in N$ and $\beta \in (N \cup \Sigma)^*$. Here α is a single non-terminal symbol. This type of grammar is also called a ***context free*** grammar (or ***CFG***).

Type 3: if in every production $\alpha \to \beta$ of P, $\alpha \in N$ and $\beta \in (N \cup \Sigma)^*$. Here α is a single non-terminal symbol and β may consist of at the most one non-terminal symbol and one or more terminal symbol. The non-terminal symbol appearing in β must be the extreme right symbol. This type of grammar is also called a ***right linear*** grammar or ***regular*** grammar (or ***RG***).

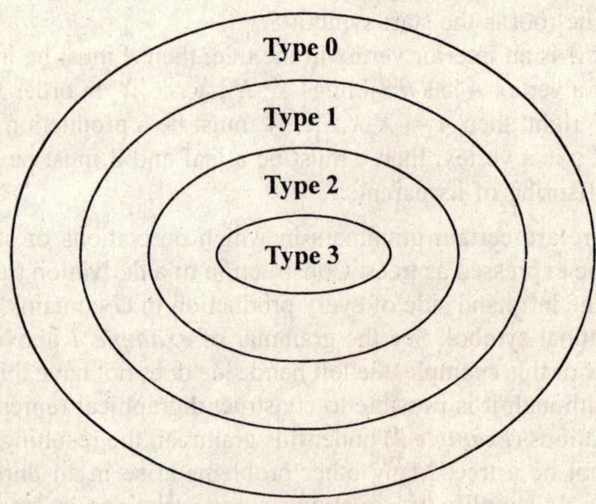

Fig. 7.1.4.

$S \to \in$

$S \to \alpha$

This classification is widely known as ***Chomsky hierarchy.*** The hierarchy can be better shown diagrammatically through the adjoining figure 7.1.4. It is obvious that type-I grammar is of type-J whenever $I > J$.

Now we shall learn how to find the language generated by a given grammar. The process involves a number of steps starting from the production with start symbol S. See the following examples.

Example 4: Find the language $L(G)$, where $G = (N, \Sigma, P, S)$ is given as: $N = \{S, W\}$; $\Sigma = [a, b, c\}$ and $P = \{S \to aW, W \to bbW, W \to c\}$.

Solution: The given grammar is of type-3. To deduce the language for this grammar we proceed step by step as follows.

Step 1: Start from one of the production $\alpha \to \beta$ in which $\alpha = S$ (start symbol). If β contains start symbol, eliminate the start symbol from the strings so obtained by using the appropriate production. In this example, we have only one production beginning with S. So, we have

$S \to aW$

The string "aW" does not contain S. So its elimination is not relevant here. Go to the step 2.

Step 2: Use other productions successively to eliminate all non-terminals presents in the string so obtained. If the a production is recursive, use it ***n*** times

$S \to aW$

$\to a(bbW)$ [Applying production 2]

$\to a(bb)^n W$ [Applying production 2, n times]

$\to a(bb)^n c$ [Applying production 3]

Step 3: If there are more than one production $\alpha \to \beta$ having the same α, apply them to get all possible alternatives. Combine all these alternatives with **OR** ($\vee$). In this case, we have two productions for W: one we have used in the step 2 & the other is $W \to c$. This we can apply without applying production 2 even once. This gives a string

$S \to aW$

$\to ac$ [Apply rule 3 once]

Step 4: Combine all the alternatives so obtained. In the case combining the two alternatives, we get

$$S \rightarrow ac \vee a(bb)^n c$$
$$= a(\varepsilon \vee (bb)^n)c$$
$$= a(bb)\ c$$

Therefore, the language of this grammar can be written in set theoretic notation as

$$L(G) = \{x \mid x = a(bb)^n c \text{ for } n \geq 0\}$$

Ans.

Example 5: Find the Language $L(G)$ where, $G = (N, \Sigma, P, S)$ is given as: $N = \{S, W\}$, $\Sigma = \{a, b, c\}$ and $P = \{S \rightarrow aSb, Sb \rightarrow bW, abW \rightarrow c\}$.

Solution: The given grammar is of type-0. To deduce the language for this grammar we process step by step as follows.

Step 1: Let us start with the production starting with start symbol S. In this example, there is one such production which is recursive. From this production, we have

$$S \rightarrow aSb$$
$$\rightarrow a^n S b^n \text{ [Apply the production 1, } n \text{ times]}$$

Now our immediate goal is to eliminate the non-terminal S from the string $a^n S b^n$. We have a production $Sb \rightarrow bW$ in P. Applying this we get

$$S \rightarrow a^n b W b^{n-1}$$

Step 2: Now eliminate W from right hand side of the above derivation. Using the production 3 $abW \rightarrow c$, we get

$$S \rightarrow a^{n-1}\ c b^{n-1}$$

Step 3: There is an alternative to the substitution in the step 1. We could have proceeded as below

$$S \rightarrow aSb$$
$$\rightarrow abW \text{ [Applying the production 2 once]}$$
$$\rightarrow c \quad \text{[Apply the production 2 once]}$$

Step 4: Combining the alternatives, we conclude that this grammar generates the language of the form $a^m\ c\ b^m$ where, $m \geq 0$. In set theoretic notation the language can be expressed as

$$L(G) = \{\ x \mid x = a^m\ c\ b^m \text{ for } m \geq 0\}$$

Ans.

Example 6: Find the Language $L(G)$ where, $G = (N, \Sigma, P, S)$ is given as: $N = \{S, A, B\}$; $\Sigma = \{(\ ), a, +\}$ and $P = \{S \to (S), S \to a + A, A \to a + B, B \to a + B, B \to a\}$.

Solution: The given grammar is of type-2. To deduce the language for this grammar, we proceed from start symbol. There are two productions beginning with S. We may start optionally with $S \to (S)$. This is a recursive production. Therefore, we have

$S \to (S)$ [Apply production 1]

$\to (^nS)^n$ [Apply production 1, n times]

Now symbol S is to be eliminated from the string obtained above. We apply production 2 to achieve this. It is important to note that the production 1 can be used any number of times, optionally. Thus, $n \geq 0$. We now have

$S \to (^na + A)^n, n \geq 0$ [Apply production 2]

$\to (^na + a + B)^n$ [Apply production 3]

One non-terminal B is yet to be eliminated from the string. There are two productions for B: one is recursive and the other ending in terminal symbol a. Combining these two for B, we can have a production for B as below

$B \to [a +]^m B, m \geq 0$ [Apply production 4 n times]

$B \to [a +]^m a, m \geq 0$ [Apply production 5]

Using this result for B in the derivation $S \to (^na + a + B)^n$, we get

$$S \to (^na + a + [a +]^m a)^n$$

In the above derivation, we may replace m with * but n cannot be replaced with *, because n stands for count of left and right parenthesis. Therefore, $L(G)$ can be expressed in set theoretic notation as

$$L(G) = \{x \mid x = (^na + a + [a +]^* a)^n \text{ for } n \geq 0\}$$

Ans.

We have seen some examples to learn how to find languages for a given grammar. It is true that for a given grammar there can be only one language. The same argument is not valid for the converse. For a given language we may find different grammars producing the same language. Therefore finding a grammar for a given language is a somewhat complicated process; though it is always possible to do that. See the following examples to know how to do it.

Example 7: Find a grammar for the language $L = \{x \mid x = a^n b^n$ for $n \geq 1\}$.

Solution: Here, language is defined over the alphabet $\Sigma = \{a, b]$. This language contains n (≥ 1) number of a and b & a's come first and all b's follows. Let S be the start symbol. Then, we have to think for a production such that whenever an "a" is added to the left side a "b" must be added to the right side. This can be achieved by the production

$$S \to aSb.$$

Next, for a sentence to be valid, it must not contain any non-terminal. This requires that there must be a production such that right side contains only terminal symbol(s). This is achieved by the production

$$S \to ab.$$

Therefore, $G = (N, \Sigma, P, S)$ is a grammar for the given language where, $N = \{S\}$, $\Sigma = \{a, b\}$, $P = \{S \to aSb, S \to ab\}$. This is a type –2 grammar.

Ans.

Example 8: Find a grammar for the language $L = \{x \mid x$ is a string of 0's and 1's with equal number n (≥ 0)$\}$.

Solution: The given language is defined over the alphabet $\Sigma = \{1, 0\}$. A string in this language contains equal number of 0 and 1. Strings 0011, 0101, 1100, 1010, 1001 and 0110 are all valid strings of length 4 in this language. Here, order of appearance of 0's and 1's is not fixed. Let S be the start symbol. A string may start with either 0 or 1. So, we should have productions $S \to 0S1$, $S \to 1S0$, $S \to 0A0$ and $S \to 1B1$ to implement this. Next A and B should be replaced in such a way that it maintains the count of 0's and 1's to be equal in a string. This is achieved by the productions $A \to 1S1$ and $B \to 0S0$. The **null** string is also an acceptable string, as it contains **zero** number of 0's and 1's. Thus $S \to \varepsilon$ should also be a production. Therefore, the required grammar G can be given by (N, Σ, P, S), where $N = \{S, A, B\}$, $\Sigma = (1, 0\}$ and $P = \{S \to 0S1, S \to 1S0, S \to 0A0, S \to 1B1, A \to 1S1, B \to 0S0, S \to e\}$.

Ans.

Example 9: Construct a phrase structure grammar g for the language

$L = \{a^m b^n \mid m, n \geq 1, m \neq \text{n}\}$

Solution: This language is defined over the alphabet $\Sigma = (a, b)$. It is important to note that

$L = L_1 \cup L_2$ where,

$L_1 = \{a^m b^n \mid m, n \geq 1, m > n\}$

$L_2 = \{a^m b^n \mid m, n \geq 1, m < n\}$

A production $P_1 = \{A \to aA, A \to aB, B \to aBb, B \to ab\}$ generates the language L_1. Similarly, production $P_2 = \{C \to Cb, C \to Db, D \to aDb, D \to ab\}$ generates the language L_2. Now, if we add two productions: $S \to A$ and $S \to C$ to the set $P_1 \cup P_2$, they together generate the language L. Therefore, the required grammar G can be given, in simplified way, by (N, Σ, P, S) where, $N = \{S, A, B, C\}$, $\Sigma = \{a, b\}$ and $P = \{S \to A, S \to C, A \to aA, A \to aB, B \to aBb, B \to ab, C \to Cb, C \to Bb\}$. Here, reader may note that non-terminal D has been combined with B.

Ans.

7.2 Regular Expression and Regular Sets

We have studied phrase structure grammar in the previous section. In type 2 grammar (which includes type 3 grammars), all productions contain a single non-terminal symbol on its left-hand side. Because of this special characteristics, type 2 and type 3 grammars can be represented by some useful alternative methods. One of them is BNF (Backus-Naur form) notation. The following steps are followed to represent productions in type 2 or type 3 grammar in BNF form.

Step 1: All non-terminals, whenever they occur, are enclosed in angle brackets (< >). For example S, A, B etc., are represented as <*S*>, <*A*>, <*B*>.

Step 2: If there is more than one production corresponding to same non-terminal, then they are listed together separated with |. ('|' stands for alternative option).

Step 3: The $\rightarrow$ symbol in production is replaced with the symbol "::=".

To make the things more clear, let us see the following examples.

Example 1: Let $G = (N, \Sigma, P, S)$ be a grammar, where $N = \{S, A\}$, $\Sigma = \{x, y, z\}$ and $P = \{S \rightarrow xS, S \rightarrow yA, A \rightarrow yA, A \rightarrow z\}$. Give BNF notation for the production of G.

Solution: The BNF representation for the productions of G can be given as:

<*S*> ::= *x* <*S*> | *y* <*A*>

<*A*> ::= *y* <*A*> | *z*

Ans.

Example 2: Let $G = (N, \Sigma, P, S)$ be a grammar, where $N = \{S, A\}$, $\Sigma = \{0, 1\}$ and $P = \{S \rightarrow 0A, A \rightarrow 11A, A \rightarrow 010A, A \rightarrow 1\}$, Give BNF notation for the productions of G.

Solution: The BNF representation for the productions of G can be given as:

<*S*> ::= 0 <*A*>

<*A*> ::= 11 <*A*> | 010 <*A*> | 1

Ans.

Note that the symbol on the left-hand side of a production may also appear in one of the strings on the right-hand side. For example, in <*S*> ::= *x* <*S*>, <*S*> appears on both the sides. Similarly, in <*A*> ::= *y* <*A*>, <*A*> appears on both the sides. When this happens, we say that the corresponding production is **recursive**. A **recursive** production is said to be **normal** if the non-terminal symbol appearing on the right-hand side is the **rightmost** symbol in the production and it appears **only once**. *Note that a recursive production that appears in a type 3 grammar normal production by definition.* A BNF notation can also be shown as a **syntax diagram**. To draw a syntax diagram we use two geometrical shapes: **circle** for terminal symbol and

rectangle for non-terminal. The step by step method to draw a syntax diagram is outlined below.

Step 1: Identify the terminals and non-terminals in a production. Represent a non-terminal by □ and a terminal by ○. Put the symbols inside the respective geometric shapes.

Step 2: Connect all the these shapes in the sequence in which the corresponding symbol appears in the production. If the production is normal, a loop is drawn from right most side to the initial state.

Step 3: If there are more than one production for a non-terminal, all of them are connected as alternatives.

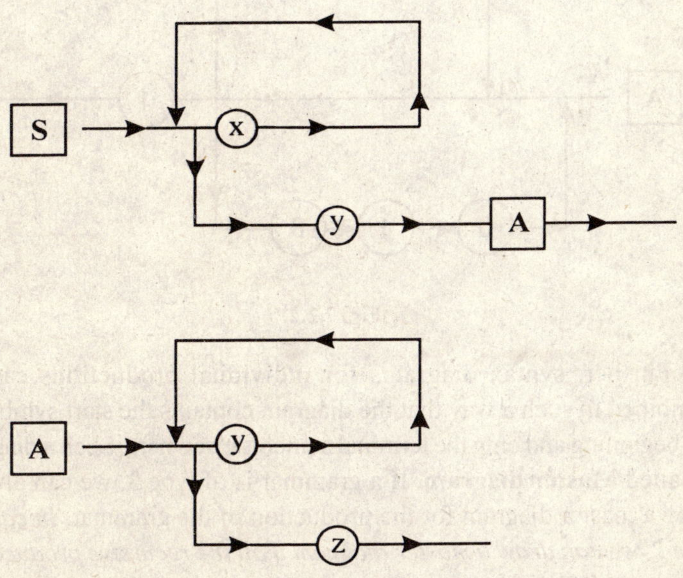

Fig. 7.2.1.

A syntax diagram for the productions of the grammar in example 1 can be given as in figure 7.2.1. The non-terminal <*S*> has two alternatives: one is a normal recursive and other is a linear one. Both these alternatives have been combined and shown as a syntax diagram for <*S*>. Similarly we can do for <*A*>.

In example 2, <*S*> has only one production and that is linear in <*A*>. This is shown in a line. The non-terminal <*A*> has three alternatives. These are connected in such a way that while moving from <*A*>, one can select any one of the three paths available at a time. The syntax diagram is shown in figure 7.2.2.

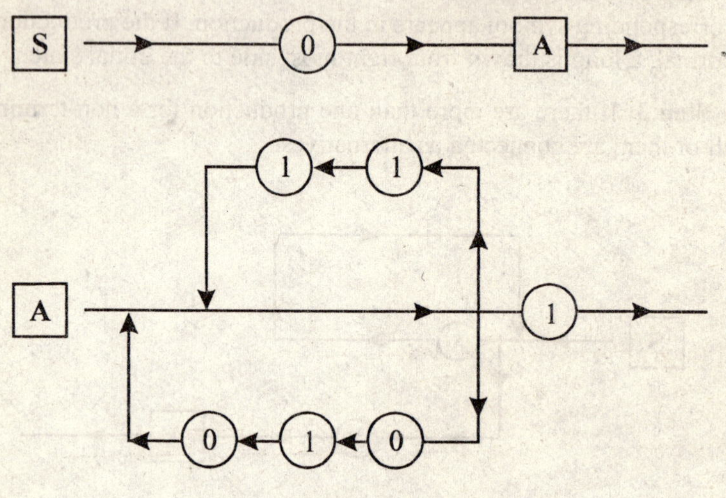

Fig. 7.2.2.

Further, syntax diagrams for individual productions can be combined in such a way that the diagram contains the start symbol in the beginning and only the terminal symbols elsewhere. Such a diagram is called **Master diagram.** If a grammar is of type 3, we can always draw a master diagram for the production of the grammar. *In case of type 2, we can draw a master diagram if all the recursive productions are normal.* This is obtained by replacing non-terminal successively by its syntax diagram. For example, see the figure 7.2.1. In the syntax diagram for <*S*>, there is one non-terminal <*A*>. If we replace this by its syntax diagram, we get a diagram as shown in figure 7.2.3. This does not contain any non-terminals.

Similarly, in example 2, a master diagram for the production of grammar can be given combining the syntax diagrams in figure 7.2.2. The diagram so obtained is shown in the figure 7.2.4.

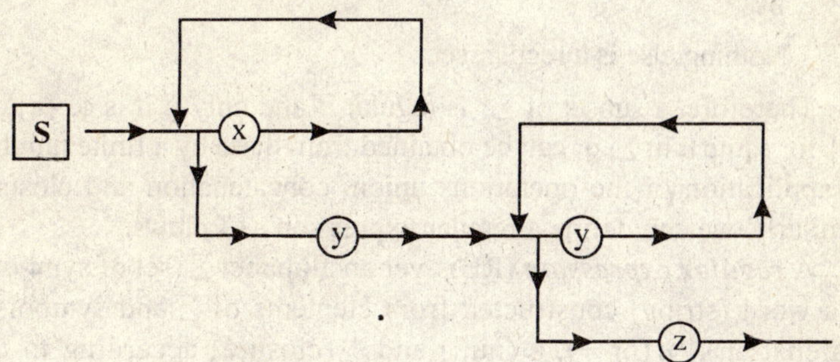

Fig. 7.2.3.

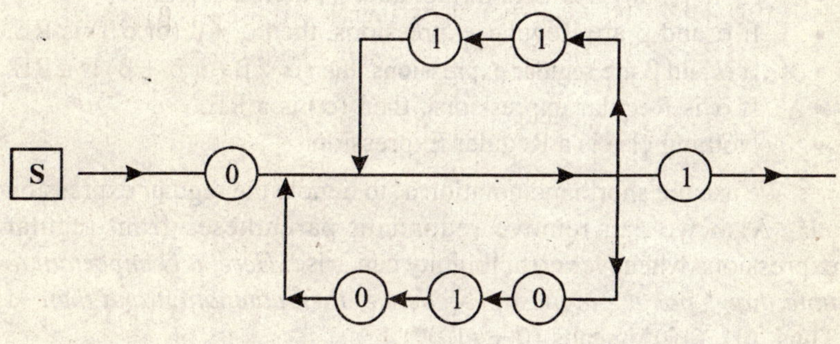

Fig. 7.2.4.

We have studied that a language generated by a regular grammar (type 3) is a regular language. From a master diagram of a regular grammar we can deduce the regular language generated by the grammar. The master diagram of figure 7.2.4 generates the expression $0\ (11 \vee 010)^*1$ and the master diagram of figure 7.2.3 generates the expression $x^*\ yy^*\ z$. Reader may wonder that how this has been done? Before responding to the query in your mind, let us first define a regular set and then a regular expression.

Let Σ be a finite alphabet. We define a regular set over Σ recursively in the following manner.

- ϕ is a regular set over Σ.
- $\{\varepsilon\}$ is a regular set over Σ.
- $\{a\}$ is a regular set over $\Sigma\ \forall\ a \in \Sigma$.

- If P and Q are regular sets over Σ the N so are $P \cup Q$, PQ and P^*.
- Nothing else is a regular set.

Therefore, a subset of Σ^* is regular if and only if it is ϕ. $\{\varepsilon\}$, or $\{a\}$ for same a in Σ, or can be obtained from these by a finite number of application of the operations union, concatenation and closure. Similarly, we can define a regular expression as follows.

A ***regular expression*** (RE) over an alphabet Σ (set of symbols) is a word (string) constructed from elements of Σ and symbols • (catenation), $\vee$ (or +), ε (null) and * (closure) according to the following definition.

- The symbol ε is a RE.
- If "a" is any symbols of alphabet Σ, then a is RE.
- If α and β are Regular expressions, then $\alpha \bullet \beta$ (or $\alpha\beta$) is RE.
- If α and β are regular expressions, then $\alpha \vee \beta$ (or $\alpha + \beta$) is a RE.
- If α is Regular expressions, then (α) is a RE.
- Nothing else is a Regular Expression.

We use the shorthand notation a^+ to denote the regular expression aa*. Also we can remove redundant parentheses from regular expressions whenever no ambiguity can arise. *Here, it is important to note that * has the highest precedence, then catenation and then +.* Thus, 0 + (10)* means (0 + (1(0*))).

Example 1: Some examples of regular expressions on alphabet $\Sigma = \{0,1\}$ are

(a) 01, denoting regular set {01}.
(b) 0*, denoting regular set {0}*.
(c) (0 + 1)*, denoting regular set {0,1}*.
(d) (0 + 1)*011, denoting the regular set of all strings of 0's and 1's ending in 011.
(e) (00 + 11)*((01 + 10)(00 + 11)*(01 + 10)(00 + 11)*)*, denoting the set of all strings of 0's and 1's containing an even number of both 0's and 1's.
(f) 0* (0 $\vee$ 1)*, denoting the set of any string in 0's and 1's.
(g) 00* (0 $\vee$ 1)*1, denoting the set of all strings in 0's and 1's that begin with 0 and ends in 1.

Example 2: $(a + b)(a + b + 0 + 1)^*$, denoting the set of all strings in $\{0, 1, a, b\}^*$ beginning with a or b.

It should be quite clear that for each regular expression we can construct the regular set denoted by that regular expression. Similarly, for each regular set, we can find at least one regular expression denoting that regular set. *Unfortunately, for each regular set, there is infinity of regular expressions denoting that set.* **Two regular expressions are equal if they denote the same regular set.** Some basic algebraic properties of regular expressions are given in the following theorem.

Theorem 1: Let α, β and γ be regular expressions, then

(a) $\alpha + \beta = \beta + \alpha$ (b) $\phi^* = \varepsilon$

(c) $\alpha + (\beta + \gamma) = (\alpha + \beta) + \gamma$ (d) $\alpha(\beta\gamma) = (\alpha\beta)\gamma$

(e) $\alpha (\beta + \gamma) = \alpha\beta + \alpha\gamma$ (f) $(\alpha + \beta)\gamma = \alpha\gamma + \beta\gamma$

(g) $\varepsilon\alpha = \alpha\varepsilon = \alpha$ (h) $\phi\alpha = \alpha\phi = \phi$

(i) $\alpha^* = \alpha + \alpha^*$ (j) $(\alpha^*)^* = \alpha^*$

(k) $\alpha + \alpha = \alpha$ (l) $a + \phi = \alpha$

Proof: (a) Let α and β denote the regular sets L_1 and L_2 respectively. Then $\alpha + \beta$ denotes the regular set $L_1 \cup L_2$ and $\beta + \alpha$ denotes the regular set $L_2 \cup L_1$. But $L_1 \cup L_2 = L_1 \cup L_2$ by the commutative property of union operation on sets. Therefore, $\alpha + \beta = \beta + \alpha$ because both the regular expressions denoted the same regular set.

Proved.

Proofs of the other parts of the theorem are left as an exercise for the reader.

Now, it is the time to answer the query, which might have come to your head while reading the regular expressions corresponding to the master diagram of figure 7.2.3 and 7.2.4. Each such diagram has sections corresponding to terminals symbols. These sections may be in *sequences* or in *parallel* or in *loop* or may be ***combination of two or more of these kinds***. The technique to find the expression lies in detecting these sections and then combining them according to the following rules.

- Each terminal symbol constitutes a segment and the symbol itself is the RE corresponding to the segment.

- If a segment D is composed of two segments D_1 and D_2, in sequence; and D_1, D_2 correspond to the RE's α_1 and α_2, respectively, then D corresponds to the regular expression $\alpha_1\alpha_2$ (concatenation).
- If a segment D is composed of alternative segments D_1 and D_2, in parallel; and D_1, D_2 correspond to the RE's α_1 and α_2, respectively, then, D corresponds to the regular expression $\alpha_1 \vee \alpha_2$ (union).
- If a segment D of master diagram is in a loop through segment D_1, and D_1 represent the RE α, then D corresponds to the regular expression α^* (closure).

Now, let us apply the above procedure on the master diagram of the figure 7.2.4. This diagram has three segment D_1, D_2, and D_3, in sequence. They are marked with dotted rectangles.

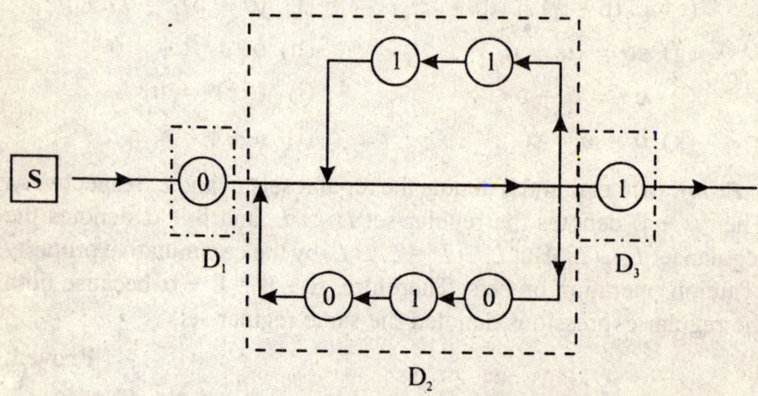

Fig. 7.2.5.

The segment D_1 represent the RE 0 and D_3 represents the RE 1. D_2 contains two segments in parallel and is in loop. The top one represents the RE 11 and bottom one corresponds to 010. Thus, segment D_2 without loop represent the RE $(11 \vee 010)$. Therefore, D_2 corresponds to the RE $(11 \vee 010)^*$. Then, the complete master diagram represents the RE corresponding to $D_1D_2D_3$, and this corresponds to $0(11 \vee 010)^*1$. I hope this must have satisfied your curiosity. Similarly, you can try to find the RE from the diagram shown in the figure 7.2.3.

A regular grammar can also be expressed as a set of equations. For example, G defined by productions

<S> ::= 0 <A> | 1 <S> | ε
<A> ::= 0 <B> | 1 <A>
<B> ::= 0 <S> | 1 <B>

can also be written as

$$S = 0A + 1S + \varepsilon$$
$$A = 0B + 1A$$
$$B = 0S + 1B$$

In dealing with languages, it is often convenient to use equations in which variables (non-terminals) and coefficient represents regular expressions. A set of equations whose coefficient are regular expressions are called ***regular expression equations***. Solving these equations mean finding the RE's for each variable satisfying the equations. By drawing master diagram for each variable and then finding RE representing the diagram a regular expression equation can be solved. We can also use Gaussian elimination method to do the same.

We shall now, show that a language is a regular expression if and only if it is defined by a right linear grammar (regular grammar). A few observations are needed to show that every regular expression has a right linear grammar.

Theorem 2: Let Σ be a finite alphabet. Then, (a) ϕ, (b) $\{\varepsilon\}$ and (c) $\{a\}$ for all a in Σ, are right linear language.

Proof: We know that a language generated by a right linear grammar is right linear. Thus, our task is show that there exist a right linear grammar for language in (a), (b) and (c) each.

- Let $G = (\{S\}, \Sigma, \phi, S)$ is a right linear grammar such that $L(G) = \phi$.
- Let $G = (\{S\}, \Sigma, \{S \rightarrow \varepsilon\}, S)$ is a right linear grammar such that $L(G) = \{\varepsilon\}$.
- Let $G = (\{S\}, \Sigma, \{S \rightarrow a\}, S)$ is a right linear grammar such that $L(G) = \{a\}$.

Proved.

Theorem 3: If L_1 and L_2 are right linear languages then, so are (a) $L_1 \cup L_2$ (b) L_1L_2 and (c) L_1*.

Proof: Since L_1 and L_2 are right linear languages, we can assume that there exist right linear grammars $G_1 = (N_1, \Sigma, P_1, S_1)$ and $G_2 = (N_2, \Sigma, P_2, S_2)$ such that $L(G_1) = \text{L}_1$ and $L(G_2) = L_2$. We may also assume that N_1 and N_2 are disjoint. This can be achieved by renaming the non-terminals arbitrarily. So, this assumption is without any loss of generality.

(a) Let G_3 be a right linear grammar given by

$$(N_1 \cup N_2 \cup \{S_3\}, \Sigma, P_1 \cup P_2 \cup \{S_3 \to S_1 \mid S_2\}, S_3)$$

Where, S_3 is a new non-terminal neither in N_1 nor in N_2. It should be clear that $L(G_3) = L(G_1) \cup L(G_2)$ because for each derivation $S_3 \Rightarrow^+ w$ in G_3, there is either a derivation $S_1 \Rightarrow^+ w$ in G_1 or $S_2 \Rightarrow^+ w$ in G_2 and conversely. Since G_3 is a right linear grammar, $L(G_3)$ is a right linear language.

(b) Let G_4 be a right linear grammar given by $(N_1 \cup N_2, \Sigma, P_4, S_1)$ where, P_4 is defined as follows:

If $A \to xB$ is in P_1, then $A \to xB$ is in P_4,

If $A \to x$ is in P_1, Then $A \to xS_2$ is in P_4,

All productions in P_2 are in P_4.

Note that, if $S_1 \Rightarrow^+ w$ in G_1then $S_1 \Rightarrow^+ w\, S_2$ in G_4. And, if $S_2 \Rightarrow^+ x$ in G_2 then $S_2 \Rightarrow^+ x$ in G_4. Thus, $L(G_1)L(G_2) \subseteq L(G_4)$. Now suppose that $S_1 \Rightarrow^+ w$ in G_4. There is no production of the form $A \to x$ in P_4 that came out of P_4. Thus, we can write a derivation in the form $S_1 \Rightarrow^+ x\, S_2 \Rightarrow^+ xy$ where, $w = xy$ and all productions used in the derivation $S_1 \Rightarrow^+ xS_2$ arose from rules (1) and (2) of the construction of P_4. Thus, we must have the derivations $S_1 \Rightarrow^+ x$ in G_1 and $S_2 \Rightarrow^+ y$ in G_2. Therefore, $L(G_4) \subseteq L(G_1)L(G_2)$. It is therefore concluded that $L(G_4) = L(G_1)L(G_2)$

(c) Let G_5, be a right linear grammar given by $(N_1 \cup \{S_2\}, \Sigma, P_5, S)$ such that S_5 is not in N_1 and P_5 is constructed as follows:

If $A \to xB$ is in P_1 then $A \to xB$ is in P_5,

If $\text{A} \to x$ is in P_1 then $A \to x\, S_5$ and $A \to x$ are in P_5,

$S_5 \to xS_1 \mid \varepsilon$ are in P_5.

Now, to prove that $L(G_5) = (L(G_1))*$, it is enough to prove that

$S_5 \Rightarrow^+ x_1S_5 \Rightarrow^+ x_1x_2S_5 \Rightarrow^+ x_1x_2x_3S_5 \ldots \Rightarrow^+ x_1x_2x_3\ldots x_{n-1}\ S_5 \Rightarrow^+ x_1x_2x_3\ldots x_{n-1}\ x_n$, if, and only if $S_1 \Rightarrow^+ x_1$, $S_1 \Rightarrow^+ x_2$, $S_1 \Rightarrow^+ x_3, \ldots, S_1 \Rightarrow^+ x_n$ are productions in P_1. This is easy to prove and is left as an exercise for the reader.

Proved.

Theorem 4: A language is a regular expression if, and only if it is a right linear language.

Proof: **Only if part:** Here, we have to prove that if a language is a regular expression then the language is a right linear language. It can be done by proving that a right linear grammar exists, which generates the given regular expression. From theorem 2 and 3, it is obvious that for any regular expression, we can find a right linear grammar and hence it is a right linear language.

If part: In this part, we have to show that if a language is a right linear language then there exists a regular expression representing this language. We know that there exists a right linear grammar for every right linear language. Let this be $G = (N, \Sigma, P, S)$. A master diagram representing P can be drawn for this grammar. From this master diagram, a regular expression can always be found. This regular expression is the representation of the language $L(G)$.

Proved.

We have seen that a regular expression can be generated (represented) by a right linear grammar. There exists a mechanism that recognizes a regular expression. Having known this, we can now construct, now, a mathematical model in such a way that set of non-terminals in a grammar G is taken as a set of states and set of terminals is taken as set of **input** symbols. Any productions $A \rightarrow xB$ is interpreted as "*while scanning a string a machine goes to state B from A after reading input symbol x*". This arrangement is shown as below:

Based on this concept, we can define a finite state machine, which is the simplest recognizer of a regular expression. The following section deals with Finite States Machines (FSM).

7.3 Finite State Machines

We take a Finite State Machine (FSM) as a recognizer of a RE. Consider the directed graph shown in the figure 7.3.1 and the adjoining table 7.3.1.

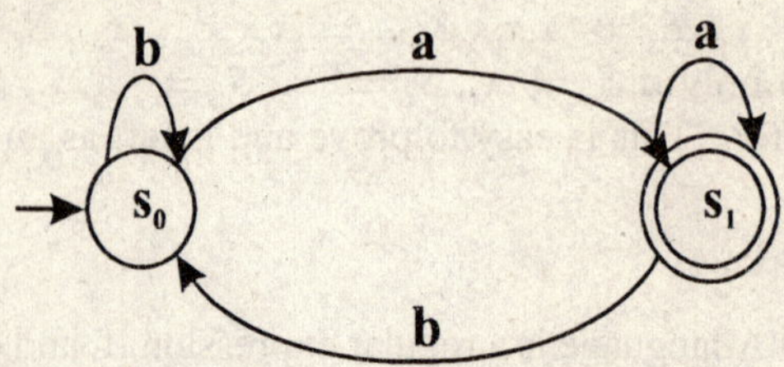

Fig. 7.3.1.

$\Im$	a	b
s_0	s_1	s_0
s_1	s_1	s_0

Table 7.3.1.

The entry in table 7.3.1 corresponding to row s_0 and column "a" is interpreted as: "*If a recognizer of a string is in state* s_0 *and reads an input symbol a, then it changes its state to* s_1." Similarly if it is on state s_1 and input symbol is b, then the recognizer goes to state s_0. Thus, entries in the table 7.3.1 are listing of possible states to which a machine may go depending upon the input symbol it reads while scanning a given string. A table like this is called **state transition table** of a machine. The same transition can be shown diagrammatically as in figure 7.3.1. If the number of states in a machine is finite it is called a finite state machine.

We define a **Finite State Machine** (FSM) as a mathematical model, $(S, I, \Im, S_0, T)$. Where, S is a non-empty, finite set of states, I is a non-empty, finite set of input symbols, S_0 is a start state. T is set of final states (also called acceptance states and $T \subseteq S$) and $\Im$ is a set of *state transition functions*, defined as:

$$\Im : S \times I \rightarrow S$$

The **FSM** shown in the figure 7.3.1 can be formalized as $M = (S, I, \Im, S_0, T)$ where, $S = \{s_0, s_1\}$, $I = \{a, b\}$, $S_0 = s_0$, $T = \{s_1\}$ and $\Im$ is given by the table 7.3.1.

Let us consider another diagram shown in the figure 7.3.2 below. This is a machine $M = (S, I, \Im, S_0, T)$.

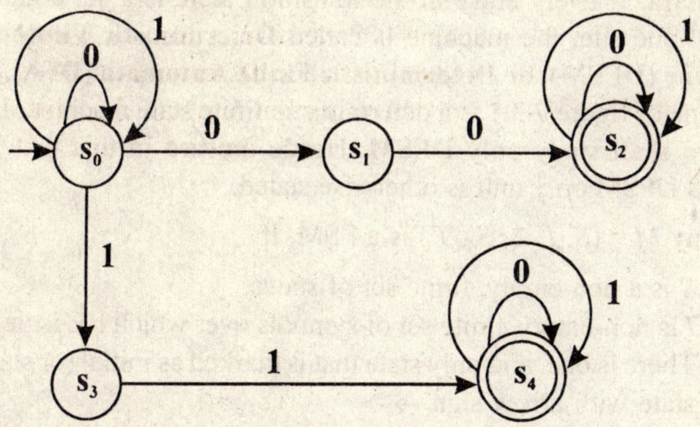

Fig. 7.3.2.

	0	1
s_0	$\{s_0, s_1\}$	$\{s_0\}$
s_1	$\{s_2\}$	ϕ
s_2	$\{s_2\}$	$\{s_2\}$
s_3	ϕ	$\{s_4\}$
s_4	$\{s_4\}$	$\{s_4\}$

Table 7.3.2.

In this machine M, $S = \{s_0, s_1, s_2, s_3, s_4\}$, $I = \{0, 1\}$, $S_0 = s_0$, $T = \{s_2, s_4\}$ and $\Im$ is given by the table 7.3.2. In this table, entries corresponding to $(s_1, 1)$ and $(s_3, 0)$ are null set i.e., once the machine M is the state s_1 and reads input symbol 1, it hangs. No transition is defined. Similarly, M hangs when it reads input symbol 0 while in state s_3. Now, notice the entry corresponding to $(s_0, 0)$. It is a set $\{s_0, s_1\}$. This implies that if the machine M is in state s_0 and reads input symbol 0, it has two states to move to. It can move to only one of them. Such a general machine is called **Non-deterministic Finite State Machine (NFSM),** or **Non-deterministic (NFA) Finite**

Automata. If every entry in the transition table is a set containing exactly one site, the machine is called **Deterministic Finite State Machine (DFSM), or Deterministic Finite Automata (DFA).** Thus Machine of figure 7.3.1 is a deterministic finite state machine. In this text we shall study only **DFSM**. Hence onward in this text, FSM implies DFSM only unless otherwise stated.

Any $M = (S, I, \Im, S_0, T)$ is a FSM, if

- S is a non-empty, finite set of states,
- I is non-empty, finite set of symbols over which RE is defined,
- There is one, and only state that is marked as initial (or starting) state with arrow sign $\rightarrow$,
- T is a non-empty subset of S. Possibly F may be equal to S i.e. all state may be a final state, and
- The transition table $\Im$ must a singleton as entry at all places. No entry should be a null set or a set having cardinality ≥ 2. Here, we represent $\{s_n\}$ by s_n without loss of any generality. Thus, we can write

  $\Im = \{f_x | f_x : S \rightarrow \ \forall \ x \in I\}$

Sometimes, some memory is provided to every state in a machine so that it can remember the input symbol it reads. In our model of consideration, a machine does not have any memory. It simply reads as input symbol, changes its state and throws the symbol. It recognizes a string as valid if the machine is in one of the final state when the string is completely scanned. Otherwise, the string is an invalid sentence. A machine scans any input strings starting for its initial state. Thus, any string $w \in I^*$ is valid for a machine $M = (S, I, \Im, S_0, T)$, if $f_w(S_0) \in T$.

Finding Language of FSM

An FSM is a recognizer of a language. The language of M is denoted as $L(M)$. We can write

$L(M) = \{w \mid w \in I^* \text{ and } f_w(S_0) \in T\}$

Given a FSM we can always find its language. Finding language of a machine M means finding all those w that satisfies the above conditions. For a given machine, there may be infinite many such strings. It is not possible to test and list all such strings. Thus, clue lies in identifying the pattern which all such strings must have. Once the pattern is identified, express the language in a set theoretic notation. Let us now see some examples.

Example 1: Find the language of FSM shown in the figure 7.3.1.

Solution: Language of the machine is defined in input symbol $I = \{a, b\}$. The I is the alphabet of $L(M)$. A string containing single a is acceptable as $f_0(s_0) = s_1 \in T$. If a string contains b, then M will remain at s_0 if it is at s_0 or it returns to s_0 if it is at s_1. This shows that any string to be accepted as valid, it must end in "a". Thus, we have

$$L(M) = \{w \mid w \in \{a, b\}^* \text{ and } w \text{ ends in "}a\text{"}\}$$

Ans.

Example 2: Find the language of FSM shown in the following figure 7.3.3.

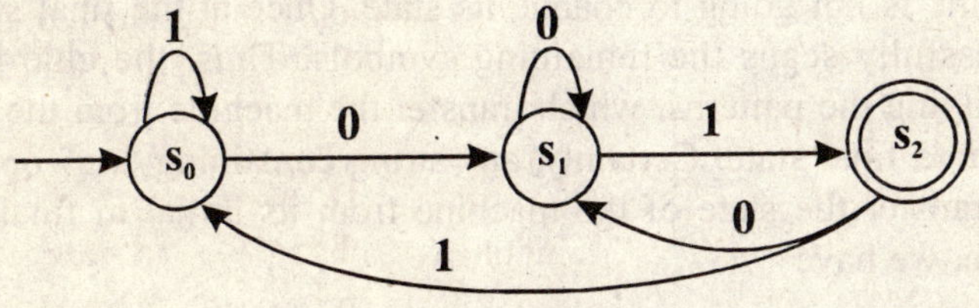

Fig. 7.3.3.

Solution: Language of the machine is defined in input symbol $I = \{0, 1\}$. The I is the alphabet of $L(M)$. A string "01" is acceptable, because we have $f_{01}(s_0) = f_1(f_0(s_0)) = f_1(s_1) = s_2 \in T$. If machine M is at state s_2 and string is not completely scanned, M changes its state to s_0 or s_1 depending upon inputs 1 and 0 respectively. After a little bit of reasoning, it can be found that any string terminating in "01" is acceptable. And a string not terminating in "01" is invalid. Thus, we have

$$L(M) = \{w \mid w \in \{0, 1\}^* \text{ and } w \text{ ends in "01"}\}$$

Ans.

Example 3: Find the language of FSM shown in the following figure 7.3.4.

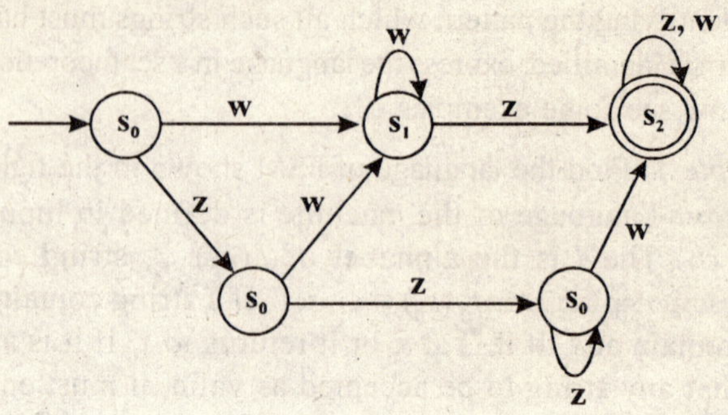

Fig. 7.3.4.

Solution: Language of the machine is defined in input symbol $I = \{w, z\}$. The I is the alphabet of $L(M)$. A string "*wz*" is acceptable. Similarly, "*zzw*" is acceptable. There is a loop at the final state on input symbol *w*, *z*. This implies that, once reached at the final state, machine is not going to change its state. Once at the final state, it successfully scans the remaining symbols. Thus, the clue lies in identifying the patterns, which transfer the machine from the initial state to a final state. Certainly, any string containing "*wz*" or "*zzw*" can transfer the state of the machine from its initial to final state. Hence, we have

$$L(M) = \{w \mid w \in \{w, z\}^* \text{ and w contains either "}wz\text{" or "}zzw\text{"}\}.$$

Ans.

Finding a Grammar from FSM

Every FSM language is a Regular language. For every regular language there exists a regular grammar. Thus, we can find a regular grammar G from a given FSM such that $L(G) = L(M)$. To obtain a grammar from FSM, we apply the following rules.

(a) Let A and B be two states in FSM such that machine transfers

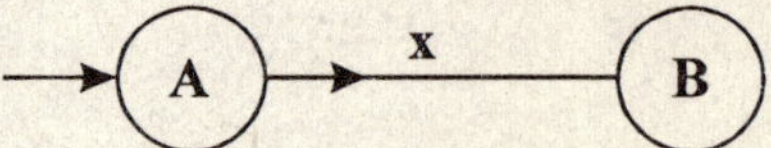

its state from A to B on input symbol x then, we have a production $A \rightarrow xB$

(b) In the above diagram, if B is a final state then it yields two productions $A \rightarrow xB$ and $A \rightarrow x$.

(c) A loop is interpreted accordingly.

See the following examples to learn the procedure of finding a Regular grammar from a given FSM.

Example 4: Find a regular grammar corresponding to the FSM shown in the figure 7.3.3.

Solution: Let $G = \{N, \Sigma, P, S\}$ be the corresponding RG. Here $\Sigma = I = \{0, 1\}$. The alphabets remain the same. The set S of M becomes N of G. thus, we have $N = \{s_0, s_1, s_2\}$. The states corresponding to the initial state of M, becomes the start symbols of G. Thus, $S = s_0$. The production P has to be found. And it is given as

1. $s_0 \rightarrow 1s_0$
2. $s_0 \rightarrow 0s_1$
3. $s_1 \rightarrow 1s_2$
4. $s_1 \rightarrow 0s_1$
5. $s_1 \rightarrow 1$ [Since on 1 machine goes to final state from s_1]
6. $s_2 \rightarrow 0s_1$
7. $s_2 \rightarrow 1s_0$

Ans.

Example 5: Find a regular grammar corresponding to the FSM shown in the figure 7.3.1.

Solution: Let $G = \{N, \Sigma, P, S)$ be the corresponding RG. Here $\Sigma = I = \{a, b\}$, $N = \{s_0, s_1\}$ and $S = s_0$. The production P has to be found. And it is given as

1. $s_0 \rightarrow bs_0$
2. $s_0 \rightarrow as_1$
3. $s_0 \rightarrow a$
4. $s_1 \rightarrow a$ [Since on a machine goes to final state from s_1]
5. $s_1 \rightarrow bs_0$
6. $s_1 \rightarrow as_1$

Ans.

Drawing a FSM for a given Language

Given a FSM, we can find $L(M)$. We can also find a RG G from FSM such that $L(G) = L(M)$. In this section we shall learn that given a language, we can draw a FSM. Every language is not a FSM language. A language is said to be a FSM language if there exist a FSM to recognize this language and rejects everything else. In brief, we can say that if the language is a regular language, there exists a FSM to recognize it. Conversely, if there is a FSM M, then the $L(M)$ is a regular language. There is a theorem by "Kleen" stating exactly the same. Procf of this theorem is beyond the scope of the intended course.

In this part of the section, we gradually learn how to design a machine that accepts the given language and rejects all strings not belonging to the language.

Example 6: Draw a finite state machine that can accept any string of {0, 1} that terminates with the symbol 1 i.e. 1 must be the last symbol in a string.

Solution: Let the FSM be M and the given language be L. Any string like 1, 11, 111, 101, 10001110001 etc must be acceptable to M. However, strings like 10, 0, 11110, 1101000000 etc must not be acceptable. Let S_0 be the initial state of M. Since "1" is a valid string in L, M must transfer its own state to a final state, say S_1, from S_0 on input symbol 1. This arrangement can be shown as below.

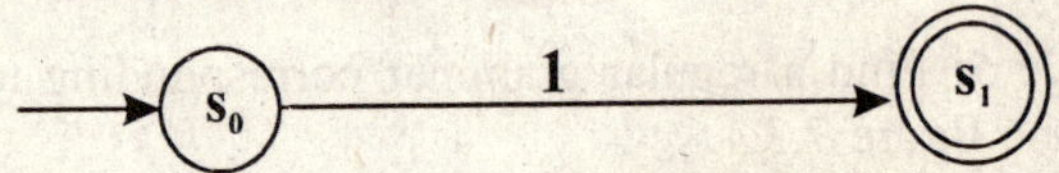

This is not a complete FSM. To complete this we have to define move from S_0 on input symbol 0 and all moves from final state S_1. Once at final state, M can scan all the following 1's in the string without changing its state. Thus, a loop on input symbol 1 will be sufficient at S_1. If M get an input 0 at state S_1, it must change it state to ensure that a valid string must terminate in 1. If string contains any number of 0's in the beginning, it may be scans without transferring state. A final arrangement is shown in figure 7.3.5.

Ans.

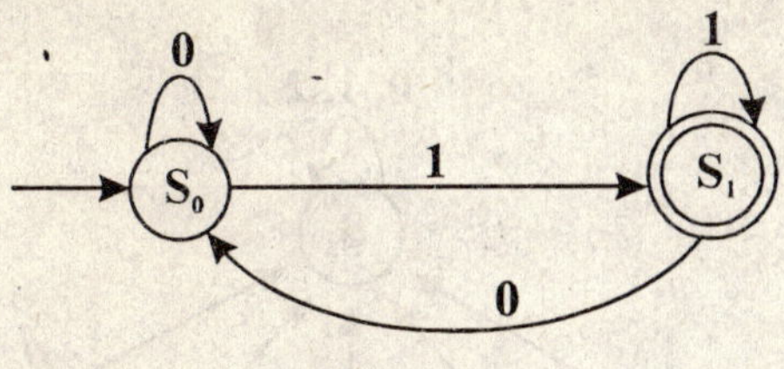

Fig. 7.3.5.

Example 7: Draw a finite state machine that can recognize only string 0120 on the input symbols {0, 1, 2}.

Solution: Let *M* is the required FSM. This accepts only string "0102". Let S_0 be the initial state of *M*. If a string contains any thing other that 0 as its first symbol, *M* must not move toward its final state after reading those symbols. The skeleton of the proposed FSM can be drawn as below.

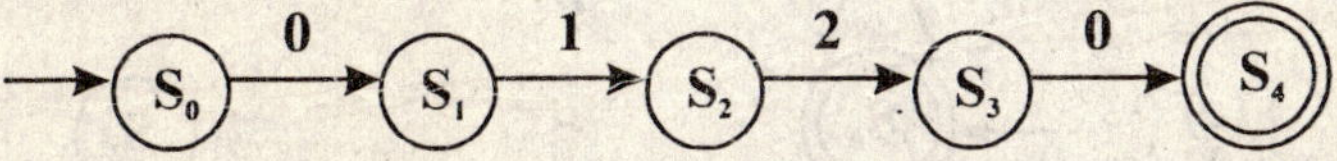

This is not a complete FSM. To complete this we have to define all the remaining moves for all states. If *M* is at S_0 and reads symbol 1 or 2 it must go to a **state of no return**, as those strings are not valid. Similarly, we can complete other moves. The final FSM is shown in the figure 7.3.6.

Ans.

In the machine of figure 7.3.6, state S_5 has a peculiar property. Once *M* has reached to this state from anywhere, there is no escape route from here. Such state is called **Trapping State**. Sometimes, it becomes necessary to introduce a trapping state in a FSM, for a language *L*. to ensure that a string having particular pattern must be rejected if does not belong to *L*. A state is said to **inaccessible** if there is no out going arc from this state to any other state. A state is said to be **redundant**, if a FSM can accept the same language even without this state. More about this, we shall learn in next section. Now, let us see some more examples on drawing a FSM for a given language.

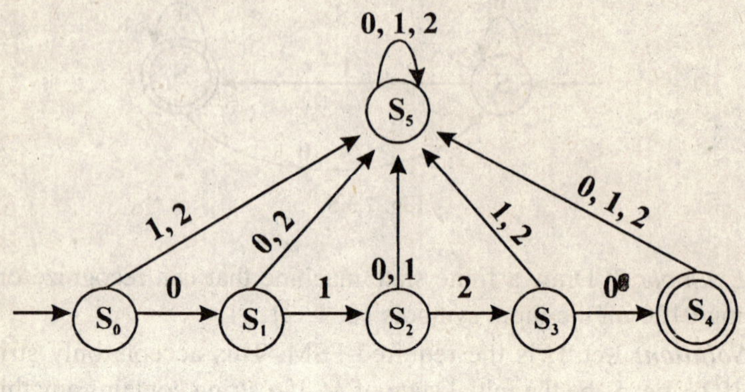

Fig. 7.3.6.

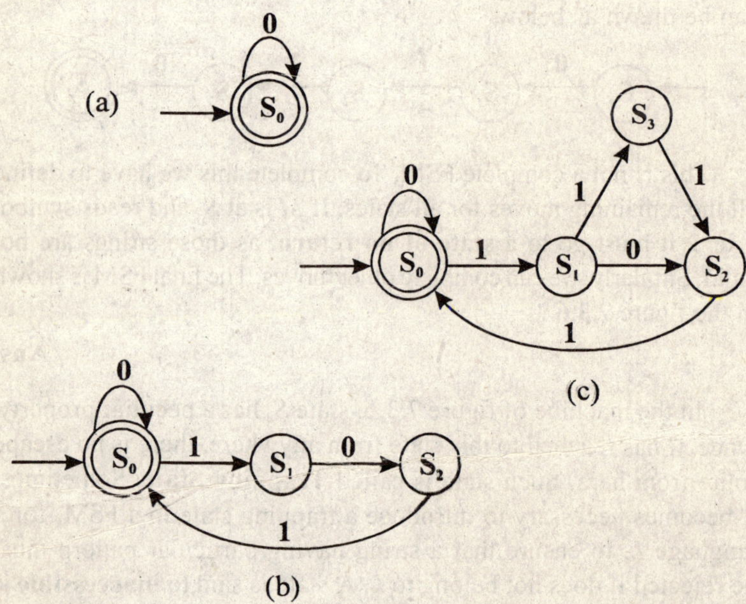

Fig. 7.3.7.

Example 8: Draw a FSM that can recognize a binary representation of any non-negative integer divisible by 5.

Solution: Let M is the required FSM and L be the language containing all bit string, the decimal equivalent of which is divisible by 5. Thus "0", "00", "00...0" all belong to L as they are divisible by 5, "101", "1010", "1111", "10100" etc are divisible by 5, thus is a valid string in L, 11, 110 etc does not belong to L. While designing a machine these patterns are very helpful. The skeleton of the proposed FSM can be drawn successively as below. Since a null string can be interpreted as decimal 0, a null string must also be acceptable by M. In order to ensure this, we have to mark initial state S_0 as final state. Skeleton in Figure 7.3.8(a) scans "000...0", that in (b) scans 101 and 1010, that in (c) scans all previous strings and 1111 also. This way, we can finally get the required FSM M.

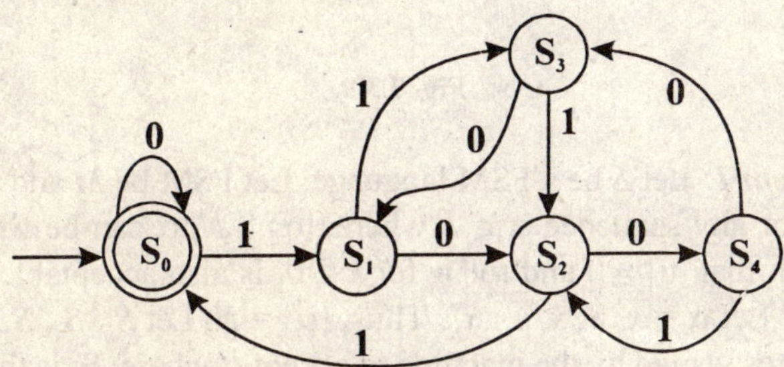

Fig. 7.3.8.

Example 9: Draw a finite state machine that can accept any string in a, *b* that contains '***abb***' or '***baa***'.

Solution: A FSM can be drawn as shown in the figure 7.3.9. How it has evolved to this stage, reader can very well reason it out. As a volunteer guide, I may suggest to reader that you never mug up the thing, which you are not able to comprehend. Try it again and again. Once you have got it, it is yours.

As we have seen that if a language is a regular language, we can find a FSM for this. But how to know that whether a language is FSM language or not? Trying to find a FSM for a language that is not regular, will be an effort in vain. Thus, it is better to ensure that given L is a regular language before proceeding to find a FSM for it. The following theorem, known as *pumping Lemma*, helps in this regard.

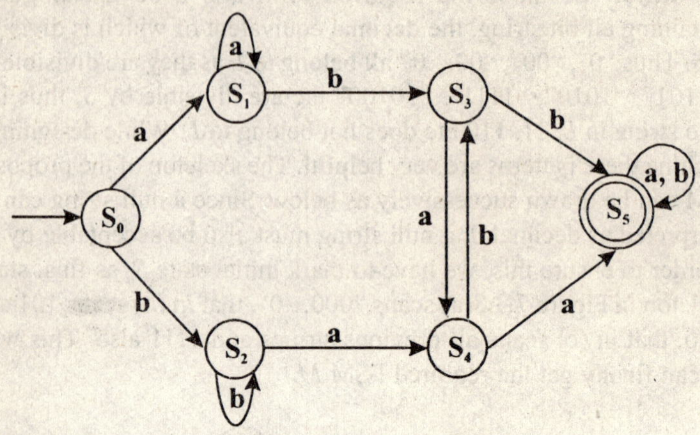

Fig. 7.3.9.

Theorem 1: Let L be a FSM language. Let FSM be M and it has N states. For any sequence $\alpha \in L$ where, $|\alpha| \geq N$, α can be written as **uvw** such that $|u| \neq 0$ and $uv^k w$ for $k \geq 0$, is also acceptable.

Proof: Let $\alpha = x_1 \, x_2 \, x_3 \, \ldots \, x_N$. Thus, $|\alpha| = N$. Let $S_0, S_1, S_2, S_3, \ldots, S_N$ be states visited by the machine to accept α where. S_0 is the initial state and S_N is the final state. This we can assume without loss of any generality. Machine M has N states, and it has visited $N + 1$ states to accept α. By Pigeonhole principle, M must have visited one state at least twice. Let us assume that state to be S_k. See the sequence depicted below.

$$\begin{bmatrix} 6\ 4\ 4\ 7\ 4\ 4\ 8 & 6\ 4\ 4\ 7\ 4\ 4\ 8 & 6\ 4\ 4\ 7\ 4\ 4\ 8 \\ a\ \ a\ \ a\ \ 0^\wedge C\ a & a\ \ a\ \ a\ \ 0^\wedge C\ a & a\ \ a\ \ a\ \ 0^\wedge C\ a \\ S_0\ S_1\ S_2\ {}^\wedge C^\wedge S_k & S_k\ \wedge\ \wedge\ \wedge\ \wedge\ S_k & S_{k+1}\ \wedge\ \wedge\ \wedge\ \wedge\ S_k \end{bmatrix}$$

The depicted sequence can be divided into three segments: u, v and w where, v can be in loop and can be scanned $k \geq 0$ times. Therefore, $uv^k w$ for $k \geq 0$, is a language accepted by M.

Proved.

Having done this theorem, let us see some examples to know how to apply this theorem to know whether a language L is FSM language or not.

Example 10: Show that the language

$$L = \{0^k \mid k = n^2, n > 0\}$$

is not a FSM language.

Proof: Let us assume that there exists a **FSM *M*** for the language L. Let there are N states in M. Let us take an integer m, large enough, such that

$(m + 1)^2 - m^2 > N$

See the sequence below which depicts $p = (m + 1)^2 - m^2$ 0's.

64447 N 448

0 0 0 0 ^0 0 0 0 0 0 0^C 0

m^2 p

Clearly, there are 0's > N between m^2 and $(m + 1)^2$. If the machine M accepts the string $0^k \mid k = (m + 1)^2$, then a certain state S_k is visited at least twice. Removal of 0's between these two visits of S_k, yields a sequence of 0's having p number of 0's and p lies between m^2 and $(m + 1)^2$ i.e.,

$(m + 1)^2 > p > m^2$

This implies that M accepts a string that does not belong to L. Hence L is not a FSM language.

Proved.

Example 11: Show that the language

$L = \{a^n b^n \mid n \geq 1\}$

is not a FSM language.

Proof: Let us assume to the contrary that the L is a FSM language. Thus, there exists a FSM, say M. Let M has N states. Let $a^N b^N$ be a string in L. To scan N a's, M changes states $N + 1$ times. Let S_0 be the initial state of M. Let us number the states as M visits it. Machine M has N states and it has visited $N + 1$ states to accept N 0's. By

Pigeonhole principle, M must have visited one state at least twice. Let us assume that state to be S_k. See the sequence depicted below.

$$\begin{bmatrix} 6\;4\;4\;4\;4\;4 & 7\;4\;4\;4\;4\;8 & 6\;4\;7^{\;N}\;4\;4 \\ a \quad a \quad a\wedge a & a\wedge a \quad a\wedge a & b \quad b \quad b\wedge b \\ S_0\; S_1\; S_2\; S_k & S_k & S_{2N} \end{bmatrix}$$

This shows that at S_k, an optional loop is created, that can be avoided. This implies that a string $x = a^{N-m} b^N$ may also be accepted for $m > 0$. Since $x \notin L$, FSM, so assumed, is not a FSM for this language. Therefore, L is not a FSM language.

Proved.

This section can be summarized by stating the following equivalence in form of a theorem without proof.

Theorem 2: The following statements are equivalent.

- L is a regular set.
- L is a right linear language
- L is a FSM language
- L is denoted by regular expression.

Important points to remember while drawing a FSM for a language.

- If a null string is in language, the initial state must be a final state.
- If a string terminates in a fixed sequence of symbols (e.g. 011, 001, 001 *az*, *aa*, etc) then, there must not be a loop on final states on symbols not mentioned in the string.
- If a string terminated in a symbol, then there must be a loop at the final state on the same symbol, but not on the other symbol $\in 1$.
- If a string starts with a particular symbol, a trapping state is required in the FSM.

These are just some tips that may help you in drawing a FSM for a given language. The best way to develop a concept is to do a rigorous practice. Try to solve problems given exercise at the end of this chapter.

7.4 Simplification of Finite State Machine

For a given FSM M we can find the smallest FSM equivalent to M by eliminating all ***inaccessible*** states in M and then ***merging*** all redundant states in M. The redundant states are determined by partitioning the set of all accessible states into equivalence classes such that each equivalence class contains indistinguishable states and is as large as possible. We then choose one representative for each equivalence class as a state for the reduced FSM. Thus we can reduce the size of M if M contains inaccessible states or two or more indistinguishable states. We shall show that this reduced machine is the smallest and the most efficient finite automation that recognizes the language defined by the original FSM M. We use the number of states in a FSM as a measure of its efficiency. Let us first learn some basic concepts needed to understand the simplification process.

Reader may recall the definition of a free semi-group and of congruence relation. For a quick reference it is reproduced here. Let $(S, *)$ be a semi-group, then an equivalence relation R defined on S is said to be ***congruence relation*** if

$$aRb \text{ and } cRd \Rightarrow (a * c)\, R\cdot(b * d)$$

Example 1: Let $(I, +)$ be a mathematical structure R be a relation defined on I as aRb iff $a \equiv_2 b$. The relation R is a congruence relation on I.

Example 2: Let $A = \{1, 0\}$ and consider the free semi-group $(A^*, \bullet)$, where $\bullet$ is concatenation operation. Define a relation R on A^* as $\alpha R \beta$ iff α and β have same number of 1's. The relation R is a congruence relation on A^*.

Let $M = (S, I, \Im, s_0, T)$ be a FSM and suppose R is equivalence relation on S, we say that R is a ***machine congruence*** on M if for any $s, t \in S$

$$sRt \Rightarrow f_x(s)\; R\; f_x(t) \text{ for all } x \text{ in } I$$

That is R-equivalent pair of states are always taken into R-equivalent pair of states by every input symbols. Since R is an equivalence relation on S, it partitions S into equivalence classes $S/R = \{[s] \mid s \in S\}$.

$$\text{Since } [s] = [t] \Rightarrow sRt$$

$$\Rightarrow f_x(s) \ \ R\, f_x(t)$$

$$\Rightarrow [f_x(s)] = [f_x(t)]$$

Thus we can define a new FSM $M/R = (S/R, I, \Im/R, [s_0], T/R)$, where

$\Im/R = \{g_x \mid g_x: S/R \to S/R$ for all x in I and $g_x([s]) = [f_x(s)]\}$,

$[s_0]$ is the class of states equivalent to initial states in M,

T/R = Set of equivalence classes containing one or more states of T of M.

This shows that this FSM is again a machine containing less number of states than that in M. This FSM M/R is called ***quotient finite state machine (QFSM)*** corresponding to relation R. Generally a QFSM is simpler and efficient than the original FSM.

Example 3: Let $M = (S, I, \Im, s_0, T)$ be a finite state machine whose state transition table is according to the table 7.4.1 Let R be an equivalence relation on S and it is defined according to the matrix M_R of table 7.4.2. It is also given that $T = \{S_2\}$ and $s_0 = S_0$. Show that R is machine congruence on M and find the corresponding QFSM.

Solution: In this FSM M, $S = \{S_0, S_1, S_2, S_3, S_4, S_5\}$, $I = \{0, 1\}$ and other things already mentioned. Based on the given states transition table, diagrammatic representation of FSM can be given as in figure 7.4.1.

$$M_R = \begin{pmatrix} 1 & 0 & 0 & 0 & 1 & 0 \\ 0 & 1 & 0 & 0 & 0 & 1 \\ 0 & 0 & 1 & 1 & 0 & 0 \\ 0 & 0 & 1 & 1 & 0 & 0 \\ 1 & 0 & 0 & 0 & 1 & 0 \\ 0 & 1 & 0 & 0 & 0 & 1 \end{pmatrix}$$

Table 7.4.2.

$\Im$ =

	0	1
S_0	S_0	S_1
S_1	S_2	S_5
S_2	S_2	S_4
S_3	S_3	S_0
S_4	S_4	S_5
S_5	S_2	S_1

Table 7.4.1.

The given relation on S is an equivalence relation, which can be easily verified. This relation partitions S into three equivalence classes.

$[S_0] = \{S_0, S_4\}$

$[S_1] = \{S_1, S_5\}$

$[S_2] = \{S_2, S_3\}$

Notice that (S_0 & S_4) are R equivalent states. Similarly, (S_1 & S_5) and (S_2 & S_3) are R equivalent pairs of states. Here, we have

$f_0(S_0) = S_0, f_0(S_4) = S_4, f_1(S_0) = S_1, f_1(S_4) = S_5$

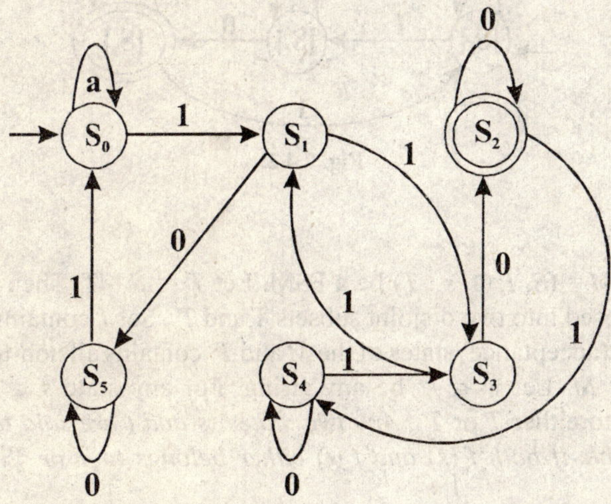

Fig.7.4.1.

This implies that on input symbol 0, equivalence class $[S_0]$ is mapped to the equivalence class $[S_0]$ and on input symbol 1 $[S_0]$ is mapped to the equivalence class $[S_1]$. Similarly we can find transitions for other equivalence classes. Therefore R is a machine congruence to M. In the QFSM, $S/R = \{[S_0], [S_1], [S_2]\}$, $I = \{0, 1\}$, $\Im/R$ is given by the table 7.4.3. The class that contains initial states of original FSM, becomes initial state in QFSM and all the classes containing at least one final state of original FSM become final states in QFSM. Therefore, in QFSM $s_0 = [S_0]$ and $T/R = \{[S_2]\}$. The diagrammatic representation of QFSM is shown in the figure 7.4.2.

	0	1
$[S_0]$	$[S_0]$	$[S_0]$
$[S_1]$	$[S_1]$	$[S_1]$
$[S_2]$	$[S_2]$	$[S_2]$

Table 7.4.3.

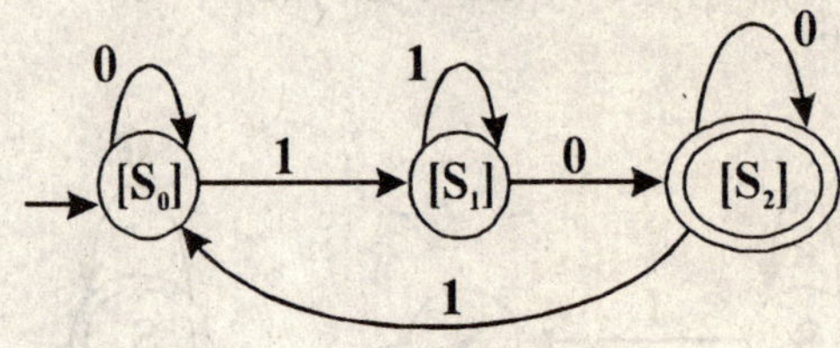

Fig. 7.4.2.

Ans.

Let $M = (S, I, \Im, s_0, T)$ be a FSM. Let $T^C = S - T$. Then set S is decomposed into two disjoint subsets T and T^C. Set T contains all the terminal (acceptance) states of the M and T^C contains all non-terminal states of M. Let $w \in I^*$ be any string. For any state $s \in S$, $f_w(s)$ belongs to either T or T^C. *Any two states: s and t are said to be **w-compatible** if both $f_w(s)$ and $f_w(t)$ either belongs to T or T^C for all $w \in I^*$.*

Theorem 1: Let $M = (S, I, \Im, s_0, T)$ be a FSM. Let R be a relation defined on S as sRt if and only if both s and t are w-compatible. Show that

(a) R is an equivalence relation.

(b) R is machine congruence to M.

Proof: (a) Let s be any state in S and w be any string in I^*. Then $f_w(s)$ is either in T or in T^C. This implies that $sRs \ \forall \ s \in S$. Hence R is reflexive on S. Let s and t be any two states in S such that sRt. Therefore, by definition of R both s and t w-compatible states. This implies that tRs. Thus, R is symmetric also. Finally, let s, t and u be any three states in S such that sRt and tRu. This implies that s t and u are w-compatible states i.e. s and u are w-compatible. Hence sRu.

Thus, R is a transitive relation. Since R is reflexive, symmetric and transitive, it is an equivalence relation.

Proved.

(b) To prove that R is machine congruence to M, we prove that every R-equivalent pair of states is taken into R-equivalent pair of states by R i.e.,

if $s, t \in S$ and $\boldsymbol{sRt}$ then $f_x(s) \mathrm{R} f_x(t)$ $\forall$ $x \in 1$

Let $w_1 \in 1^*$ and $w = w_1 \bullet x$, where $\bullet$ is a concatenation operation and x is any input symbol from 1.

Since s & t are R related, s & t are w-compatible i.e. $f_w(s)$ and $f_w(t)$ both are either in T to in T^C. But $w = w_1 \bullet x$, thus

$f_w(s) = f_{w_1} \bullet_x(s) = f_{w_1}(f_x(s))$ and
$f_w(t) = f_{w_1} \bullet_x(t) = f_{w_1}(f_x(t))$

The above two results show that $f_x(s)$ and $f_x(t)$ are w_1-compatible. Since w_1 is any string in I^*, we say that $f_x(s)$ and $f_x(t)$ are w-compatible. Thus,

$f_x(\mathrm{s})\ R\ f_x(\mathrm{t})$ $\forall$ $x \in I$ [Since x is any symbol of I]

Thus we have proved that $sRt \Rightarrow f_x(s)\ R\ f_x(t)$ $\forall$ $x \in I$. Therefore R is machine congruence to M.

Proved.

We have discussed a quotient machine, which is obtained from a given FSM. In the following theorem, we shall show that a language of a quotient machine is same as that of original FSM. Therefore, the process of simplification of FSM only makes the machine more efficient without losing any generality of the nature of language acceptable to the machine.

Theorem 2: Let $M = (S, I, \Im, s_0, t)$ be a FSM and R be a relation of w-compatibility on S. If $M/R = (S/R, I, \Im/R, [s_0], T/R)$ be the corresponding QFSM then $L(M) = L(M/R)$.

Proof: Let $\Im/R = \{g_x \mid g_x\text{: } S/R \rightarrow S/R$ for all x in $I\}$ and $\Im = \{f_x \mid f_x\text{: } S \rightarrow S$ for all x in $I\}$. Let $w \in L(M)$ then $f_w(s_0) \in T$. The equivalence class $[s_0]$ contains all states of S that are R-equivalent to s_0. Since R is machine congruence to M (as proved in theorem 1(b)), all R-equivalent states are mapped to R-equivalent pair of states. Thus $f_w(s_0) \in T \Rightarrow [f_w(s_0)] \in T/R$. Also $g_w([s_0]) = [f_w(s_0)]$ because M/R is QFSM. This shows that $g_w([s_0] \in T/R$. Therefore, $w \in L(M/R)$. Hence from the result

$w \in L(M) \Rightarrow w \in L(M/R)$

we conclude that $L(M) \subseteq L(M/R)$ ______________________ (A)

Conversely, let $w \in L(M/R)$. Then $g_w([s_0]) \in T/R$. Since $g_w([s_0]) = [f_w(s_0)]$, $[f_w(s_0] \in T/R$. This implies that $\exists$ some state $t \in T$ such that $t\,R\,f_w(s_0)$. That is t and $f_w(s_0)$ are w-compatible states $\forall\ w \in 1^*$. If we take $w_1 = \varepsilon$ then $f_{w_1}(t)$ and $f_{w_1}(f_w(s_0))$ both either belong to T or to T^C. Since $f_{w_1}(t) = t \in T$, $f_{w_1}(f_w(s_0)) = f_w(s_0)$ also belongs to T. Thus w is acceptable by M and hence $w \in L(M)$. This shows that

$L(M/R) \subseteq L(M)$ ________________________________ (B)

From (A) and (B), we have $L(M) = L(M/R)$

Proved.

Simplification of a FSM is an evolutionary process. It is achieved through step-by step approximation. Let us define a relation R_k on S of $M = (S, I, \Im, s_0, T)$ as for any two states s and t, sRt iff s and t are w-compatible for all $w \in 1^*$ with $|w| \leq k$. The following theorems give the basic concepts involved in simplifying a given FSM.

Theorem 3: Let R_k be relation on S of $M = (S, I, \Im, s_0, T)$ as defined above. Then

(a) $R_{k+1} \subseteq R_k\ \forall\ k \geq 0$.
(b) Each R_k is an equivalence relation.
(c) $R \subseteq R_k\ \forall\ k \geq 0$.

Proof: (a) Let $(s, t) \in R_{k+1} \Rightarrow s\ R_{k+1}\ t$

$\Rightarrow s$ and t are w-compatible $\forall\ w \in I^*$ with $|w| \leq k+1$

$\Rightarrow s$ and t are w-compatible $\forall\ w \in I^*$ with $|w| \leq k$

$\Rightarrow s\ R_k\ t$

$\Rightarrow (s, t) \in R_k$

Therefore, $R_{k+1} \subseteq R_k$

Proved.

(b) It can be proved according to the proof of the theorem 1(a).
(c) Prove as in part(a).

Theorem 4: For any FSM $M = (S, I, \Im, s_0, T)$, and relation R_k as defined above the following are true.

(a) $S/R_0 = \{T, T^C\}$
(b) For $k \geq 0$ R $s, t \in S$, $s\ R_{k+1}\ t$ if and only if
 (i) $s\ R_k\ t$ and
 (ii) $f_x(s)\ R_k\ f_w(t)\ \forall\ x \in I$.

Proof: Since $|\varepsilon| = 0$, no input symbol is there for state transition in M. Thus a state is either in T or in T^C. Therefore, $S/R_0 = \{T, T^C\}$.

(b) Let $w \in I^*$ and $|w| \leq k + 1$. Then $w = vx$ for some $x \in I$ and $v \in I^*$ with $|v| \leq k$. Similarly, if $x \in I$ and $v \in I^*$ with $|v| \leq k$ then $w = vx \in I^*$ with $|w| \leq k + 1$.

Now for any state $s, t \in S$, we have

$$f_w(s) = f_{vx}(s) = f_v(f_x(s))$$

and

$$f_w(t) = f_{vx}(t) = f_v(f_x(t))$$

Now s and t are w-compatible if and only if both $f_x(s)$ and $f_x(t)$ $\forall$ $x \in I$ are v-compatible i.e. sR_{k+1} iff $f_x(s)\ R_k\ f_x(t)\ \forall\ x \in I$.

Next, since $R_{k+1} \subseteq R_k\ \forall\ k \geq 0$ (Theorem 3(a)), we have $sR_{k+1}t \Rightarrow sR_kt$.

Proved.

Theorem 5: If $R_k = R_{k+1}$ for any non-negative integer k, then $R_k = R$.
Proof: From the theorem 4, for any two states s and t of S

$$sR_{k+2}t \quad \textbf{iff} \quad f_x(s)\ R_{k+1} f_x(t)\ \forall\ x \in I$$

$$\textbf{iff} \quad f_x(s)\ R_k f_x(t)\ \forall\ x \in I \quad [\text{Since } R_k = R_{k-1}]$$

$$\textbf{iff} \quad sR_{k+1}t$$

This implies that $R_{k+2} = R_{k+1} = R_k$, similarly by induction, $R_k = R_n\ \forall$ $n \geq k$.

Since $R_0 \supseteq R_1 \supseteq R_2 \ldots \supseteq R_n \ldots$, we have

$$R = \bigcap_{n=0}^{\infty} R_n = R_k$$

Therefore, $R = R_k$.

Proved.

Having seen all the basic theorems, it is the time to learn the working method to find QFSM for a given FSM. The algorithm is outlined below.

Step 1: Start with $P_0 = \{T, T^C\}$ of S corresponding to relation R_0.

Step 2: Construct successive partitions $P_1, P_2, P_3, \ldots P_n$ corresponding to the equivalence relations $R_1, R_2, R_3, \ldots R_n$ respectively, using

2.1 Let $P_k = \{A_1, A_2, A_3, \ldots A_m\}$, where each A_i is an equivalence class of S obtained by equivalence relation R_k. Examine each A_i and decompose it into further subclass where two elements s & t of A_i fall into same subclasses if a input $x \in I$ both s & t into same class A_i.

2.2 Call this new partition of S P_{k-1}.

Step 3: Repeat Step 2 until $P_k = P_{k+1}$.

Step 4: The resulting $P_k = P$ corresponds to the relation R. The QFSM (S/R, I,$\Im/R$ $[s_0]$, T/R of M obtained by R is the simplified machine.

We shall demonstrate the working of this algorithm in the following examples.

Example 4: Simplify the FSM shown in the figures 7.4.3.

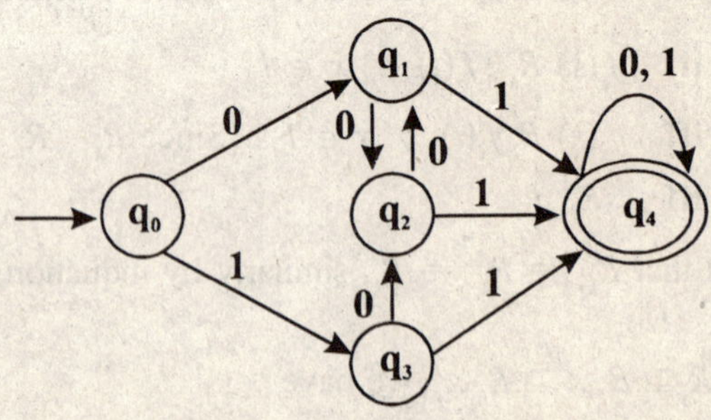

Fig. 7.4.3.

Solution: Here, in the given FSM $M = (S, I, \Im, s_0, T)$, we have $S = \{q_0, q_1, q_2, q_3, q_4\}$, $I = \{0, 1\}$, $s_0 = q_0$, $T = \{q_0\}$ and $\Im$ can be derived from the diagram of M. The initial partition P_0 is obtained on empty string ε i.e. without making a move. Thus, we have

$P_0 = \{T, T^C\}$, where $T = \{q_4\}$ and $T^C = \{q_0, q_1, q_2, q_3\}$. We can now write different partitions as collection of subsets of S only like

$$P_0 = \{\{q_4\}, \{q_0, q_1, q_2, q_3\}\}$$

This completes the starting steps i.e. step 1. Now we have to decompose the subclasses so obtained. The subclass $\{q_4\}$ contains only one element, so it cannot further decomposed. We, therefore, consider $\{q_0, q_1, q_2, q_3\}$ for further decomposition, if possible.

On input symbol 0 all states in T^C goes to some states belonging to T^C only, but on input symbol 1, q_0 goes to q_1 in T^C and other states q_1, q_2 and q_3 goes to q_4 in T. This implies that on input symbol 0 q_1, q_2 and q_3 go to states in T^C and on input symbol 1 they go to a state in T, so they form a class different from the class to which q_0 belongs. Thus we have a new partition

$$P_1 = \{\{q_4\}, \{q_0\} \{q_1, q_2, q_3\}\}$$

We repeat the above procedure on P_1 to get P_2. Only subclass that can be, if at all, partitioned is $\{q_1, q_2, q_3\}$. On input symbol 0 all these states go to some state of $\{q_1, q_2, q_3\}$ and on input symbol 1 they go to $\{q_4\}$, so no further partition is possible. Therefore, we have

$$P_2 = \{\{q_4\}, \{q_0\} \{q_1, q_2, q_3\}\}$$

Now $P_1 = P_2$: so we stop here and it the P we are looking for. Thus $P = \{[q_0], [q_1], [q_4]\}$. The relation corresponding to P_0 is R_0, P_1 is R_1, and so on. Finally P corresponds to the machine congruence relation R. The QFSM $= (S/R, I, \Im/R, [s_0], T/R)$, where

$S/R = \{[q_0], [q_1], [q_4]\}$,

$I = \{0, 1\}$,

$[s_0] = [q_0]$,

$T/R = \{[q_4]\}$ and $\Im$/R is given by the following table 7.4.4.

	0	1
$[q_0]$	$[q_1]$	$[q_0]$
$[q_0]$	$[q_1]$	$[q_0]$
$[q_0]$	$[q_1]$	$[q_0]$

Table 7.4.4.

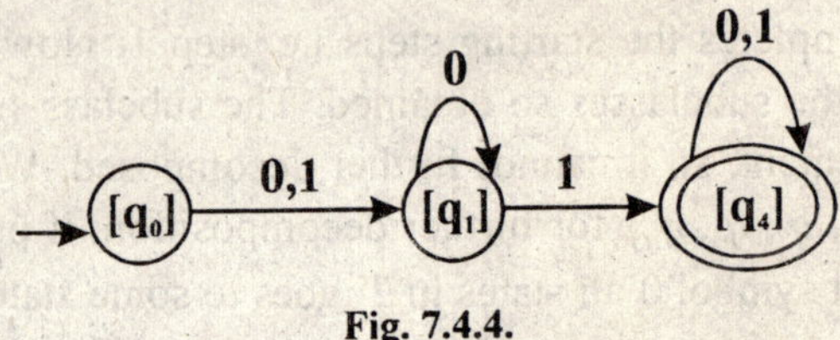

Fig. 7.4.4.

The final simplified QFSM is shown in the figure 7.4.4

Ans.

Example 5: Simplify the FSM shown in the following figure.

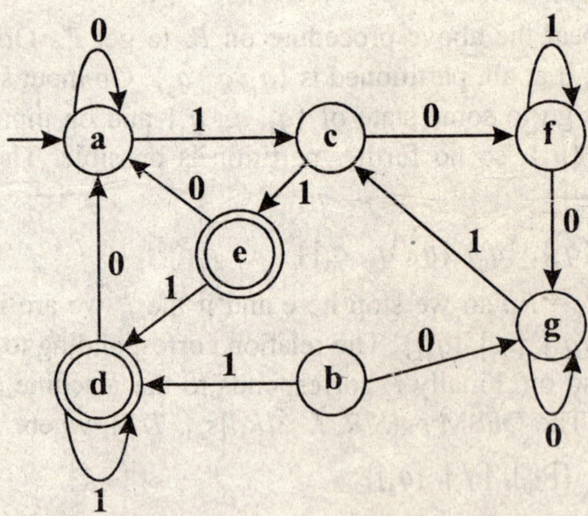

Fig. 7.4.5.

Solution: In the given FSM $M = (S, I, \Im, s_0, T)$, we have $S = \{a, b, c, d, e, f, g\}$, $I = \{0, 1\}$, $s_0 = a$, $T = \{d, e\}$ and $\Im$ is obvious from the diagram 7.4.5.

Here, $T = \{d, e\}$ and $T^C = \{a, b, c, f, g\}$. Thus, the initial partition P_0 is given by

$P_0 = \{\{d, e\}, \{a, b, c, f, g\}\}$

Notice that, state b has no incoming arc and hence it is an inaccessible state. It can be eliminated from the machine. Now decomposition procedure is applied on T and T^C. The set $\{d, e\}$ cannot be decomposed further as on input symbol 1, d and e go to states in $\{d, e\}$ and on 1 it goes to T^C. In T^C, on input symbol 1, c goes to states in T where as a, g and f go to states in T^C. On input symbol 0, c goes to states in $\{a, g, f\}$ and states in $\{a, g, f\}$ get transferred to some states in $\{a, g, f\}$. Thus $\{a, c, f, g\}$ has been decomposed into two subclasses: $\{a, g, f\}$ and $\{c\}$. Hence, we have

$P_1 = \{\{d, e\}, \{a, f, g\}\{c\}\}$

Now to find P_2, we consider each class of P_1. Again $\{d, e\}$ cannot be decomposed further. In class $\{a, f, g\}$, states a and g go to states c in $\{c\}$ on input symbol I, whereas, state f goes to f itself. This implies that $\{a, f, g\}$ can be decomposed into two subclasses $\{a, g\}$ and $\{f\}$. Class $\{c\}$ contains single state, no further decomposition is possible. Therefore, we have

$P_2 = \{\{d, e\}, \{a, g\}, \{f\}, \{c\}$

It can be easily verified that

$P_3 = \{\{d, e\}, \{a, g\}, \{f\}, \{c\}\}$

	0	1
[a]	[a]	[c]
[c]	[f]	[d]
[d]	[a]	[d]
[f]	[a]	[f]

Table 7.4.6.

Since $P_2 = P_3$, we have $P = P_2 = \{[a], [c], [d], [f]\}$. Therefore, in the QFSM $M/R = (S, R, I, \Im/R, [s_0], T/R)$, we have $S/R = P_0$ $1 = \{0, 1\}$, $s_0 = [a]$, $T = \{[d]\}$ and $\Im/R$ is given by the following table 7.4.5. The simplified machine is shown in the figure 7.4.6.

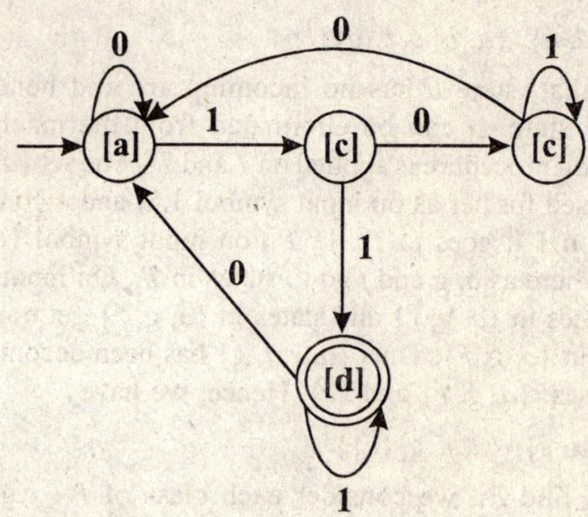

Fig. 7.4.6.

Example 6: Simplify the FSM shown in the following diagram.

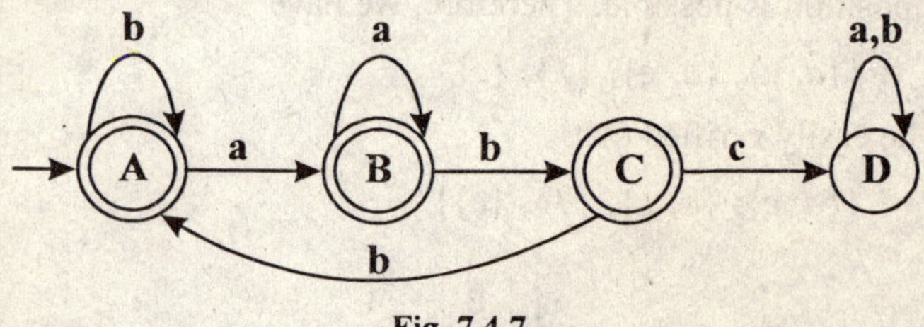

Fig. 7.4.7.

Solution: In the given FSM $M = (S, I, \Im, s_0, T)$, we have $S = \{A, B, C, D\}$, $I = \{a, b\}$, $s_0 = A$, $T = \{A, B, C\}$ and $\Im$ is obvious from the diagram 7.4.7.

Here, $T = \{A, B, C\}$ and $T^C = \{D\}$. Thus, the initial partition P_0 is given by

$P_0 = \{\{A, B, C\}, \{D\}\}$

Notice that, there is no inaccessible state in this machine. Now decomposition procedure is applied on T and T^C. The set $\{D\}$ cannot be decomposed further as it contains only one state. In T, on input symbol a, C goes to states in T^C whereas A and B go to states in T. Thus $\{A, B, C\}$ has been decomposed into two subclasses: $\{A, B\}$ and $\{C\}$. Hence, we have

$$P_1 = [\{A, B\}, \{C\}, \{D\}\}$$

Now to find P_2, we consider each class of P_1. Again $\{C\}$ and $\{D\}$ cannot be decomposed further. In class $\{A, B\}$, state A goes to state A in $\{A, B\}$ on input symbol b, whereas, state B goes to C on the same input symbol. This implies that $\{A, B\}$ can be decomposed into two subclasses: $\{A\}$ and $\{B\}$. Therefore, we have

$$P_2 = \{\{A\}, \{B\}, \{C\}, \{D\}\} = P$$

This shows that the given FSM is already in simplified form. Therefore, it cannot be simplified further.

Ans.

Exercise

1. Consider right-linear grammar for
   - Identifiers which can be of arbitrary length and must start with a letter and may contain any characters from roman alphabet and any digits.
   - Identifiers which can be one to six symbols in length and must start with I, J, K, L, M or N and may contain nay lower/ uppercase roman alphabet and digits.
   - All strings of 0's and 1's having both an odd number of 0's and an odd number of 1's.
2. Construct context free grammars that generate
   - All strings of 0's and 1's having equal number of 0's and 1's.
   - Well formed statements in prepositional calculus.
   - $\{0^i j^j \mid i \neq j \text{ and } i.j \geq 0\}$.
   - All possible sequence of balanced parentheses.

3. Describe the language generated by the productions $S \to bSS \mid a$. Observe that it is not always easy to describe what language a grammar generates.
4. What class of language can be generated by grammars with only *left context*, i.e. grammar in which each production is of the form $\alpha A \to \alpha\beta$, where α and β belongs to $(N \cup \Sigma)^*$?
5. Show that every context free languages can be generated by a grammar $G = (N, \Sigma, P, S)$ in which each production is of either the form $A \to \alpha$, α in N^* or $A \to w$, w in Σ^*.
6. Let G be a grammar defined by the productions:

   1. $S \to aSBC \mid abC$ 2. $CB \to BC$ 3. $bB \to bb$
   4. $bC \to bc$ 5. $cC \to cc$

   Prove that $L(G) = \{a^n b^n c^n \mid n \geq 1\}$.
7. In an unrestricted grammar G there are many ways of deriving a given sentence that are essentially the same, differing only in order in which productions are applied. If G is context-free, then we can represent these essentially similar derivations by means of a derivation tree. However, if G is a context sensitive or unrestricted, we can define equivalence classes of derivations in the following manner.

   Let $G = (N, \Sigma, P, S)$ be an unrestricted grammar. Let D be the set of all derivations of the form $S \Rightarrow^* w$. That means elements of D are sequences of the form $(\alpha_0, \alpha_1, \alpha_2, \ldots \alpha_n)$ such that $\alpha_0 = S$, $\alpha_n \in \Sigma^*$ and $\alpha_{i-1} \to a_i$ for $1 \leq i \leq n$.

   Define a relation R_0 on D by $(\alpha_0, \alpha_1, \alpha_2, \ldots \alpha_n)$ R_0 $(\beta_0, \beta_1, \beta_2, \ldots \beta_n)$ if and only if there exists i between 1 and $(n - 1)$ such that

   1. $\alpha_j = \beta_j$ for $1 \leq j \leq n$ such that $j \neq I$.
   2. We can write $a_{i-1} = \gamma_1\gamma_2\gamma_3\gamma_4\gamma_5$ and $\alpha_{i+1} = \gamma_1\delta\gamma_3\varepsilon\gamma_5$ such that $\gamma_2 \to \delta$ and $\gamma_4 \to \varepsilon$ are in P, and either $\alpha_i = \gamma_1\delta\gamma_3\gamma_4\gamma_5$ and $\beta_i = \gamma_1\gamma_2\gamma_3\varepsilon\gamma_5$ or conversely. Let R be the equivalence closure of the relation R_0.

   A grammar G is said to be ***unambiguous*** if each w in $L(G)$ appears as the last component of a derivation in one and only one equivalence class under R as defined above. Show that every right-linear language has an ***unambiguous right linear grammar.***

8. Which of the following are regular sets? Give regular expressions for these regular sets.
   - The set of words in $\{0,1\}^*$ having equal number of 0's and 1's.
   - The set of words in $\{0,1\}^*$ with an even number of 0's and an odd number of 1's.
   - The set of words in Σ^* whose length is divisible by 3.
   - The set of words in $\{0,1\}^*$ with no sub-string 101.
9. Show that if L is any regular set, then there is infinite number of regular expressions denoting L.
10. Show that the following identities for regular expressions α, β and γ:

    (a) $\alpha(\beta + \gamma) = \alpha\beta + \alpha\gamma$ (b) $\alpha + (\beta + \gamma) = (\alpha + \beta) + \gamma$

    (c) $\alpha(\alpha\beta) = (\alpha\beta)\gamma$ (d) $\alpha\,\varepsilon = \varepsilon\alpha = \alpha$

    (e) $(\alpha + \beta)\,\gamma = \alpha\gamma + \beta\gamma$ (f) $\alpha + \alpha = \alpha$

    (g) $\phi^* = \varepsilon$ (h) $\alpha^* + \alpha = \alpha^*$

    (i) $(\alpha^*)^* = \alpha^*$ (j) $\alpha + \phi = \alpha$

    (k) $(\alpha + \beta)^* = (\alpha^*\beta^*)^*$
11. Solve the following set of regular expressions:

$$A_1 = (01^* + 1)\,A_1 + A_2$$

$$A_2 = 11 + 1A_1 + 00A_3$$

$$A_3 = \varepsilon + A_1 + A_2$$

12. A ***right linear grammar*** $G = (N, \Sigma, P, S)$ is called a ***regular*** grammar when
    - All productions with the possible exception of $S \to \varepsilon$ are of the form $A \to aB$ or $A \to a$ where A and B are in N and a is in Σ.
    - If $S \to \varepsilon$ is in P, then S does not appear on the right of any production.

Show that every regular set has a regular grammar. Next construct a regular grammar for the regular set generated by the right linear grammar

$A \to B \mid C$ $\quad$ $B \to 0B \mid 1B \mid 011$

$C \to 0D \mid 1C \mid e$ $\quad$ $D \to 0C \mid 1D$

13. Find the regular expressions accepted by the FSM shown in the figures *a*, *b*, *c*, and *d*.

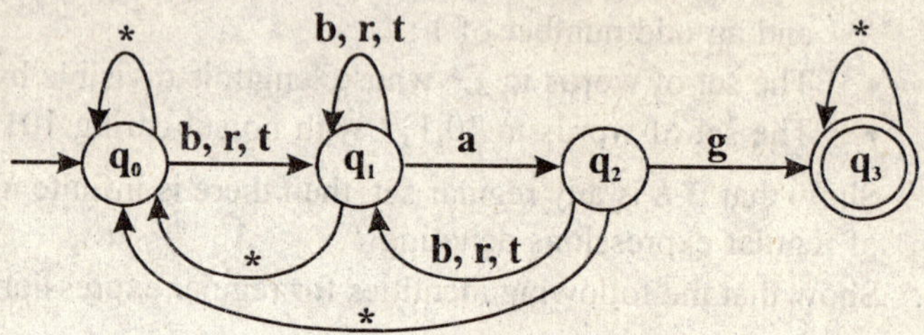

Fig. 7.13ex (a).

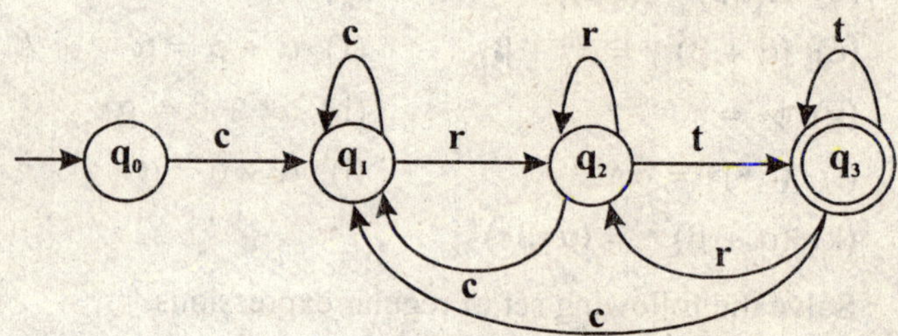

Fig. 7.13ex (b).

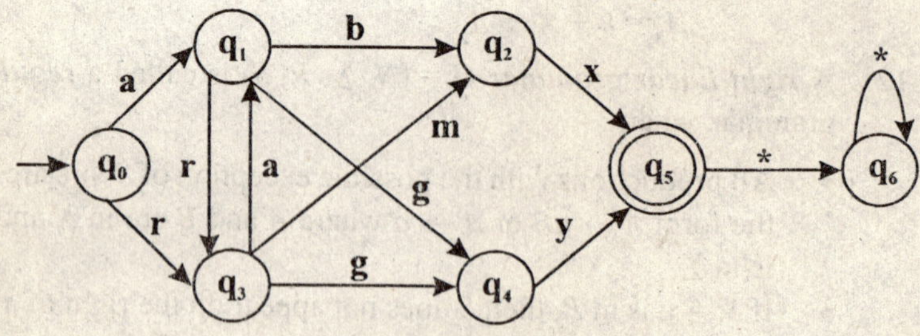

Fig. 7.13ex (c).

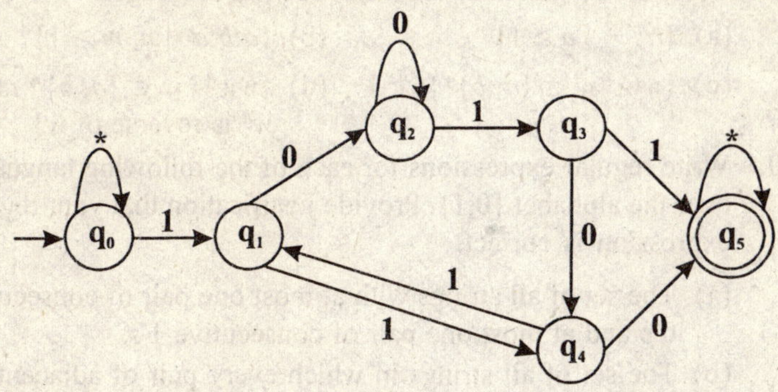

Fig. 7.12ex (d).

14. Draw FSM corresponding to the following regular expressions:

    (a) $a + bb + bab^*a$
    (b) $(ab^*a + b)a^*b$
    (c) $a\,(ab^* + ba)^*$
    (d) $(b + a\,(a + bb)^*ba)^*(\varepsilon + (b\,(ba + ab)^*\,a)$.

15. The regular expression a^* represents the set $\{\varepsilon, a, aa, aaa, \ldots\}$. Find the regular expression representing the regular set $\{a, aa, aaa, \ldots\}$.
16. Find the set of strings on $T = \{a, b\}$ produced by the regular expression $b(a + b)^*ab$.
17. Draw a FSM to recognize the regular expression $b(a + b)^*ab$.
18. Let $G = (N, \Sigma, P, S)$, where $N = \{S, C, D\}$, $\Sigma = \{a, b\}$ and P is given below:

    1. $S \to aCDa$
    2. $C \to baCDb$
    3. $D \to Cab$
    4. $aC \to baa$
    5. $bDb \to abab$

    (a) Show that $baabbabaaabbaba \in L(G)$.
    (b) Show that $L(G)$ contains infinitely many elements (words).
    (c) Show that if a is any string generated from S by P and includes some non-terminal symbols then it is always possible to extend the generation from α to a string that involves only terminals. That is, any generation process in G can be forced to terminate.

19. Determine a grammar for the following languages.

    (a) $\{a^{n^2} + \mid n > 1\}$ (b) $\{a^n b^n a^m \mid n, m > 1\}$

    (c) $\{ww \mid w \in \{a, b\}^*\}$ (d) $\{ww^R \mid w \in \{a, b\}^*$ *and* w^R is reverse of $w\}$.

20. Write regular expressions for each of the following languages over the alphabet {0,1}. Provide justification that your regular expression is correct.

    (a) The set of all strings with at most one pair of consecutive 0's and at most one pair of consecutive 1's.
    (b) The set of all strings in which every pair of adjacent 0's appears before any pair of adjacent 1's.
    (c) The set of all strings not containing 101 as a sub-string.

21. Describe in statement the sets denoted by the following regular expressions.

    (a) (11 + 0)*(00 + 1)* (*b*) (1 + 01 + 001)*(ε + 0 + 00)

    (*c*) [00 + 11 + (01 + 10)(00 + 11)*(01 + 10)]*.

22. Construct finite automata equivalent to the following regular expressions.

    (a) 10 + (0 + 11)0*1 (*b*) 01[((10)* + 111)* + 0]*1

    (*c*) ((0 + 1)(0 + 1))* + ((0 + 1)() + 1)(0 + 1)]*

23. Construct regular expressions corresponding to the state diagram given in the following figures.

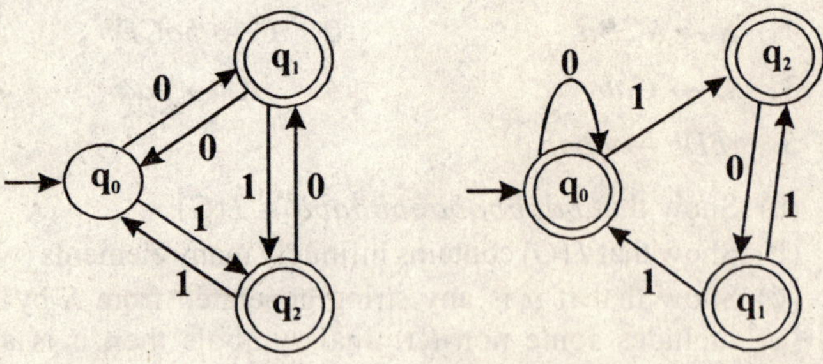

Fig. 7.23ex.

24. Consider the language L specified by the grammar $G = (N, \Sigma, P, S)$, where $N = \{S, A, B\}$, $\Sigma = \{a, b, c\}$ and P is set containing following productions:

1. $S \to AB$
2. $A \to ab$
3. $A \to aAB$
4. $B \to c$
5. $B \to Bc$

(a) Determine whether each of the following strings is a sentence in the language.

$aabb$ $\quad$ $aaabbc$ $\quad$ $aaabbbccc$ $\quad$ $ababcc$

(b) Describe the language L in set-theoretic notations.

25. Give a grammar that specifies each of the following languages:

(a) $L = \{a^{2m}\, b^{2n} \mid m \geq 1, n \geq 1\}$

(b) $L = \{(ab)^m\, c^{2n} \mid m \geq 1, n \geq 1\}$

(c) $L = \{a^m\, b^n \mid 1 \leq m < n, n \geq 1\}$

(d) $L = \{a^m\, b^n \mid m \leq n \leq 2m, m \geq 1\}$

(e) $L = \{a^m\, b^m\, c^n \mid m \geq 1, n \geq 1)$

(f) $L = \{a^m\, b^n\, c^q \mid m \geq 1, n \geq 1, m + n = q\}$

26. (a) Give a type-3 grammar that generates the language $L = \{x \mid x \in \{a, b\}^*$ and x does not contain two consecutive a's .}

(b) Give a type-2 grammar that generates the language $L = \{x \mid x \in \{a, b)^*$ and x contains twice as many a's as b's.}

27. For each of the following grammar state whether it is type-1, 2 or 3. Find $L(G)$ for each of them. Whenever grammar is type-2 or 3, give its BNF representation and syntax diagram.

- $G = (N, \Sigma, P, S)$, where $N = \{S, A\}$, $\Sigma = \{x, y, z\}$ and $P = \{S \to xS, S \to yA, A \to yA, A \to z\}$
- $G = (N, \Sigma, P, S)$, where $N = \{S\}$, $\Sigma = \{a\}$ and $P = \{S \to aaS, S \to aa\}$.
- $G = (N, \Sigma, P, S)$, where $N = \{S\}$, $\Sigma = \{a, b\}$ and $P = \{S \to aaS, S \to a, S \to a\}$

- $G = (N, \Sigma, P, S)$, where $N = \{S\}$, $\Sigma = \{a, b, c\}$ and $P = \{S \to aS, S \to bS, S \to c\}$
- $G = (N, \Sigma, P, S)$, where $N = \{S, A, B\}$, $\Sigma = \{a, +, (.)\}$ and $P = \{S \to (S), S \to a + A, A \to a + B, B \to a + b, B \to a\}$
- $G = (N, \Sigma, P, S)$, where $N = \{S, A\}$, $\Sigma = \{a, b\}$ and $P = \{S \to aA, A \to bS\ A \to a\}$
- $G = (N, \Sigma, P, S)$, where n $= \{S, A, B\}$, $\Sigma = \{x, y, z\}$ and $P = \{S \to SA, SA \to BS, BS \to xy, B \to x, A \to z\}$

28. Give two distinct derivation (sequences of substitution that start at S) for the string xyz of $L(G)$, where G is the grammar of exercise 27(g).
29. Let G be the grammar of exercise 27(e). Can you give two distinct derivations of the string $((a + a + a))$?
30. Find the regular expression corresponding to the grammar of exercise 27(a), 27(b), 27(c), 27(d), 27(e) and 27(f).
31. Find the regular expression corresponding to the syntax diagram shown below.

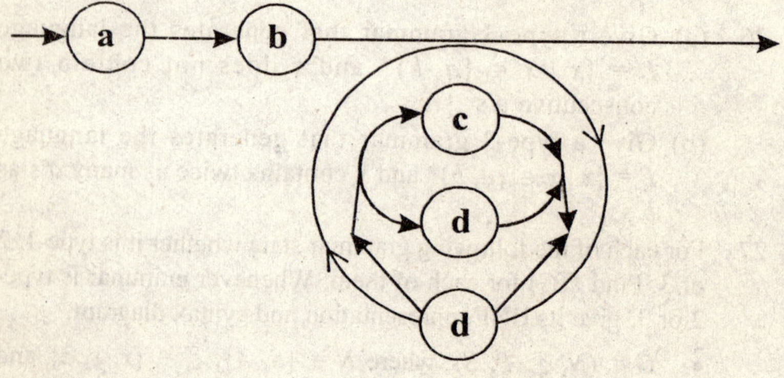

Fig. 7.31ex.

32. Draw the digraph of the machine whose state transition table is shown below.

	0	1	2
S_0	S_1	S_0	S_2
S_1	S_0	S_0	S_1
S_2	S_2	S_0	S_2

	a	b
S_0	S_1	S_0
S_1	S_2	S_1
S_2	S_3	S_2
S_3	S_3	S_3

Table 7.32ex.

33. Construct the state transition table of the finite state machine whose digraph is shown in the figure 7.33ex.

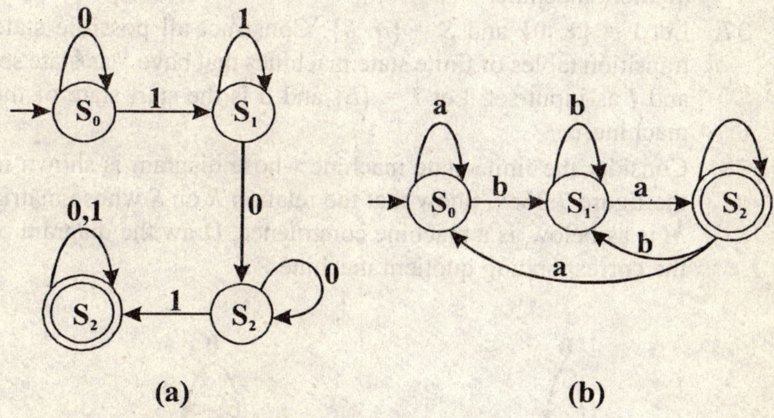

Fig. 7.33ex.

34. Let $M = (S, I, \Im, s_0, T)$ be a finite state machine. Define a relation R on I as follows $x_1 R x_2$ iff $f_{x_1}(s) = f_{x_2}(s)$ for every s in S. Show that R is an equivalence relation.
35. Let $(S, *)$ be a finite semi group. Then we may consider the machine $(S, S, \Im, s_0, T)$, where $\Im = \{f_x \mid x \in S\}$, and $f_x(y) = x * y$ for all $x, y, \in S$. Define a relation R on S as follows: $x\,R\,y$ iff there is some $z \in S$ such that $f_z(x) = y$. Show that R is transitive.
36. Consider the machine whose state transition diagram is given as below:

	0	1
1	1	4
2	3	2
3	2	3
4	4	1

Table 7.36ex.

Here $S = \{1, 2, 3, 4\}$. (a) Show that $R = [(1, 1), (1, 4)\ (4, 1), (4, 4), (2, 2), (2, 3)\ (3, 2), (3, 3)]$ is a machine congruence. (b) Construct the state transition table for the corresponding quotient machine.

37. Let I = $\{1, 0\}$ and $S = \{a, b\}$. Construct all possible state transition tables of finite state machines that have S as state set and I as input set. Let $T = \{b\}$ and a is the start state of the machine.

38. Consider the finite state machine whose diagram is shown in the figure 7.36ex. Show that the relation R on S whose matrix M is as below, is a machine congruence. Draw the diagram of the corresponding quotient machine.

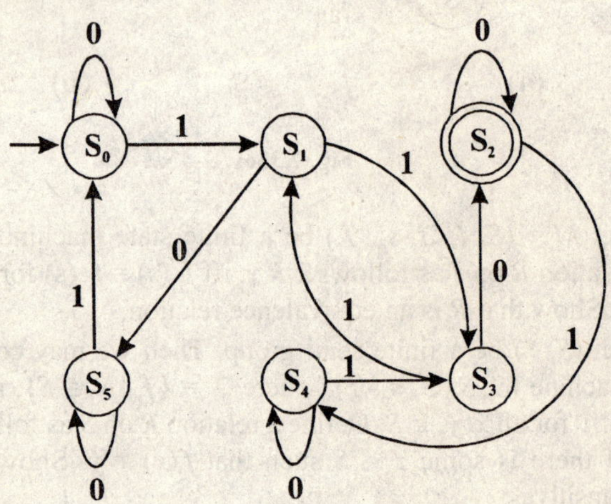

$$M_R = \begin{pmatrix} 1 & 0 & 0 & 0 & 1 & 0 \\ 0 & 1 & 0 & 0 & 0 & 1 \\ 0 & 0 & 1 & 1 & 0 & 0 \\ 0 & 0 & 1 & 1 & 0 & 0 \\ 1 & 0 & 0 & 0 & 1 & 0 \\ 0 & 1 & 0 & 0 & 0 & 1 \end{pmatrix}$$

Fig. 7.38ex.

39. Draw the finite state machine whose state transition table is given as below and answer the following:

	0	1
S_0	S_0	S_1
S_1	S_1	S_2
S_2	S_2	S_3
S_3	S_3	S_0

Table 7.39ex.

(a) List the values of transition function f_w for w = 01001.

(b) List the values of transition function f_w for w = 11100.

(c) Describe the set of binary words w having the property that $f_w = f_{010}$.

(d) Describe the set of binary words w having the property that $f_w(s_0) = s_2$.

40. See the finite state machine shown in the figure 7.40ex and answer the following questions.

(a) List the values of transition function f_w for w = *abba.*

(b) List the values of transition function f_w for w = *babab.*

(c) Describe the set of binary words w having the property that $f_w(s_0) = s_2$.

(d) Describe the set of binary words w having the property that $f_w(s_0) = s_0$.

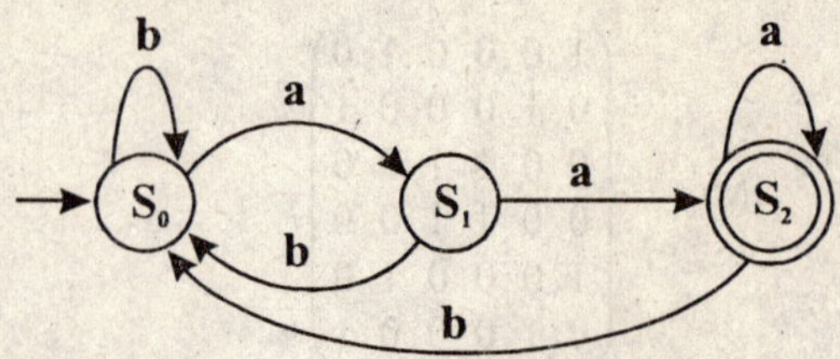

Fig. 7.40ex.

41. Describe the language accepted by the finite state machines shown in the figures 7.41ex.

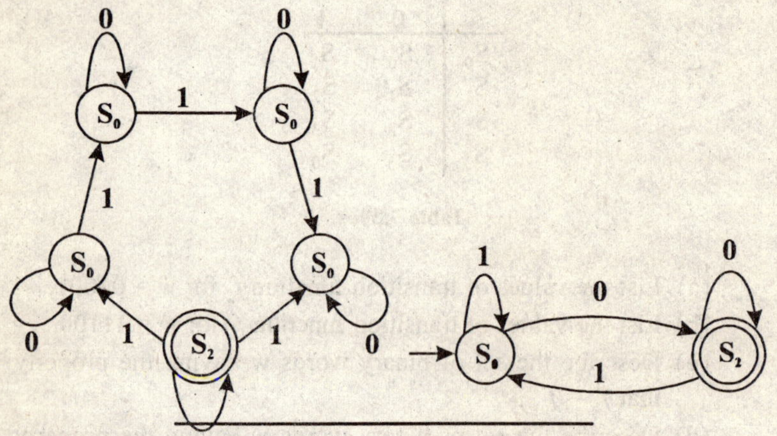

Figure 7.41ex(a).

Figure 7.41ex(b).

Fig. 7.41ex (c).

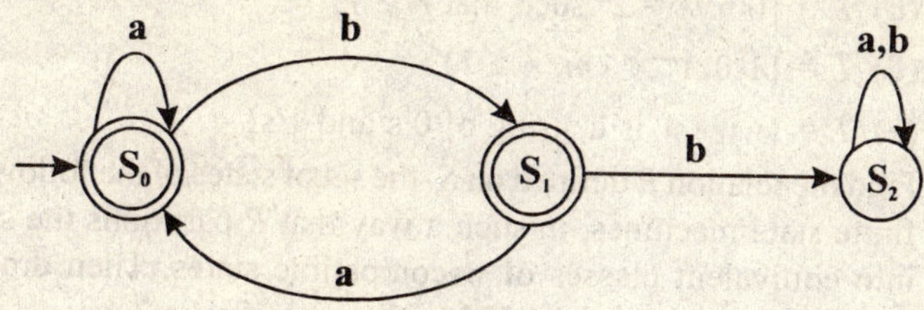

Fig. 7.41ex (d).

42. Let $M = \{S, I, \Im, s_0, T)$ be a finite state machine. Suppose that if s is in T and w in I^*, then $f_w(s)$ is in T. Prove that $L(M)$ is a sub-semi group of $(I^*, \bullet)$, where $\bullet$ is a concatenation operation.
43. Construct type 3 grammar for the regular expression accepted by the finite state machines shown in exercise 41.
44. For each of the following input string construct a finite state machine that accepts it and no others. Mark the **trapping** state, if any.

    (a) Inputs a, b; strings where the number of b's is divisible by 3.
    (b) Inputs x, y; strings where the number of y's is even.
    (c) Inputs 0, 1; strings that contain 0011.
    (d) Inputs a, b; strings that contain ab and end in bbb.
    (e) Inputs 0, 1; strings that end with 0011.
    (f) Inputs w, z; strings that contain wz or zzw.
    (g) Inputs w, z; strings that end in wz or zzw.
    (h) Input 0, 1, 2; string 0120 is the only string recognized.
    (i) Inputs x, y, z; strings xzx or yx or zyx are to be recognized.
    (j) Inputs 0, 1; strings ending in 0101.
    (k) Inputs x, y; strings having exactly two x's.
    (l) Inputs a, b; strings that do not have two consecutive b's.

45. Show that each of the following language is not a finite state language.

    (a) $L = \{0^m 1^n \mid m \geq n\}$
    (b) $L = \{0^m 1^n \mid m \leq n\}$

(c) $L = \{0^m \mid m = 2^n \text{ such that } n \geq 1\}$

(d) $L = \{1^n 0^m 1^{n+m} \mid m, n \geq 1\}$

(e) $L = \{ww \mid w \text{ is a string of 0's and 1's}\}$

46. Find the relation R defined on S -the set of states of the following finite state machines, in such a way that R partitions the set S into equivalent classes of w-compatible states. Then draw a simplified version of the FSMs shown in figure 7.4ex.

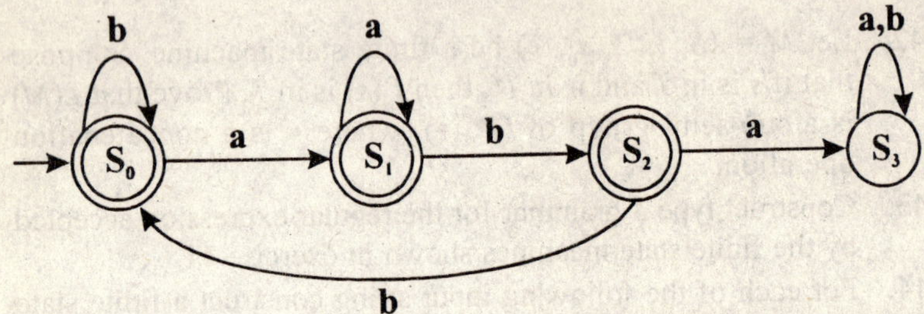

Fig. 7.46ex (a).

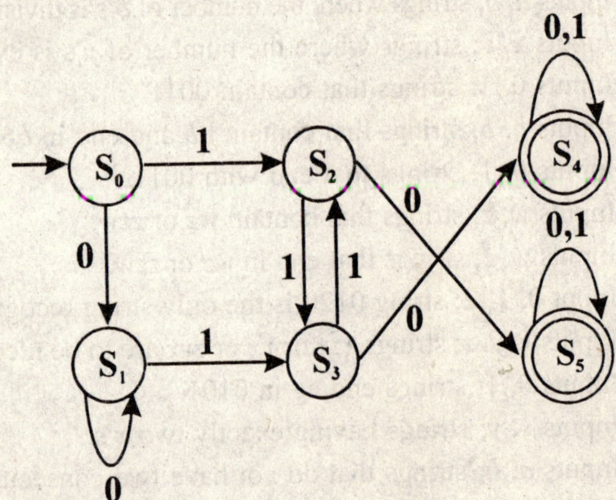

Fig. 7.46ex (b).

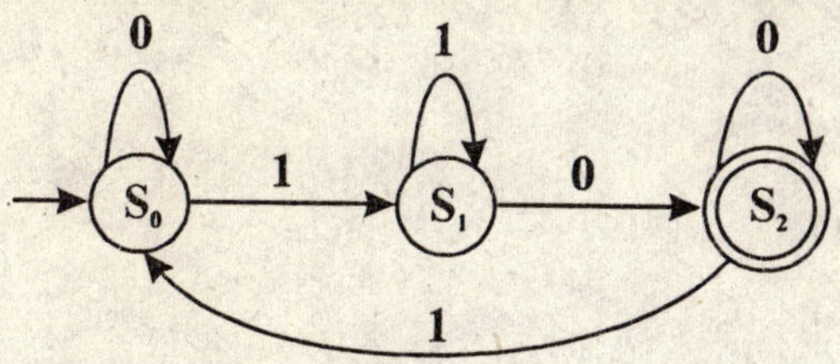

Fig. 7.46ex (c).